Advancements in Nanomaterials for Energy Conversion and Storage

In this book, readers will find an exhaustive examination of the latest advancements in nanomaterials, covering their synthesis, characterization, and utilization in energy storage and conversion. Additionally, the text delves into the diverse applications of these nanomaterials across various fields such as supercapacitors, fuel cells, biofuel cells, solar cells, batteries, and organic electronics. The discussion also encompasses the challenges faced, historical context, and future outlooks within this rapidly evolving domain.

Features:

- Cutting-edge insights: Stays abreast of the latest breakthroughs in nanomaterial science, with a succinct review of advanced materials tailored for energy storage applications.
- Developmental journey: Traces the evolution of energy storage materials, from their inception to their current state of the art.
- Versatile applications: Explores the diverse applications of nanomaterials in energy storage, spanning supercapacitors, fuel cells, biofuel cells, solar cells, batteries, and beyond.
- Visual aids: Enhances readers' understanding, with key figures and tables spotlighting the intricate applications of various nanomaterials.

Geared toward researchers and graduate students in chemical engineering, chemical sciences, nanomaterials, and energy engineering/conversion, this book serves as an indispensable resource for those seeking to push the boundaries of nanotechnology in the pursuit of sustainable energy solutions.

Emerging Materials and Technologies

Series Editor:
Boris I. Kharissov

The *Emerging Materials and Technologies* series is devoted to highlighting publications centered on emerging advanced materials and novel technologies. Attention is paid to those newly discovered or applied materials with potential to solve pressing societal problems and improve quality of life, corresponding to environmental protection, medicine, communications, energy, transportation, advanced manufacturing, and related areas.

The series takes into account that, under present strong demands for energy, material, and cost savings, as well as heavy contamination problems and worldwide pandemic conditions, the area of emerging materials and related scalable technologies is a highly interdisciplinary field, with the need for researchers, professionals, and academics across the spectrum of engineering and technological disciplines. The main objective of this book series is to attract more attention to these materials and technologies and invite conversation among the international R&D community.

MXenes
From Research to Emerging Applications
Edited by Subhendu Chakroborty

Biodegradable Polymers, Blends and Biocomposites
Trends and Applications
Edited by A. Arun, Kunyu Zhang, Sudhakar Muniyasamy, and Rathinam Raja

Bioinspired Materials and Metamaterials
A New Look at the Materials Science
Edward Bormashenko

Computational Studies
From Molecules to Materials
Edited by Ambrish Kumar Srivastava

2D Semiconductors for Environmental Remediation
Edited by Honey John, Nisha T Padmanabhan, Sona Stanly, and Jith C Janardhanan

Materials from Natural Sources
Structure, Properties, and Applications
Edited by Ramesh Gardas, Neha Patni, and Amita Chaudhary

Dielectric Materials for Capacitive Energy Storage
Edited By Haibo Zhang and Hua Tan

Multifunctional Coordination Materials for Green Energy Technologies
Edited by Ghulam Yasin, Anuj Kumar, Sajjad Ali, Tuan Anh Nguyen, and Saira Ajmal

Advancements in Nanomaterials for Energy Conversion and Storage
Edited by Piyush Kumar Sonkar and Vellaichamy Ganesan

For more information about this series, please visit:
www.routledge.com/Emerging-Materials-and-Technologies/book-series/CRCEMT

Advancements in Nanomaterials for Energy Conversion and Storage

Edited by
Piyush Kumar Sonkar and Vellaichamy Ganesan

CRC Press
Taylor & Francis Group
Boca Raton London New York

CRC Press is an imprint of the
Taylor & Francis Group, an **informa** business

Designed cover image: © Shutterstock Images

First edition published 2025
by CRC Press
2385 NW Executive Center Drive, Suite 320, Boca Raton FL 33431

and by CRC Press
4 Park Square, Milton Park, Abingdon, Oxon, OX14 4RN

CRC Press is an imprint of Taylor & Francis Group, LLC

ISBN: 9781032768403 (hbk)
ISBN: 9781032770819 (pbk)
ISBN: 9781003481157 (ebk)

DOI: 10.1201/9781003481157

Typeset in Times
by codeMantra

Dedicated to our Authors and

Collaborators

&

A heartfelt tribute to

Late Prof. S. Abraham John

Who immensely contributed to Science

(one of the authors of this book)

Centre for Nanoscience and Nanotechnology

Department of Chemistry

Gandhigram Rural Institute-Deemed to be University

Gandhigram-624302, Tamil Nadu, India

Contents

About the Editors

Piyush Kumar Sonkar received his B.Sc. (Chemistry), M.Sc. (Chemistry), and Ph.D. (Chemistry) degrees from Banaras Hindu University, India. He is presently working as an assistant professor in the Department of Chemistry, MMV, Banaras Hindu University, Varanasi, India. His research interests include nanomaterials, nanocomposites, fuel cells, electrochemical devices, supercapacitors, biosensors, chemical sensors, and new materials. He has published more than 45 international and national research papers in various reputed journals. He has published six book chapters and one book in national/international publications. He has presented his research work in various international/national seminars and conferences.

Vellaichamy Ganesan completed his B.Sc. from Ayya Nadar Janaki Ammal College, Sivakasi; M.Sc. from Bharathidasan University, Tiruchirappalli; and Ph.D. from Madurai Kamaraj University, Madurai, Tamil Nadu, India. Later, he visited the University of Houston, Houston, USA, and Université de Lorraine, Nancy, France, for postdoctoral research. At present, he is working as a professor in the Department of Chemistry, Institute of Science, Banaras Hindu University (BHU), Varanasi. BHU recognized his research in 2020 by awarding him the most productive researcher award of the year. He is a recipient of the Indo-US Science and Technology Forum Fellowship (IUSSTF Fellow), USA; Commonwealth Fellowship, UK; and DAAD Fellowship, Germany. His current areas of research interest include materials electrochemistry, energy conversion and storage, catalytic materials for fuel cells, electrocatalysis, metal-organic frameworks, electrochemical sensors, photocatalysis, etc. He has published more than 120 papers in reputed national and international journals.

Contributors

Nasreen Amin
Department of Botany
Banaras Hindu University
Varanasi, India

Uday Pratap Azad
Department of Chemistry
Guru Ghasidas Vishwavidyalaya
Bilaspur, India

Shahid Bashir
Department of Physics, Faculty of Science, Centre for Ionics Universiti Malaya
Universiti Malaya
Kuala Lumpur, Malaysia

Wei Sea Chang
School of Engineering
Monash University
Selangor, Malaysia

Alok Kumar Chaudhari
Department of Chemistry
DDU Gorakhpur University
Gorakhpur, India

N. R. Elizer
Department of Physics, Faculty of Science, Centre for Ionics
Universiti Malaya
Kuala Lumpur, Malaysia

Ameer Farithkhan
Department of Chemistry, Centre for Nanoscience and Nanotechnology
The Gandhigram Rural Institute (deemed to be university)
Gandhigram, India

Somenath Garai
Department of Chemistry, Institute of Science
Banaras Hindu University
Varanasi, India

Z. L. Goh
Department of Physics, Faculty of Science, Centre for Ionics Universiti Malaya
Universiti Malaya
Kuala Lumpur, Malaysia

N. S. K. Gowthaman
School of Engineering
Monash University
Selangor, Malaysia

Mandakini Gupta
Department of Chemistry
Sunbeam Women's College
Varuna, Varanasi, Uttar Pradesh, India

Saral Kumar Gupta
Department of Physical Sciences
Banasthali Vidyapith
Banasthali, India

Tanu Gupta
Department of Chemistry
Rajendra College
Chapra, India

Amit Jaiswal
Department of Chemistry
C.M.P. Degree College
Prayagraj, India

S. Abraham John
Department of Chemistry, Centre for Nanoscience and Nanotechnology
The Gandhigram Rural Institute (deemed to be university)
Gandhigram, India

Vinod K. Kannaujiya
Department of Botany, MMV
Banaras Hindu University
Varanasi, India

Ranjeet Kumar
Department of Chemistry
C.M.P. Degree College
Prayagraj, India

Ranjeet Kumar
Department of Chemistry
Vikramajit Singh Sanatan Dharma College
Kanpur Nagar, India

Ravendra Kumar
Department of Chemistry, MMV
Banaras Hindu University
Varanasi, India

Upendra Kumar
Advanced Functional Materials Laboratory, Department of Applied Sciences, IIIT Allahabad
Prayagraj, India

Sajith Kurian
Department of Chemistry
Mar Ivanios College (Autonomous)
Thiruvananthapuram, India

Divya Kushwaha
Department of Chemistry, MMV
Banaras Hindu University
Varanasi, India

Dasnur Nanjappa Madhusudan
Department of Chemistry, Institute of Science
Banaras Hindu University
Varanasi, India

Pershaanaa Manogran
Department of Physics, Faculty of Science, Centre for Ionics
Universiti Malaya
Kuala Lumpur, Malaysia

Vijay Lakshmi Mishra
Post doctoral Fellow
National Institute for Materials Science
Tsukuba, Ibaraki, Japan

Saumi Pandey
Department of Botany, MMV
Banaras Hindu University
Varanasi, India

S. K. S. Patel
Department of Chemistry, MMV
Banaras Hindu University
Varanasi, India

Shankab J. Phukan
Department of Chemistry, Institute of Science
Banaras Hindu University
Varanasi, India

Ramasamy Ramaraj
Department of Physical Chemistry, School of Chemistry
Madurai Kamaraj University
Madurai, India

K. Ramesh
Department of Physics, Faculty of Science, Centre for Ionics Universiti Malaya
Universiti Malaya
Kuala Lumpur, Malaysia

S. Ramesh
Department of Physics, Faculty of Science, Centre for Ionics Universiti Malaya
Universiti Malaya
Kuala Lumpur, Malaysia

Shobhana Ramteke
Integrated Regional Office, Ministry of Environment, Forest and Climate Change
Government of India
Naya Raipur, India

Shreya Rastogi
Department of Chemistry, Institute of Science
Banaras Hindu University
Varanasi, India

Akshay K. Ray
Department of Chemistry, Institute of Science
Banaras Hindu University
Varanasi, India

Manas Roy
Department of Chemistry
National Institute of Technology
Agartala, India

Neeraj K. Sah
Department of Chemistry, Institute of Science
Banaras Hindu University
Varanasi, India

Bharat Lal Sahu
Department of Chemistry
Guru Ghasidas Vishwavidyalaya
Bilaspur, India

Kamatchi Sankaranarayanan
Physical Sciences Division
Institute of Advanced Study in Science and Technology
Guwahati, India

Ashish Kumar Singh
Department of Chemistry
Guru Ghasidas Vishwavidyalaya
Bilaspur, India

Divya Pratap Singh
Department of Chemistry
Rajkiya Engineering College
Azamgarh, India

Nandita Singh
Department of Chemistry
Guru Ghasidas Vishwavidyalaya
Bilaspur, India

Shalini Singh
Department of Chemistry, Institute of Science
Banaras Hindu University
Varanasi, India

Sunil Kumar Singh
Department of Chemistry
Guru Ghasidas Vishwavidyalaya
Bilaspur, India

V. B. Singh
Department of Chemistry, Center of Advanced Study, Institute of Science
Banaras Hindu University
Varanasi, India

Pankaj Srivastava
Department of Chemistry, Institute of Science
Banaras Hindu University
Varanasi, India

Varsha Yadav
School of Applied Sciences
Shri Venkateshwara University
Gajraula, India

Introduction

Nanomaterials is one of the promising areas for researchers to explore new dimensions in the field of material science, energy, medical, engineering, and industries. They have a wide range of applications in energy storage, drug delivery, nanomedicine, chemical and biological sensors, and so on. Recently, consumption of energy is continuously increasing, and consequently, fossil fuels are unremittingly exhausting. Therefore, it is highly essential to uncover sustainable energy sources with low-cost materials to replace the high-cost commercial materials/devices for the development of cost-effective alternative renewable energy devices. Nanomaterials offer a wide range of constituents that have the capability to replace high-cost materials with high efficiency and durability. It is very essential to understand the recently reported materials to explore effectively new (variety of) materials for sustainable development. In this regard, detailed information on various nanomaterials would be required. Of the various uses of nanomaterials, this book is specially focused on energy storage applications. A brief description of nanomaterials (introduction, synthesis, characterizations, and their energy storage applications) is discussed here.

In this book, Chapters 1 and 2 discuss the synthetic aspects of the nanomaterials and the mode of preparations. Chapters 3–10 describe different nanomaterials reported recently such as carbon-based nanomaterials, mesoporous carbon materials, halide perovskite, organic nanomaterials, and their energy storage applications. Chapter 11 deliberates on organic nanomaterials. In Chapter 12, specially designed materials for battery applications are discussed. Currently, a very large number of literature are available for energy storage applications using nanomaterials. Major challenges are also provided in some particular chapters. Thus, this book will be a unique one for beginners and researchers working in this field.

THE OBJECTIVES OF THIS BOOK

- To provide the historical, current, and future prospective of nanomaterials for energy storage applications.
- To provide a brief overview of recent advancements in nanomaterials for energy storage applications.
- To discuss the major challenges in the development of nanomaterials regarding energy storage applications.
- To highlight the major challenges in the development of advanced nanomaterials.
- To provide a special emphasis on synthesis and characterization of individual/group of nanomaterials and their composites for energy applications.

SCOPE OF THIS BOOK

This book provides brief descriptions of nanomaterials and their energy storage applications. It deals with the synthesis of nanomaterials and designing strategies. In addition, special emphasis is provided on solar cells; photoelectrochemical (PEC), photovoltaic, and third-generation solar cells are discussed in three chapters. Besides this, perovskite, mesoporous materials, metal oxide-based composites, carbon-based materials, and poly(amidoamine) dendrimers are discussed in separate chapters. A specially dedicated chapter on organic nanomaterials is also discussed. So, detailed descriptions of material syntheses and their energy storage applications make this book unique for the readers. This book would be helpful for the postgraduate and research students studying/working in the field of nanomaterials. In addition, it would be highly beneficial for young scientists, industries, and engineers who work in the development of novel materials in the field of energy storage devices.

SALIENT FEATURES OF THIS BOOK

- A detailed description of nanomaterials for energy storage applications with recent information is provided.
- Brief details of different energy storage materials, synthesis, and major challenges are highlighted.
- Individual chapters are dedicated to specific nanomaterials such as carbon materials, perovskite, metal oxide-based composites, and organic nanomaterials.
- Key figures (e.g. charts, flow diagrams, bar graphs) and tables (comparison with the literature) focusing on the detailed applications of particular nanomaterials are provided.
- Standard abbreviations, formulas, chapter summary, and key attractive figures are included in each chapter.
- A few references (predominantly review articles) are provided at the end of each chapter for further reading.
- Detailed reactions and schemes to increase the interest and better understanding of the readers are provided.

1 Synthesis of Mono-, Bi-, and Trimetallic Nanostructures by Anti-Galvanic Displacement Reaction

Ameer Farithkhan, N. S. K. Gowthaman, Wei Sea Chang, S. Abraham John, and Ramasamy Ramaraj

1.1 INTRODUCTION

Fabrication of atomically accurate metallic nanoparticles (MNPs) has been a primary aim for chemists and material scientists owing to their importance in achieving structure-composition synergism and ease of in-depth investigation of active-site chemistry (Qian et al., 2012; Boettcher et al., 2007; Parker et al., 2020; Jadzinsky et al., 2007; Verma et al., 2008; Lu and Chen, 2012; Liu et al., 2012). Although this long desire has been partially fulfilled *via* wet-chemical routes, achieving fully homogeneous atomic level nanoparticles is still a significant challenge (Tan and Zang, 2015). The discovery of galvanic reaction (GR) by Italian scientist Luigi Galvani (1737–1798) entails the reduction of a noble metal cation (of high oxidation potential) by a less noble metal (of high reduction potential) with an appropriate driving force originated from the different electrochemical potentials in the solution advantaged the molecular engineering of diverse nanomaterials in the nanoworld (da Silva et al., 2017; Skrabalak et al., 2008; Zhao et al., 2019). With a history of more than 240 years, the standard galvanic reduction also termed "galvanic replacement reaction (GRR)" has evidenced an immense engineering breakthrough in the designing and fabrication of advanced metal nanostructures (Chee et al., 2017; Xia et al., 2013; Cobley and Xia, 2010). However, at ultimate atomic level, the accurate control of the reaction process is still unachievable through GR, due to its rapid and spontaneous nature. A new spectacle, anti-galvanic reaction (AGR), which deals with the reduction of less noble metal ions by more noble metals (Figure 1.1), not only violates the classical galvanic principle but also promises a novel approach for tuning the structure, composition, and properties of metal nanostructures (Liu and Astruc, 2017). Since AGR defies the typical GR, it was thought to be a near-impossibility until the research

DOI: 10.1201/9781003481157-1

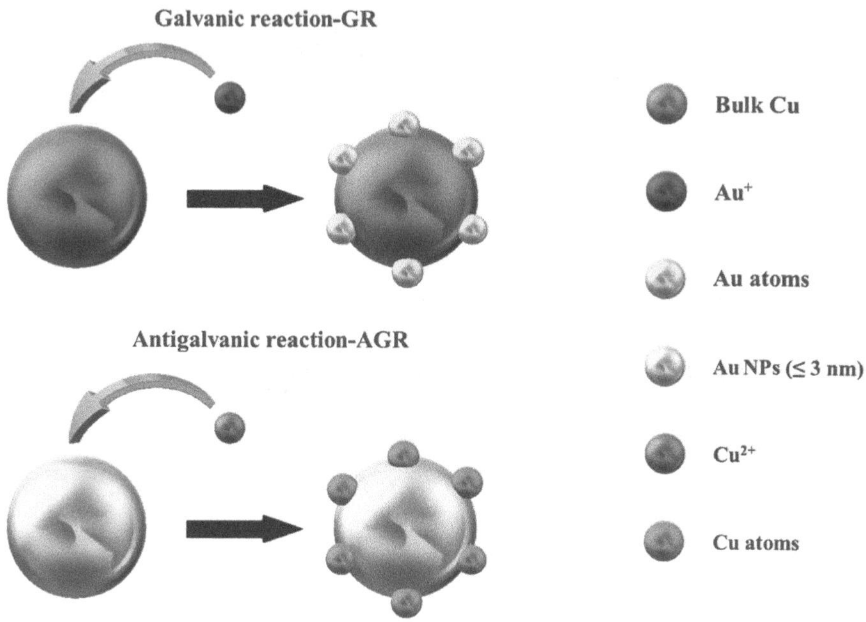

FIGURE 1.1 GR and AGR reactions.

endeavours of the past two decades confirmed the association of quantum confinement effect with ultrasmall nanoparticles (metal nanoclusters (MNCs)), when their size shrinks to 3 nm or less. Indeed, the AGR phenomenon was first evidenced by the reaction of core-shell $[Au_{25}(SC_2H_4Ph)_{18}]$ (L = SC_2H_4Ph = phenylethanethiolate) MNCs with highly reactive Ag^+ ions, which yielded the diverse $Au_{25}Ag(SC_2H_4Ph)_{18}$, $Au_{24}Ag_1(SC_2H_4Ph)_{18}$, and $Au_{23}Ag_2(SC_2H_4Ph)_{18}$ adducts utilizing the optimized $Ag^+/Au_{25}(SC_2H_4Ph)_{18}^-$ ratio, which was further confirmed by voltammetry and mass spectrometry (MS) (Choi et al., 2010). However, Wang and Wu first hypothesized that the thiolate ligands of $[Au_{25}(SC_2H_4Ph)_{18}]$ MNCs might engage in the AGR phenomenon by providing a negative charge to Au and Ag nanocrystals as the reaction proceeds (Wu, 2012). Eventually, this hypothesis was disclaimed and proved that AGR deviates from the stereotypical free-energy-based systems and relies upon the size effect of MNPs and not with the accord of reductive ligands (Wang et al., 2014). Later, the reaction of MNCs with the same metal ions was also found, which was termed as pseudo-anti-galvanic reaction (Fengming et al., 2020). Various researchers have been actively engaged in the advancement of this frontier field, which will be discussed below.

1.2 PRINCIPLE AND FACTORS INFLUENCING AGR

Since AGR becomes a ground-breaking research arena, its principle and influencing factors earn significant consideration. Obviously, AGR is a surprising behaviour of ultrasmall MNPs of size 3 nm or less, triggered by their quantum size effects.

Besides, this size-dependent redox behaviour of ultrasmall nanoparticles was also defined by equation (1.1) (Plieth et al., 1985).

$$\mu_d = \mu_b - (2M/zF\rho)\gamma/r \tag{1.1}$$

where, μ_d and μ_b, respectively, stand for reduction potential of MNPs and electrode, M stands for molar mass, Z is the charge transfer number, ρ is the specific mass, F is the Faraday's constant, and γ and r denote free energy and radius of particles employed. Clearly, according to the above equation, the reduction potential of MNPs diminishes as the size decreases and even drops off compared to the noble metals (Chaki et al., 2004). For instance, Jin and co-workers employed cyclic voltammetry (CV) to assess the redox behaviour of drop-dried glassy carbon electrode (GCE) fabricated with distinct MNPs.

The GCE with Au NPs (15–20 nm) displayed a typical redox peak at ~0.95 V, whereas Au nanoclusters (2 nm) demonstrated a dramatic decline of redox peak from 0.95 to ~0.45 V, which clearly validates that the small size of Au NCs facilitated its easier oxidation to Au(I) while reducing the Ag^+ ions (Figure 1.2) (Sun et al., 2014).

In addition, Wu synthesized stabilizer (ligand and surfactant)-free Au NPs and explored the unusual reduction of Ag^+ by Au NPs when its oxidation potential (~ −0.75 V) is

$$Au_{<3\text{ nm}} + Ag^+_{(aq)} \rightarrow Ag_{(s)} + Au^+(aq) \tag{1.2}$$

lower than the reduction potential of Ag^+ ions (~ −0.21 V), and it is unfavourable when the stereotypical standard redox electrode potentials is considered (Figure 1.3; Wang et al., 2014). Despite the fact that the product diversity of AGR is also influenced by other factors, Wu et al. showed the solvent effect on AGR by engaging the reaction between Ag^+ and $Au_{25}(SC_2H_4Ph)_{18}$ precursors and validated that the atomically monodispersed $Au_{25}Ag_2(SC_2H_4Ph)_{18}$ MNCs were only obtained in acetonitrile, while a polydispersed $Au_{24-x}Ag_x(SC_2H_4Ph)_{18}$ is produced under toluene or methane (Yao et al., 2015). Later, Wu and his team discovered dependency of AGR on ion-dose and ion-precursors because dissimilar species own unique redox potentials

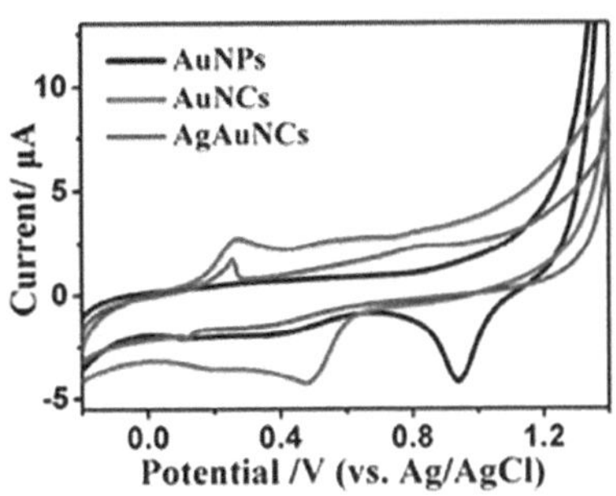

FIGURE 1.2 CVs of respective 11-MUA-AuNPs, 11-MUA-AuNCs, and 11-MUA-Ag/AuNCs (1:2) in 0.5 M H_2SO_4 on GCE. Reproduced with permission from Sun et al. (2014). Copyright 2014, The Royal Society of Chemistry.

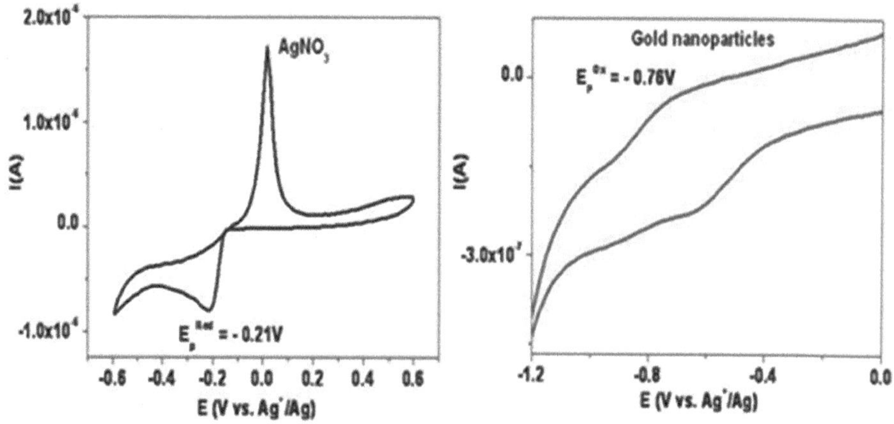

FIGURE 1.3 CVs of 4.5 mM $AgNO_3$ (left) and ablated gold nanoparticles (right) in 0.1 M Bu_4NPF_6/CH_3CN-THF. Reproduced with permission from Wang et al. (2014). Copyright 2014, Wiley-VCH.

and coordination style (Tian et al., 2015). As a consequence, they examined the $Au_{25}L_{18}$ reactions with four distinct kinds of Ag^+ precursors ($AgNO_3$, Ag-EDTA, Ag-PET, and Ag-DTZ), where ethylenediaminetetraacetic acid (EDTA) and dithizone (DTZ), respectively, and with Ag-DTZ in three various dosages, which leads to different products, were examined by thin-layer chromatography (TLC) technique (Figure 1.4).

The nature of the ligand plays a vital role in yielding the desired AGR product. Accordingly, Sahu and Prasad constructed a core-shell architecture of Ag(shell)@Au(core) NPs and examined whether the reduction of Ag^+ ions was carried out on the surface of Au or by ligand dodecyl amine (DDA) (Sahu and Prasad, 2015). For reference, dodecanethiol (DDT) was also used as a capping agent. When DDA-capped Au

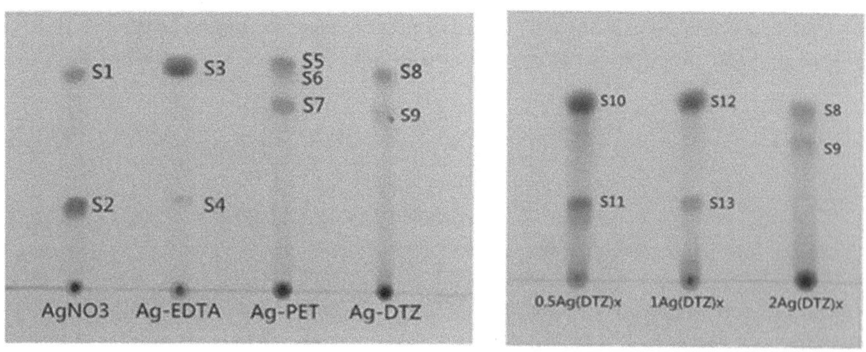

FIGURE 1.4 Photograph of the TLC plate, starting from left to right: $AgNO_3$, Ag–EDTA, Ag–PET, and Ag–DTZ were utilized as the precursor (left). Photograph of the TLC plate, from left to right: 0.5 eq., 1 eq., and 2 eq. Ag–DTZ were employed in the AGR reaction (right). Reproduced with permission from Tian et al. (2015). Copyright 2015, The Royal Society of Chemistry.

NPs were reacted, core-shell Ag@Au nanostructures were formed, but when DDT was used, individual Au and Ag parts were produced. In this case, DDT prevents the reaction between Au NPs and Ag^+ ions since it has a strong affinity towards them. However, the development of Ag@Au architecture was observed because of the detachment of DDA from Au NP surface under reflux conditions and its zero-binding affinity for Ag.

1.3 APPLICATIONS OF AGR

Evolving as an outstanding research streamline, the AGR phenomenon starts making its impact in diverse research areas. With a futuristic vision for constructing proficient electrochemical systems, the mechanism of AGR has been employed in the engineering and fabrication of monometallic NPs, dimetallic NPs, trimetallic NPs, metal ion sensing, and antioxidation.

1.3.1 Preparation of Monometallic NCs

1.3.1.1 Au-based Nanomaterials

In recent years, atomically accurate gold nanoparticles have attracted momentous attention for both basic research and industrial applications. The recent achievement in the fabrication of atomically well-defined nanoparticles has paved the way for new possibilities of studying the structural, catalytic, and stability characteristics of Au nanoparticles at a basic level (Dykman and Khlebtsov, 2011; Thangavel and Ramaraj, 2008; Liu et al., 2008; Hong et al., 2015; Zhang et al., 2014; Jin et al., 2010). Starting from the pseudo-anti-galvanic reaction, which involves the reaction of MNCs with the same metal ions, Wu and his research team demonstrated the formation of fluorescent $Au_{24}(SC_2H_4Ph)_{20}$ NC by the reaction of $Au_{25}(SC_2H_4Ph)_{18}$ with $Au(SC_2H_4Ph)$ ligand and displayed the diverse reaction products $Au_{25}(SC_2H_4Ph)_{18}$, $Au_{38}(SC_2H_4Ph)_{24}$, and $Au_{144}(SC_2H_4Ph)_{60}$ NCs under different wavelengths of the electromagnetic spectrum (Figure 1.5) (Yao et al., 2015). It is noteworthy that when $Au_{25}(SC_2H_4Ph)_{18}$ is

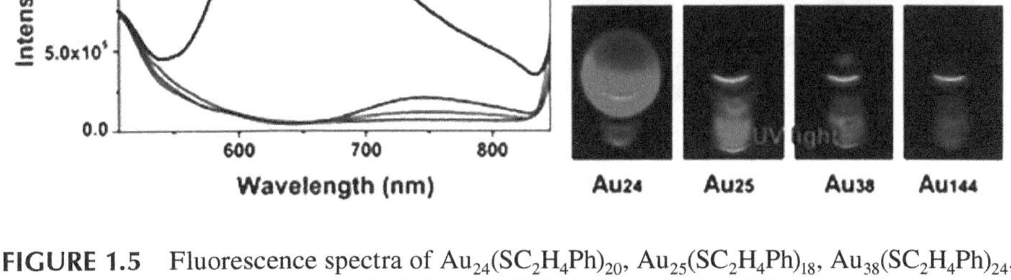

FIGURE 1.5 Fluorescence spectra of $Au_{24}(SC_2H_4Ph)_{20}$, $Au_{25}(SC_2H_4Ph)_{18}$, $Au_{38}(SC_2H_4Ph)_{24}$, and $Au_{144}(SC_2H_4Ph)_{60}$, respectively (left), and digital photographs of four nanoclusters under visible and 365 nm UV light irradiation (right). Reproduced with permission from Yao et al. (2015). Copyright 2015, The Royal Society of Chemistry.

coupled with the reference $AuCl_4^-$, owing to the greater oxidation capacity of $AuCl_4^-$ compared to the Au-PET complex, only large gold NPs (>3 nm) were produced. Since AGR is also reliant on ion-precursors, it takes part in governing the product diversity, stability, and catalytic efficiency of final outcomes. Accordingly, Wu et al. executed the transformation of Au_{23}(S-*c*$C_6H_{11})_{16}^-$ to Au_{28}(S-*c*-$C_6H_{11})_{20}$ and $Au_{20}Cd_4$(SH)(S-*c*-$C_6H_{11})_{19}$ NCs by the aid of $Cd(NO_3)_2$ and Cd(S-*c*-$C_6H_{11})_2$, respectively; then, they analysed the impact of ligands in detail (Zhu et al., 2018). As a result, $Cd(NO_3)_2$ acts as an oxidant in the transformation of Au_{23}(S-*c*$C_6H_{11})_{16}^-$ to Au_{28}(S-*c*-$C_6H_{11})_{20}$, whereas the ion-precursor Cd(S-*c*-$C_6H_{11})_2$ acts as both an oxidant and a doping reagent in the transformation of Au_{23}(S-*c*$C_6H_{11})_{16}^-$ to Au_{20}Cd4(SH)(S-*c*$C_6H_{11})_{19}$. Doping of Cd to the parent nanocluster improves the catalytic performance and robust stability of the product. Further, the same group prepared ultrafine $Au_{44}(SC_2H_4Ph)_{32}$ NC using the AGR method and witnessed that Cu^{2+} ions play a pivotal role in converting $Au_{25}(SC_2H_4Ph)_{18}^-$ anion into the cationic $Au_{25}(SC_2H_4Ph)_{18}^+$ and neutral $Au_{22}(SC_2H_4Ph)_{16}$ and thus the final product by dividing the entire reaction into four stages: oxidation, reduction, decomposition, and recombination (Li et al., 2015). The sequence of the aforementioned reactions is illustrated in Figure 1.6. Then, they investigated its catalytic efficiency for the reduction of 4-nitrophenol in detail, in which $Au_{44}(SC_2H_4Ph)_{32}$ NC exhibited momentous catalytic activity in both ambient and low temperatures.

1.3.2 Preparation of Bimetallic NCs

In contrast with monometallic NCs, bimetallic (or) alloy NCs encompassing two unique electron-rich metal atoms promise the enrichment of catalytic activity, stability, and selectivity, owing to their distinct engineering in geometric and electronic architecture (Mazhar et al., 2017; Roopan et al., 2014; Tiwari et al., 2020; Ren et al., 2019). Bimetallic NCs attract more attention as they exhibit surprising chemical conversion

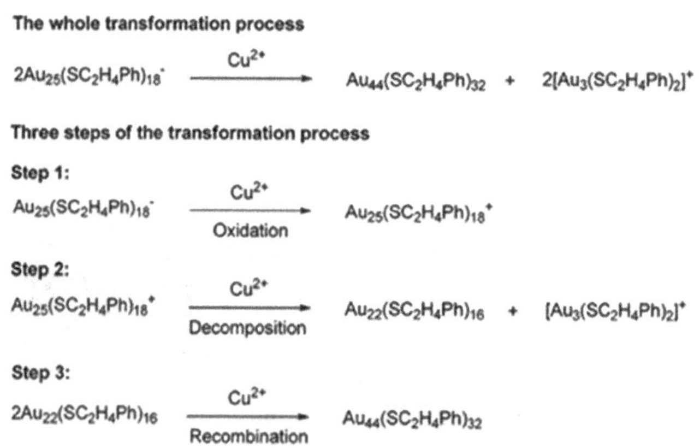

FIGURE 1.6 Evolutionary process from $Au_{25}(SC_2H_4Ph)_{18}^-$ to $Au_{44}(SC_2H_4Ph)_{32}$. Reproduced with permission from Li et al. (2015). Copyright 2015, The Royal Society of Chemistry.

kinetics, which surpass their monometallic counterparts (Kim et al., 2014; Wang et al., 2013). The diversity of alloying in AGR is one of its most intriguing features. Until now, there have been four alloying modes recorded: (a) Insertion of heteroatom while the mother NC remains intact, (b) replacement of atoms for mother NC by heteroatom without altering the overall structure, (c) heteroatoms replace metal atoms of mother NC with the transformation of its framework, and (d) structural change takes place; however, foreign atoms do not replace metal atoms of the parent NC. Incorporating these alloying modes, the development of various alloy NCs is discussed below.

1.3.2.1 Au-Ag Alloy NC

An interesting finding from Jin and co-workers revealed a shuttling mechanism of heteroatom into the parent $Au_{24}(PPh_3)_{10}(SC_2H_4Ph)_5Cl_2$ NC, which leads to the formation of $Au_{24}Ag_1(PPh_3)_{10}(SC_2H_4Ph)_5Cl_2$ (Wang et al., 2017). In addition to alloying, a foreign Ag atom pushes an Au atom into the pre-existing void in the parent nanocluster without altering its composition, yielding the desired product. Moreover, Jin et al. utilized the hollowing-refilling method to convert the obtained $Au_{24}Ag_1(PPh_3)_{10}(SC_2H_4Ph)_5Cl_2$ into $Au_{23}Ag_2(PPh_3)_{10}(SC_2H_4Ph)_5Cl_2$ NC. In a previous study, Ag atoms squeezed into the parent Au NC. Wu et al. decorated $Au_{25}(SC_2H_4Ph)_{18}$ NC by just depositing Ag atoms on its surface without sacrificing its structural scaffold.

Supported by the UV-vis, inductively coupled plasma-atomic emission (ICP-AES)-MS and theoretical studies, a simple mix of $Au_{25}(SC_2H_4Ph)_{18}$ with Ag^+ ions under acetonitrile yields $Au_{25}Ag_2(SC_2H_4Ph)_{18}$ NC (yield 89%), in which the doped Ag atoms did not replace any Au atoms. However, $Au_{25}Ag_2(SC_2H_4Ph)_{18}$ was discovered to have unique optical functionalities, high ambient stability, and outstanding catalytic activity. In the presence of 0.2 mol% of $Au_{25}Ag_2(SC_2H_4Ph)_{18}$, the hydrolysis of 1,3-diphenylprop-2-ynyl acetate achieved the intended product with a 52% yield, while other related catalysts suffered. Recently, Jin et al. reported a novel technique for the fabrication of $Au_{25-x}Ag_x(Sc\text{-}C_6H_{11})_{18}$-NCs heavily doped by Ag atoms (X = ~19), in which heteroatoms not only replaced parent metal atoms but also transformed the framework of $Au_{23}(SC_6H_{11})_{16}^-$ (Li et al., 2016). Further, this structural transformation of $Au_{25-x}Ag_x(ScC_6H_{11})_{16}^-$ NC facilitated the introduction of Ag atoms in the staple motifs of NC, which would otherwise be impossible by typical ways.

1.3.2.2 Au-Cd Alloy NC

In the continuation of aforesaid research endeavours, a unique Cd-based Au alloy NC, $Au_{26}Cd_5$ ($[Au_{13}Cd_2(PPh_3)_6(SC_2H_4Ph)_6(NO_3)_2]_2Cd(NO_3)_4$), was prepared by peeling and doping (Figure 1.7) of $Au_{25}(Au_{25}(SCH_2CH_2Ph)_{18})$ NC to improve the catalytic performance (Li et al., 2017). The obtained $Au_{26}Cd_5$ NC is an assembly of two $Au_{13}Cd_2$ icosahedrons, with the two Cd dopants coordinated to the Au_3 atoms of individual Au_{13} moieties, as determined by single-crystal X-ray crystallography (SCXC). Despite the fact that Au_{25} has no catalytic activity, the synergistic effect of Cd and Au atoms in $Au_{26}Cd_5$ NC resulted in outstanding catalytic performance towards the A^3-coupling reaction.

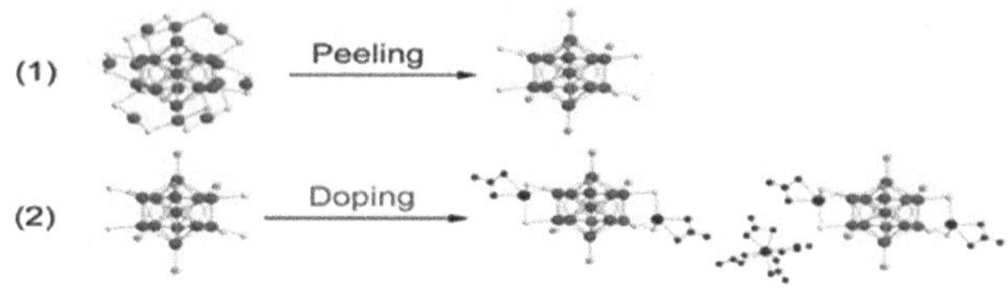

FIGURE 1.7 Structural transformation from Au_{25} to $Au_{26}Cd_5$. Reproduced with permission from Li et al. (2017). Copyright 2017, Wiley-VCH.

1.3.2.3 Au-Hg Alloy NC

Wu et al. adopted the alloying mode, which entails replacing metal atoms in parent NC without creating structural changes. An atomically monodispersed $Au_{24}Hg_1(SC_2H_4Ph)_{18}$ NC was fabricated from the $Au_{25}(SC_2H_4Ph)_{18}$ NC by replacing the outer-shell Au atom by invading the Hg atom. Supported by the thermogravimetric analysis (TGA) and matrix-assisted laser desorption/ionization time-of-flight mass spectrometry (MALDI-TOF-MS), instead of replacing core or inner-shell Au atom, Hg atoms substituted themselves on outer-shell by declining HOMO/LUMO energies of the parent NC (Liao et al., 2015).

1.3.3 Preparation of Trimetallic NCs

1.3.3.1 Au-Ag-M (M = Pt or Pd) Alloy NC

Trimetallic NPs have maintained a higher level of catalytic performance and selectivity than monometallic and bimetallic nanoparticles (BNPs) due to the innovative physicochemical characteristics (e.g., catalytic, optical, magnetic, and electronic) generated from the synergistic effects of their monometallic counterparts (Nasrollahzadeh et al., 2020; Park and Ahn, 2020; Nie et al., 2020; Cai et al., 2017). Negishi et al. created Ag- and Pt/Pd-doped trimetallic, $Au_{20}Ag_4Pd(SC_2H_4Ph)_{18}$ and $Au_{20}Ag_4Pt(SC_2H_4Ph)_{18}$ NCs with finely adjusted chemical composition and metallic positions (Hossain et al., 2018). Initially, while the reaction between $[Au_{24}Pd(SC_2H_4Ph)_{18}]^0$/$[Au_{24}Pt(SC_2H_4Ph)_{18}]^0$ with Ag-SC_2H_4Ph proceeds, the replacement of Au by Ag progressed with up to five atoms after six hours. Then the strength of the original clusters was dramatically lowered when the reaction time was increased to 12 h, yielding $Au_{20}Ag_4Pd(SC_2H_4Ph)_{18}$/$Au_{20}Ag_4Pt(SC_2H_4Ph)_{18}$ NCs. The incorporation of Pd or Pt as the core atom triggered the deformation of the geometric architecture of the cluster, which facilitated positioning of the Ag atom at the cluster surface and fine tunability of the chemical composition.

1.3.3.2 Au-Ag-Hg Alloy NC

Adopting the bi-anti-galvanic method, Wu et al. prepared $Au_{16.8}Ag_{7.2}Hg_1(SC_2H_4Ph)_{18}$ NC by doping Ag and Hg atoms into $Au_{25}(PET)_{18}$ (Yan et al., 2016). SCXC and theoretical studies demonstrated that Hg and Ag have taken the position of $Au_{25}(SC_2H_4Ph)_{18}$'s outer-shell and inner-shell Au atoms, respectively. The first oxidation/reduction potential shift of $Au_{25}(SC_2H_4Ph)_{18}$ after being doped with both Hg and Ag was investigated using differential pulse voltammetry (DPV). Surprisingly, Hg doping reduces the electrochemical gap of $Au_{25}(SC_2H_4Ph)_{18}$ by 0.12 V; however, Ag doping enhances the electrochemical gap of Au_{25} by 0.14 V, bringing the final bandgap of product NC to 1.64 V, which is comparable to the parent NC's 1.63 V. Moreover, the coexistence of synergism between trimetallic atoms was also confirmed by the proficient reduction of 4-nitrobenzene.

1.3.4 Metal Ion Sensing by the AGR

Physical chemistry has long been fascinated by the optical properties of MNPs, dating back to Faraday's research of colloidal gold in the mid-nineteenth century (Goia and Matijević, 1999). An optical probe is an essential component of a standard optical sensor, and the optical probes chosen can affect the sensor's sensing characteristics. Even though the quantum dots and organic dyes have all been exploited as optical probes in sensor development over the last couple of decades, owing to the confined surface plasmon with resonance wavelength in the visible range, the optical characteristics of MNPs play a pivotal role in optical sensing platform (Choi et al., 2016; Resch-Genger et al., 2008; Li et al., 2015; Yuan et al., 2016). MNCs have already emerged as a new era of intriguing optical sensor probes ascribed to their tunable fluorescence, outstanding photostability, and ultrafine size. Zhu et al. detected Cu^{2+} ions using $[Ag_{62}S_{13}(SBut)_{32}]^{4+}$ (SBut = tert-butyl thiolate) NC as a fluorescent probe through AGR mechanism. When a Cu^{2+} salt was added to a solution of $[Ag_{62}S_{13}(SBut)_{32}]^{4+}$, the absorption peak at 543 nm progressively migrated to 534 nm, resulting in a colour change from red to yellow (Wang et al., 2014). The mechanism behind this fluorescent quenching was studied via MALDI-TOF MS by analysing the molecular weight variations of $[Ag_{62}S_{13}(SBut)_{32}]^{4+}$ NCs upon addition of Cu^{2+} ions. The shift of broad peak to the higher mass after adding one equivalent of Cu^{2+} indicates that Cu is attaching to the silver core rather than replacing the Ag atoms in the NCs. Interference investigation revealed that foreign metal ions other than Cu^{2+} (Co^{2+}, Cd^{2+}, Mn^{2+}, Hg^{2+}, Fe^{2+}, and Ni^{2+}) failed to quench the fluorescent $[Ag_{62}S_{13}(SBut)_{32}]^{4+}$, confirming the great selectivity of $[Ag_{62}S_{13}(SBut)_{32}]^{4+}$ for Cu^{2+}. Similarly, Wu and co-workers utilized the AGR mechanism in sensing Ag^+ ions by $Au_{25}(SG)_{18}$ (SG = glutathionate) NC probe (Wu et al., 2012). Oxidation of $Au_{25}(SG)_{18}$ by Ag^+ ions increases the intensity of photoluminescence, which is primarily used to detect the silver ion quantitatively, with a detection limit of about 20 ppb. The pronounced selectivity of $Au_{25}(SG)_{18}$ towards Ag^+ ions was confirmed by the interference analysis, which specified that the other metal ions encompassing Pb^{2+}, Cd^{2+}, Hg^{2+}, Cu^{2+}, Zn^{2+}, Ni^{2+}, Co^{3+}, Tb^{3+}, Eu^{3+}, Pd^{2+}, Fe^{2+}, Fe^{3+}, Mg^{2+}, K^+, Na^+, Ca^{2+}, Cr^{3+}, Mn^{2+}, and Au^{3+} ion only quenched the photoluminescence of $Au_{25}(SG)_{18}$ to the minor level.

For the first time, the utilization of Ag NCs templated with ssDNA as the ultrasensitive nano-biosensor for the detection of Cu^{2+} ions was demonstrated by Liu et al. (2013). The ssDNA-Ag NCs reduce Cu^{2+} to Cu^0 when copper (II) ions are introduced, resulting in non-fluorescent Ag/Cu alloy nanoparticles and an anti-galvanic replacement reaction (Figure 1.8). Because the size of the DNA-Ag NC undergoes progressive expansion after reacting with Cu^{2+} ions through the AGR, it was discovered that 5×10^{-9} to 7.5×10^{-7} M Cu^{2+} ions are readily detected by resonant light scattering (RLS) with a detection limit of 2×10^{-9} M.

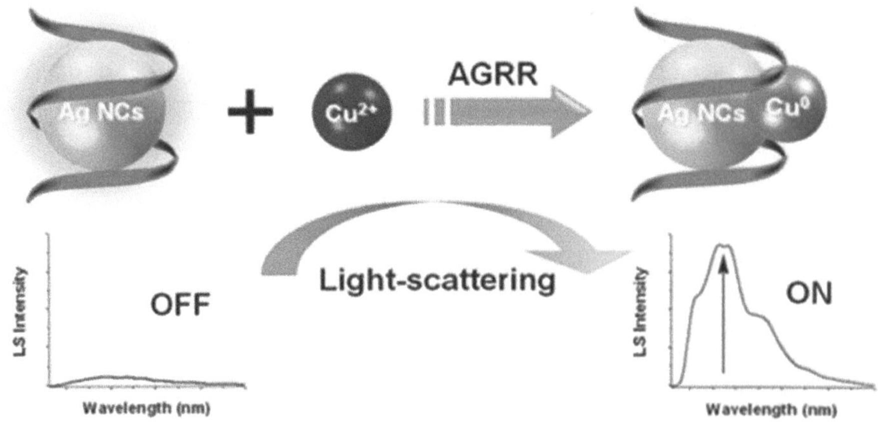

FIGURE 1.8 Schematic illustration of the anti-galvanic replacement reaction of DNA–Ag NCs monitored by the light-scattering (LS) technique. Reproduced with permission from Liu et al. (2013). Copyright 2013, The Royal Society of Chemistry.

Wu and colleagues developed Ag_{30}(captopril)$_{18}$ NC as a novel sensor for detecting Hg^{2+} levels through the AGR, which has excellent solubility in water and different polar organic solvents (Xia et al., 2015). The as-prepared Ag_{30}(captopril)$_{18}$ colorimetric probe displayed elevated selectivity and sensitivity towards Hg^{2+} ion, indicating that even low levels (~6 ppb) can be sensed in environmental samples. After introducing Hg^{2+}, even in tiny amounts from lake water samples, there is an instantaneous drop in the absorbance of Ag_{30}(captopril)$_{18}$ in the UV-vis-NIR spectrum; however, the addition of other metal ions such as Cu^{2+} also illustrates the analogous reaction. Because the oxidation-reduction reaction between Ag_{30}(captopril)$_{18}$ and Cu^{2+} is slow at low temperatures, the interference of Hg^{2+} might be eliminated by allowing the reaction to take place at the aforesaid temperature.

1.3.5 Antioxidation of Cu(0)

CuNPs catalyse a wide range of reactions ascribed to their low cost, environmental pleasantness, elevated efficiency, and plentiful availability (Ponce et al., 2005; Gawande et al., 2016). In many cases, the air oxidation of CuNPs reduces their catalytic activity (Gan et al., 2018). As a result, the protection of air oxidation of Cu(0) atoms is gaining attention in material science and catalysis. Astruc and his team exploited the AGR technique for realizing the aforementioned purpose (Liu et al., 2017). They investigated the antioxidation of Cu(0) by fabricating Au/Cu and Ag/Cu BNPs and discovered that the Ag/Cu BNPs are unstable and confirmed by the rapid colour change of the solution from orange to dark green (Cu_2O) (Figure 1.9) after 24 h. However, the Au/Cu alloy NPs demonstrated concrete stability even after two months. In general, Au is more stable than silver; thus, it is difficult to oxidize Cu in Au/Cu alloy than Ag/Cu NPs.

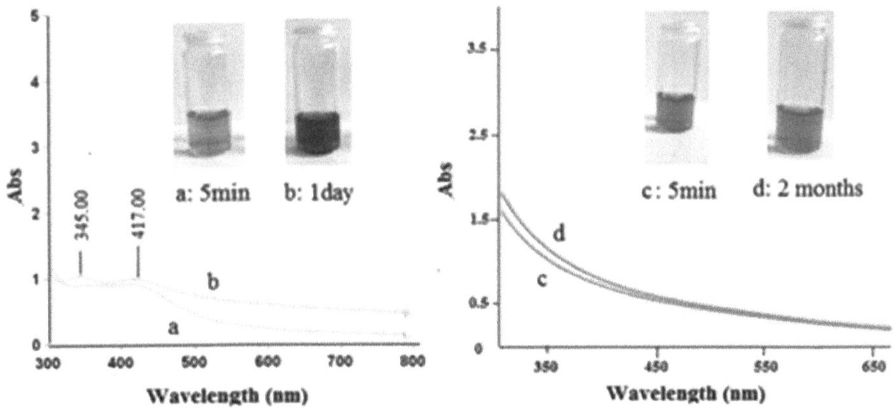

FIGURE 1.9 Photographs and UV-vis spectra of BNPs: (a) Ag/Cu BNPs after 5 min; (b) Ag/Cu BNPs after one day in air; (c) Au/Cu BNPs after 5 min; and (d) Au/Cu BNPs after two months in air. Reproduced with permission from Liu et al. (2017). Copyright 2017, The Royal Society of Chemistry.

1.4 CONCLUSIONS

This chapter details the mechanistic understanding, influencing factors, research endeavours, and advanced applications of AGR in MNC synthesis, metal ion sensing, and antioxidation. In addition, substantial experiments carried out to understand the fundamental aspects of the AGR in various research fields are highlighted in this chapter. AGR deals with the reduction of less noble metal ions by more noble metals, which not only defies the standard galvanic principle but also offers a revolutionary method for modifying the structure, composition, and characteristics of metal nanostructures. Through AGR, the creation of atomically accurate and monodispersed MNCs with distinct chemical compositions and structures is realized. Although significant progress has been made since the general concept of AGR was proposed in 2012, AGR research is still in its early stages, and more work has to be done in the future.

ACKNOWLEDGEMENTS

We thank the Department of Science and Technology (DST/IDP/MED/04/2017), New Delhi, India, for the financial support. Ramaraj acknowledges the financial support from the Raja Ramanna Fellowship Scheme, Department of Atomic Energy, India.

REFERENCES

Boettcher, S. W., Strandwitz, N. C., Schierhorn, M., Lock, N., Lonergan, M. C., and Stucky, G. D. (2007). Tunable electronic interfaces between bulk semiconductors and ligand-stabilized nanoparticle assemblies, *Nat. Mater.* 6, 592–596.

Cai, X.-L., C. H., Liu, J., Lu, Y., Zhong, Y. N., Nie, K. Q., Xu, J. L., Gao, X., Sun, X.-H., and Wang, S.-D. (2017). Synergistic effects in CNTs-PdAu/Pt trimetallic nanoparticles with high electrocatalytic activity and stability, *Nano-Micro Lett.* 9, 1–10.

Chaki, N. K., Sharma, J., Mandle, A. B., Mulla, I. S., Pasricha, R., and Vijayamohanan, K. (2004). Size dependent redox behaviour of monolayer protected silver nanoparticles (2-7 nm) in aqueous medium, *Phys. Chem. Chem. Phys.* 6, 1304–1309.

Chee, S. W., Tan, S. F., Baraissov, Z., Bosman, M., and Mirsaidov, U. (2017). Direct observation of the nanoscale Kirkendall effect during galvanic replacement reactions, *Nat. Commun.* 8, 1–8.

Choi, J., Shin, D.-M., Song, H., Lee, D., and Kim, K. (2016). Current achievements of nanoparticle applications in developing optical sensing and imaging techniques, *Nano Converg.* 3, 1–13.

Choi, J.-P., Fields-Zinna, C. A., Stiles, R. L., Balasubramanian, R., Douglas, A. D., Crowe, M. C., and Murray, R. W. (2010). Reactivity of $[Au_{25} (SCH_2CH_2Ph)_{18}]-$ nanoparticles with metal ions, *J. Phys. Chem. C*, 114, 15890–15896.

Cobley, C. M., and Xia, Y. (2010). Engineering the properties of metal nanostructures via galvanic replacement reactions, *Mater. Sci. Eng. R Rep.* 70, 44–62.

da Silva, A. G. M., Rodrigues, T. S., Haigh, S. J., and Camargo, P. H. C. (2017). Galvanic replacement reaction: Recent developments for engineering metal nanostructures towards catalytic applications, *Chem. Commun.* 53, 7135–7148.

Dykman, L. A., and Khlebtsov, N. G. (2011). Gold nanoparticles in biology and medicine: Recent advances and prospects, *Acta Naturae* 3, 34–55.

Fengming, J., Hongwei, D., Yan, Z., Shengli, Z., Lingwen, L., Nan, Y., and Wanmiao, G. (2020). Module replacement of gold nanoparticles by a pseudo-AGR process, *Acta Chim. Sin.* 78, 407–411.

Gan, Z., Xia, N., and Wu, Z. (2018). Discovery, mechanism, and application of antigalvanic reaction, *Acc. Chem. Res.* 51, 2774–2783.

Gawande, M. B., Goswami, A., Felpin, F.-X., Asefa, T., Huang, X., Silva, R., Zou, X., Zboril, R., and Varma, R. S. (2016). Cu and Cu-based nanoparticles: Synthesis and applications in catalysis, *Chem. Rev.* 116, 3722–3811.

Goia, D., and Matijević, E. (1999). Tailoring the particle size of monodispersed colloidal gold, *Colloids Surf. A: Physicochem. Eng. Asp.* 146, 139–152.

Hong, X., Tan, C., Chen, J., Xu, Z., and Zhang, H. (2015). Synthesis, properties and applications of one-and two-dimensional gold nanostructures, *Nano Res.* 8, 40–55.

Hossain, S., Ono, T., Yoshioka, M., Hu, G., Hosoi, M., Chen, Z., and Nair, L. V. (2018). Thiolate-protected trimetallic $Au_{\sim 20}Ag_{\sim 4}Pd$ and $Au_{\sim 20}Ag_{\sim 4}Pt$ alloy clusters with controlled chemical composition and metal positions, *J. Phys. Chem. Lett.* 9, 2590–2594.

Jadzinsky, P. D., Calero, G., Ackerson, C. J., Bushnell, D. A., and Kornberg, R. D. (2007). Structure of a thiol monolayer-protected gold nanoparticle at 1.1 Å resolution, *Science* 318, 430–433.

Jin, R., Qian, H., Wu, Z., Zhu, Y., Zhu, M., Mohanty, A., and Garg, N. (2010). Size focusing: A methodology for synthesizing atomically precise gold nanoclusters, *J. Phys. Chem.* Lett. 1, 2903–2910.

Kim, D., Resasco, J., Yu, Y., Asiri, A. M., and Yang, P. (2014). Synergistic geometric and electronic effects for electrochemical reduction of carbon dioxide using gold-copper bimetallic nanoparticles, *Nat. Commun.* 5, 1–8.

Li, M., Cushing, S. K., and Wu, N. (2015). Plasmon-enhanced optical sensors: A review, *Analyst* 140, 386–406.

Li, M.-B., Tian, S.-K., and Wu, Z. (2017). Improving the catalytic activity of Au_{25} nanocluster by peeling and doping, *Chin. J. Chem.* 35, 567–571.

Li, M.-B., Tian, S.-K., Wu, Z., and Jin, R. (2015). Cu^{2+} induced formation of Au_{44} $(SC_2H_4Ph)32$ and its high catalytic activity for the reduction of 4-nitrophenol at low temperature, *Chem. Commun.* 51, 4433–4436.

Li, Q., Wang, S., Kirschbaum, K., Lambright, K. J., Das, A., and Jin, R. (2016). Heavily doped $Au_{25-x}Ag_x(SC_6H_{11})_{18-}$ nanoclusters: Silver goes from the core to the surface, *Chem. Commun.* 52, 5194–5197.
Liao, L., Zhou, S., Dai, Y., Liu, L., Yao, C., Fu, C., Yang, J., and Wu, Z. (2015). Mono-mercury doping of Au_{25} and the HOMO/LUMO energies evaluation employing differential pulse voltammetry, *J. Am. Chem. Soc.* 137, 9511–9514.
Liu, G., Feng, D.-Q., Zheng, W., Chen, T., and Li, D. (2013). An anti-galvanic replacement reaction of DNA templated silver nanoclusters monitored by the light-scattering technique, *Chem. Commun.* 49, 7941–7943.
Liu, H., Tian, Y., and Xia, P. (2008). Pyramidal, rodlike, spherical gold nanostructures for direct electron transfer of copper, zinc-superoxide dismutase: Application to superoxide anion biosensors, *Langmuir* 24, 6359–6366.
Liu, X., and Astruc, D. (2017). From galvanic to anti-galvanic synthesis of bimetallic nanoparticles and applications in catalysis, sensing, and materials science, *Adv. Mater.* 29, 1605305.
Liu, X., Ruiz, J., and Astruc, D. (2017). Prevention of aerobic oxidation of copper nanoparticles by anti-galvanic alloying: Gold versus silver, *Chem. Commun.* 53, 11134–11137.
Liu, X., Yu, M., Kim, H., Mameli, M., and Stellacci, F. (2012). Determination of monolayer-protected gold nanoparticle ligand-shell morphology using NMR, *Nat. Commun.* 3, 1–9.
Lu, Y., and Chen, Y. (2012). Sub-nanometre sized metal clusters: From synthetic challenges to the unique property discoveries, *Chem. Soc. Rev.* 41, 3594–3623.
Mazhar, T., Shrivastava, V., and Tomar, R. S. (2017). Green synthesis of bimetallic nanoparticles and its applications: A review, *J. Pharm. Sci. Res.* 9, 102–110
Nasrollahzadeh, M., Sajjadi, M., Iravani, S., and Varma, R. S. (2020). Trimetallic nanoparticles: Greener synthesis and their applications, *Nanomaterials* 10, 1784–1803.
Nie, F., Ga, L., Ai, J., and Wang, Y. (2020). Trimetallic PdCuAu nanoparticles for temperature sensing and fluorescence detection of H_2O_2 and glucose, *Front. Chem.* 8, 244–253.
Park, J. H., and Ahn, H. S. (2020). Electrochemical synthesis of multimetallic nanoparticles and their application in alkaline oxygen reduction catalysis, *Appl. Surf. Sci.* 504, 144517–14424.
Parker, J. F., Fields-Zinna, C. A., and Murray, R. W. (2020). The story of a monodisperse gold nanoparticle: $Au_{25}L_{18}$, *Acc. Chem. Res.* 43, 1289–1296.
Plieth, W. J. (1985). The work function of small metal particles and its relation to electrochemical properties, *Surf. Sci.* 156, 530–535.
Ponce, A. A., and Klabunde, K. J. (2005). Chemical and catalytic activity of copper nanoparticles prepared via metal vapor synthesis, *J. Mol. Catal.* 225, 1–6.
Qian, H., Zhu, M., Wu, Z., and Jin, R. (2012). Quantum sized gold nanoclusters with atomic precision, *Acc. Chem. Res.* 45, 1470–1479.
Ren, W., Zang, W., Zhang, H., Bian, J., Chen, Z., Guan, C., and Cheng, C. (2019). PtCo bimetallic nanoparticles encapsulated in N-doped carbon nanorod arrays for efficient electrocatalysis, *Carbon* 142, 206–216.
Resch-Genger, U., Grabolle, M., Cavaliere-Jaricot, S., Nitschke, R., and Nann, T. (2008). Quantum dots versus organic dyes as fluorescent labels, *Nat. Methods.* 5, 763–775.
Roopan, S. M., Surendra, T. V., Elango, G., and Kumar, S. H. H. (2014). Biosynthetic trends and future aspects of bimetallic nanoparticles and its medicinal applications, *Appl. Microbiol. Biotechnol.* 98, 5289–5300.
Sahu, P., and Prasad, B. L. V. (2015). Preparation of Ag (Shell)-Au (core) nanoparticles by anti-galvanic reactions: Are capping agents the "real heroes" of reduction?" *Colloids Surf. A: Physicochem. Eng. Asp.* 478, 30–35.
Skrabalak, S. E., Chen, J., Sun, Y., Lu, X., Au, Cobley, C. M., and Xia, Y. (2008). Gold nanocages: Synthesis, properties, and applications, *Acc. Chem. Res.* 41, 1587–1595.

Sun, J., Wu, H., and Jin, Y. (2014). Synthesis of thiolated Ag/Au bimetallic nanoclusters exhibiting an anti-galvanic reduction mechanism and composition-dependent fluorescence, *Nanoscale* 6, 5449–5457.

Tan, C., and Zhang, H. (2015). Wet-chemical synthesis and applications of non-layer structured two-dimensional nanomaterials, *Nat. Commun.* 6, 1–13.

Thangavel, S., and Ramaraj, R. (2008). Polymer membrane stabilized gold nanostructures modified electrode and its application in nitric oxide detection, *J. Phys. Chem. C*, 112, 19825–19830.

Tian, S., Yao, C., Liao, L., Xia, N., and Wu, Z. (2015). Ion-precursor and ion-dose dependent anti-galvanic reduction, *Chem. Commun.* 51, 11773–11776.

Tiwari, P., Das, G. M., and Dantham, V. R. (2020). Optical properties of au-ag bimetallic nanoparticles of different shapes for making efficient bimetallic-photonic whispering gallery mode hybrid microresonators, *Plasmonics* 15, 1251–1260.

Verma, A., Uzun, O., Hu, Y., Hu, Y., Han, H.-S., Watson, N., Chen, S. Irvine, D. J., and Stellacci, F. (2008). Surface-structure-regulated cell-membrane penetration by monolayer-protected nanoparticles, *Nat. Mater.* 7, 588–595.

Wang, M., Wu, Z., Chu, Z., Yang, J., and Yao, C. (2014). Chemico-physical synthesis of surfactant-and ligand-free gold nanoparticles and their anti-galvanic reduction property, *Chem. Asian J.* 9, 1006–1010.

Wang, R., Wu, Z., Chen, C., Qin, Z., Zhu, H., Wang, G., and Wang, H. (2013). Graphene-supported Au-Pd bimetallic nanoparticles with excellent catalytic performance in selective oxidation of methanol to methyl formate, *Chem. Commun.* 49, 8250–8252.

Wang, S., Abroshan, H., Liu, C., Luo, T.-Y., Zhu, M., Kim, H. J., Rosi, N. L., and Jin, R. (2017). Shuttling single metal atom into and out of a metal nanoparticle, *Nat. Commun.* 8, 1–7.

Wu, Z. (2012). Anti-galvanic reduction of thiolate-protected gold and silver nanoparticles, *Angew. Chem. Int. Ed.* 51, 2934–2938.

Wu, Z., Wang, M., Yang, J., Zheng, X., Cai, W., Meng, G., Qian, H., Wang, H., and Jin, R. (2012). Well-defined nanoclusters as fluorescent nanosensors: A case study on $Au_{25}(SG)_{18}$, *Small* 8, 2028–2035.

Xia, N., Yang, J., and Wu, Z. (2015). Fast, high-yield synthesis of amphiphilic Ag nanoclusters and the sensing of Hg^{2+} in environmental samples, *Nanoscale* 7, 10013–10020.

Xia, X., Wang, Y., Ruditskiy, A., and Xia, Y. (2013). 25th Anniversary Article: Galvanic replacement: A simple and versatile route to hollow nanostructures with tunable and well-controlled properties, *Adv. Mater.* 25, 6313–6333.

Yan, N., Liao, L., Yuan, J., Lin, Y.-J., Weng, L.-H., Yang, J., and Wu, Z. (2016). Bimetal doping in nanoclusters: Synergistic or counteractive? *Chem. Mater.* 28, 8240–8247.

Yao, C., Chen, J., Li, M.-B., Liu, L., Yang, J., and Wu, Z. (2015). Adding two active silver atoms on Au_{25} nanoparticle, *Nano Lett.* 15, 1281–1287.

Yao, C., Tian, S., Liao, L., Liu, X., Xia, N., Yan, N., Gan, Z., and Wu, Z. (2015). Synthesis of fluorescent phenylethanethiolated gold nanoclusters via pseudo-AGR method, *Nanoscale* 7, 16200–16203.

Yuan, Z., Hu, C.-C., Chang, H.-T., and Lu, C. (2016). Gold nanoparticles as sensitive optical probes, *Analyst* 141, 1611–1626.

Zhang, Y., Chu, W., Foroushani, A. D., Wang, H., Li, D., Liu, J., Barrow, C. J., Wang, X., and Yang, W. (2014). New gold nanostructures for sensor applications: A review, *Materials* 7, 5169–5201.

Zhao, Y., Sun, Y., Jiang, Y., Song, S., Zhao, T., Zhao, Y., Wang, X., Li, B., Yang, B., and Lin, Q. (2019). Fluorescent probe gold nanodots to quick detect Cr (VI) via oxidoreduction quenching process, *Sci. China Chem.* 62, 133–141.

Zhu, M., Wang, P., Yan, N., Chai, X., He, X., Zhao, Y., and Xia, N. (2018). The fourth alloying mode by way of anti-galvanic reaction, *Angew. Chem. Int. Ed.* 57, 4500–4504.

2 Nanomaterials in Electrochemical Synthesis

Alok Kumar Chaudhari and V. B. Singh

2.1 INTRODUCTION

Electrolytic synthesis can be considered to be originated with the production of electricity through chemical means by Alessandro Volta in 1799, followed by its application for energy storage as the first modern electrical battery in 1800. Electrolysis was mainly considered as merely scientific curiosity until about 1839 when the advantages of electrodeposition for the reproduction of surfaces and objects were announced by several researchers at about the same time. During electrolysis, the electric current passes through the electrolyte solution, causing the reduction of an electroactive species at the cathode and oxidation at the anode. This interconversion between electrical energy and chemical energy accomplished by the flow of ionic current in an electrolyte solution between two electrodes produces a net chemical change using electrical energy. The use of electrochemistry for effective transformations has some obvious advantages such as low cost, high level of control, least production of waste and no contaminating reagents (Schotten et al., 2020). Presently, electrochemical synthesis has several exciting applications in the areas of inorganic and organic chemistry. It is not only helpful for mechanistic studies but also helpful for precise control of oxidation/reduction process, electron transfer stoichiometry of organometallic compounds, etc. As a result, electrochemical synthesis is extremely useful in the production of wonderful oxidation states in inorganic compounds, highly specific organic compounds and species enabled to initiate polymerization process (electroinitiated polymerization).

Although large-scale productions still take place through chemical synthesis, for energy storage applications where purity of the material has prime importance, electrochemical synthetic routes are favored. Advanced electrochemical synthesis methods have recently experienced growth because they have been found useful for a wide range of applications where conventional methods of electrochemical synthesis are inadequate (Schotten et al., 2020). It not only offers several advantages over the conventional chemical oxidizing and reducing agents but also has been considered as a green chemical process (Chen et al., 2020) (Figure 2.1).

Nanomaterials are not only different from conventional materials on the basis of their characteristic dimensions, but they are also able to show entirely different properties in addition to the bulk properties. When the particle size of a material reduces to nanoscale, the fraction of surface atoms not only increases dramatically but also reduces the traveling space from the core to the surface for all the wandering ions. This variation in traveling distance has excellent applicability for lithium-ion

DOI: 10.1201/9781003481157-2

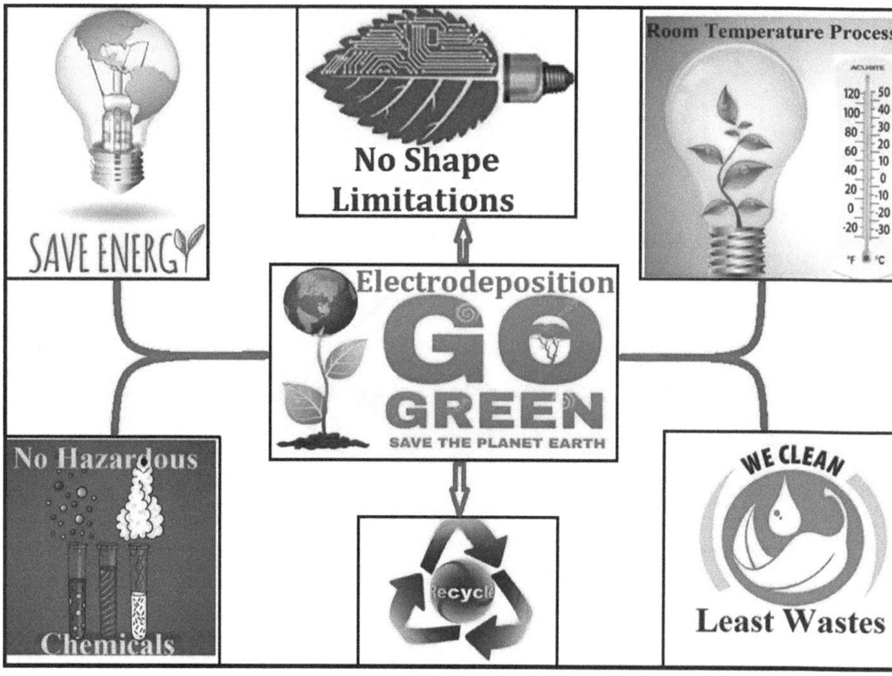

FIGURE 2.1 Schematic representation of advantages of electrochemical synthesis.

systems in which the diffusion rate of lithium ions in the electrodes controls the charge and discharge rates. Quantum confinement effect results in increased bandgap when the dimensions of a semiconductor are reduced below a certain size. The bandgap of the material can be simply tuned by varying the size of the material, so the optical properties can also be controlled.

Because of the abovementioned special characteristics of the nanomaterials, they play a vital role in electrochemical synthesis and are widely used in the preparation of hard, mesoporous or composite electrodes and other energy storage materials. During electrochemical synthesis, nanomaterials are highly influenced by the colloidal properties of the reaction solution such as van der Waals, electrostatic and solvent interaction forces. They significantly influence the structure and properties of the electrode materials and can facilitate the transfer of charge and mass inside the electrodes. Particularly, with the emergence of molecular electronics and nanoelectronics where single molecules are expected to control the transport of electrons, nanomaterials can now be explored for a vast variety of energy storage applications.

Hydrogen can be an ideal energy source for the future, and it is accounted as a renewable and green energy option which enhances the energy conversion efficiency and reduces the global demand for fossil fuels and thus the emission of greenhouse gases (Gholami and Salavati-Niasari, 2017). Although hydrogen is regarded as a convenient, versatile and safe energy source, its storage is relatively complex because of its low storage density and high decomposition temperature. Several efforts have been made to overcome these challenges, and various hydrogen storage systems have been explored (Felderhoff et al., 2007; Eberle et al., 2009). Electrochemical storage of hydrogen using nanostructures has attracted a lot of attention of the investigators in the past few decades. These nanomaterials seem to provide the highest probability of high storage density, moderate/low decomposition temperatures, fast kinetics, good reversibility and low manufacturing cost.

In the 21st century, the development of nanotechnology and the enhanced availability of nanomaterials of various shapes, sizes and properties have offered unparalleled solutions to the energy storage complications. Thus, the use of nanomaterials for energy storage can provide an extra dimension of size variation (in addition to composition, pressure and temperature) for tailoring the device applicability and performances. This chapter gives a general introduction of electrochemically synthesized nanomaterials for energy storage applications and their advantages and challenges. Separate sections have been dedicated to cover the fundamentals of electrochemical synthesis, basic requirements, influencing factors and applications of electrochemically synthesized nanomaterials for energy storage. Finally, we have summarized several important aspects and challenges covered in this chapter and the future aspects of electrochemical synthetic methods for energy storage nanomaterials.

2.2 PRINCIPLES OF ELECTROCHEMICAL SYNTHESIS

A simple setup of electrochemical synthesis consists of a reaction vessel containing a reaction solution and two electrodes connected with a power supply (Figure 2.2). The reaction solution or the bath solution consists of the desired electroactive species, solvent, substrate and sometimes additives and supporting electrolytes for facilitated electrochemical reactions and lower cell resistance, respectively. Oxidation takes place at the anode, while reduction takes place at the cathode simultaneously. The electrode at which the reaction of interest takes place is denoted as the working

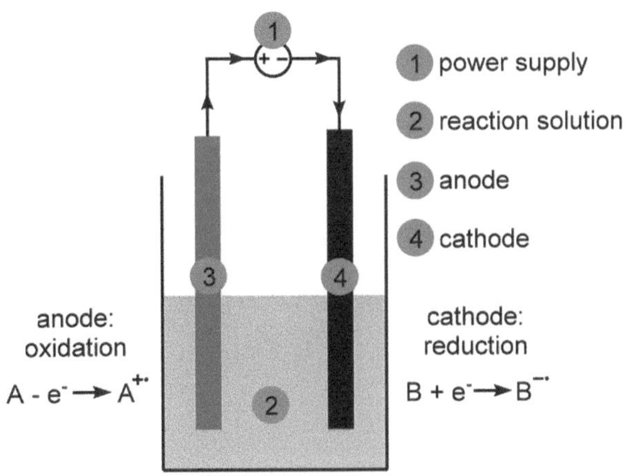

FIGURE 2.2 Simple setup for electrochemical synthesis. Adapted from Schotten et al. (2020). Copyright 2020 Royal Chemical Society.

electrode, and the other electrode is denoted as the counter electrode. The reaction taking place on the counter electrode, called counter reaction, is vital to consider during the optimization of an electrochemical reaction because it may be the rate-limiting factor. The migration of charged electroactive species through the bath solution completes the electrical circuit. Electrochemical reactions are considered to be initiated by the formation of radical species via single electron transfer to the substrate, which eventually forms the end product. The supply of electrons was maintained using a power supply, causing the movement of electrons from the anode to the cathode developing an electrochemical potential. A chemical reaction takes place when the electrochemical potential is higher than the redox potential of the substrate (Figure 2.3a). The current describes the flow of electrons and thus the rate of electrochemical reactions taking place at the electrodes.

An electrochemical double layer, i.e. the Helmholtz layer, is formed at the surface of the electrode consisting of two layers of the oppositely charged species. One layer is the electrode, i.e. the electric conductor, and the other one is the ionic conductor in contact with the electrode such as an electrolyte. For example, the anodic double layer comprises a positively charged layer on the surface of anode and a negatively charged layer of the solvated electrolytes in contact with the anode surface (Figure 2.3b). Usually, the thickness of the electrochemical double layer is of few nanometers and is determined by the electrolyte concentration and the applied voltage. The concentration of the substrate near the electrode is described as the Nernst diffusion layer (Figure 2.3c). When the substrate is produced, the substrate concentration is higher at the electrode in comparison with the bulk solution and vice versa.

The current (I) and the potential (U) are the most important electrochemical reaction parameters. The flow of electrons is expressed in terms of the current, while the potential specifies the energy by which electrons are moved. The charge (Q) is defined as the current per unit time. The rate of the electrochemical reaction is dependent on the number of electrons supplied to the reaction medium, which is

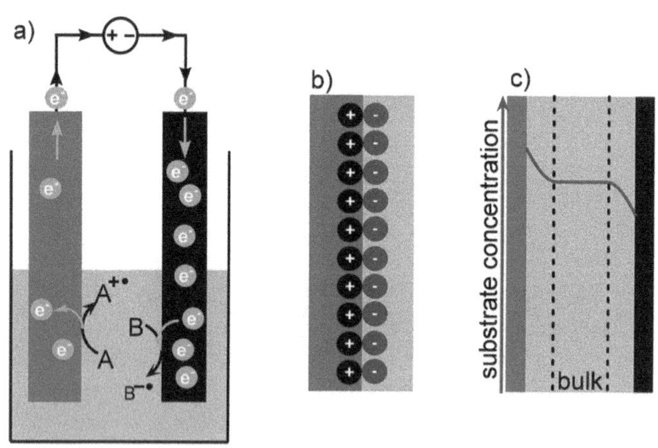

FIGURE 2.3 (a) Movement of electrons in an electrochemical reaction, (b) Helmholtz double layer and (c) Nernst diffusion layer; normal: anode, bold: cathode and gray: reaction solution. Adapted from Schotten et al. (2020). Copyright 2020 Royal Chemical Society.

controlled by the current and the charge. Both potential and current are related to each other by resistance (R), and both of them cannot be changed separately without varying the resistance of the electrochemical system. Resistance of the system is determined by various factors such as electrodes, electrolytes, solvent, electrochemical reaction, wiring and additives. Current, potential and resistance all three parameters are related to each other and can be expressed as follows:

$$U = R \times I$$

where U is potential in volt, R is resistance in ohm and I is current in ampere. There are two sets of conditions in which the electrochemical synthesis can be carried out.

2.2.1 Galvanostatic Synthesis

Those electrochemical synthesis reactions which run under the constant current are called galvanostatic reactions, and they simply require working and counter electrodes. Potential of the reaction system is not controlled in this condition, and it varies according to the resistance to maintain the constant current. A low potential is required for a system having low resistance and vice versa. Thus, current density becomes an important parameter and will be proportional to the rate of reaction. Variation in the rate of reaction and the reaction time by varying current density is the most important and thoroughly investigated process parameter. When the electrode material is oxidized or reduced, a competing reaction may occur, which will consume electrons for an undesired side reaction and decrease the faradaic efficiency. This can, for example, happen in an oxidation reaction when easily oxidized metals such as iron or copper are used as the anode. The total charge and the amount of reaction products can be controlled by controlling the current (I) and the time (t) for which it is flowing (Schotten et al., 2020).

$$Q = I \times t$$

2.2.2 Potentiostatic Synthesis

Potential is also attributed as voltage and is the amount of energy involved during the movement of electrons from the anode to the cathode, causing a potential difference. Higher potential difference causes transfer of electrons from the cathode to the reaction solution. Simultaneously, there will be removal of electrons at the anode from the reaction solution, resulting in the flow of current. In potentiostatic conditions, constant potential is maintained by variation in current and thus does not allow direct electron equivalence calculations. For electron equivalence calculations, current over reaction time has been recorded using a multimeter. In addition to working electrode and counter electrode, potentiostatic setup also requires a reference electrode, which precisely measures the potential at the working electrode. Thus, it indicates the potential involved in the electrochemical synthesis excluding any type of setup resistance. However, almost always, a higher electrode potential compared to its standard potential is required for an electrochemical reaction to occur, which is referred to as overpotential. It arises because most of the reactions are not performed under standard conditions, and thus, the redox potential is influenced by the concentration. For reversible reactions, it is explained by the Nernst equation (Schotten et al., 2020).

2.3 BASIC REQUIREMENTS AND FACTORS INFLUENCING ELECTROCHEMICAL SYNTHESIS

There are several factors which influence the electrochemical synthesis such as (a) variables of bath composition, for example, electrolyte concentration, electrode material, additives and solvent; (b) variables of bath operation, for example, current density and temperature; and (c) miscellaneous variables such as current type, electrode shape and current efficiency. Some of them have been discussed in the above section, and other important ones will be discussed in this section along with the basic setup.

Day-by-day improvement in technology and increasing demands of nanomaterials of specific shape, size and surface area in the field of energy storage resulted in the requirement of some high-end equipment for electrochemical synthesis. However, the basic instrumental requirements for the electrochemical synthesis of energy storage nanomaterials remain very simple. The most important ones are mentioned briefly.

2.3.1 Bath Reactor Vessel

The electrochemical cell or reactor vessel has several possible setups. It can be as simple as both electrodes can be fitted into a beaker sharing the reaction solution (Figure 2.4a). As both electrodes fit into the same compartment, they are also called undivided electrochemical cell. They are suitable for the electrochemical conditions in which stable reaction products are formed. But for the reactions which produce

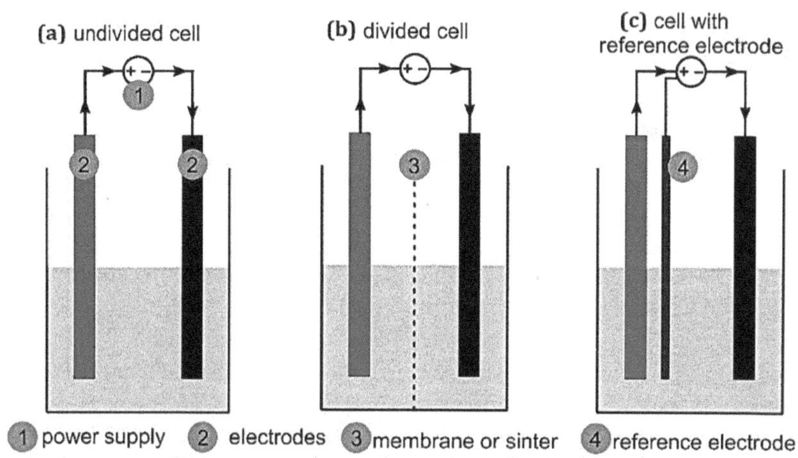

FIGURE 2.4 Electrochemical reactor vessels: (a) undivided cell, (b) divided cell and (c) cell with reference electrode. Adapted from Schotten et al. (2020). Copyright 2020 Royal Chemical Society.

electrochemically unstable products, i.e. the product may decompose at the counter electrode or it is incompatible with the counter reaction, a divided cell can be used. The divided cell has anodic and cathodic reaction chambers divided by a membrane, usually an ion-exchange membrane to allow the charge transfer (Figure 2.4b). Sometimes, they are made in-house for specific synthetic applications.

2.3.2 Power Supply Unit

Power supply pumps the electrons in the system causing the flow of electrons from the anode to the cathode. Thus, it drives the electrochemical synthesis reaction. Potentiostats are most widely used for this purpose because they not only precisely control the supplied current or the voltage during electrolysis but also allow precise potential control at the working electrode with the help of a reference electrode. A simple bench-top power supply can be used for the electrolysis, which does not require a reference electrode.

2.3.3 Solvent

The selection of a solvent is crucial for an electrochemical synthesis. Several factors should be considered during the selection of a solvent such as reagent solubility, intermediate stabilization, conductivity, polarity, promotion of counter reaction and electrochemical window. Degradation of solvent may become the primary reaction because of its significantly higher concentration, and thus, it must be checked. A large electrochemical window of the solvent denotes a wide potential range over which the solvent is resistant to oxidation or reduction. Commonly used protic solvents are fluorinated alcohols, ethanol, methanol and sometimes water, while typical aprotic solvents are DMF, DMSO, EG, PEG, THF, DMA, DCM and MeCN.

2.3.4 Electrolytes

A supporting electrolyte is added to all the non-conductive solvents as well as solvents having poor conductivity during the electrochemical reactions to ensure that the electrolysis occurs by lowering the potential and decreasing the resistance of the system. The supporting electrolyte must be fully soluble in the reaction solution and must form solvent-separated ion pairs which enhance the conductivity of the solution. Additionally, it should not show any reactivity or any redox change under the applied potential of electrolysis. Some of the most commonly used electrolytes in electrochemical synthesis are tetrabutylammonium (TBA) salts, alkali metal salts and halides.

2.3.5 Electrodes

Principally, every conducting material can be used as an electrode. As the whole electrochemical synthetic process takes place on the surface of electrode, the shape, size and composition of the electrodes can significantly vary the reaction outcome. Porous electrodes have a significantly higher surface area compared to rods or plates;

thus, they not only increase the reaction rate but also provide several benefits for the synthesis of nanomaterials. Particularly, mesoporous electrodes are extremely important for energy storage nanomaterials. The electrodes must be chemically and electrochemically inert to avoid any undesired competing reaction or side reaction. For example, if hydrogen evolution reaction is competing with the desired reduction reaction, it can be avoided by using electrodes having large hydrogen overpotential. A counter electrode having larger surface area than the working electrode will promote the counter reaction, and the overall reaction rate is not limited by the counter reaction rate. A reference electrode is situated very close to the working electrode to minimize the resistance when voltage is applied between them by the power supply (Figure 2.4c). Thus, the reference electrode accurately measures the voltage applied on the working electrode, for example, saturated calomel electrode (SCE), silver chloride electrode and mercury-mercurous sulfate electrode.

2.3.6 Additives

Additives are used to improve the reaction performance, and they can significantly influence the structure and properties of the end product either by stabilizing the intermediate products or by promoting the counter reaction. They can reduce the crystallite growth by blocking the growth surface or by varying the cathodic overpotential during electrodeposition of metals (Mockute et al., 2000; Szeptycka et al., 2001). Addition of acids was found beneficial for oxidation reactions because they cater protons for hydrogen reduction, which is the counter reaction. In a similar way, addition of metal salts results in reduction and deposition on the cathode; fluorinated alcohols are used to stabilize the radicle intermediates and boric acid is used in electroplating solutions to improve the coating appearance and to reduce the coating brittleness.

2.4 ELECTROCHEMICAL SYNTHESIS OF NANOMATERIALS FOR ENERGY STORAGE APPLICATIONS

In this section, we will discuss the energy storage applications of electrochemically synthesized nanomaterials. Several operating parameters of electrochemical synthesis provide precise control over the size and surface morphology of the products. Preparation methods and applications of metals, alloys, metal oxides and carbon-based materials have been explored for energy storage applications.

2.4.1 Electrochemically Synthesized Nanomaterials for Lithium-Ion Battery Applications

Nanomaterials provide high surface-to-volume ratio, altered physical properties and short diffusion path for electronic conduction and ionic transport. They can be used as electrode materials to improve the energy density of lithium-ion batteries. Alloy anodes such as Sn, Si and Ge have much higher theoretical capacity than the conventional graphite anodes (372 mAh g^{-1}). Particularly, silicon anodes have almost ten times higher theoretical capacity (4,200 mAh g^{-1}) than the graphite anodes, and they can increase the energy density of the fuel cell up to 40% (Zhang, 2011). Apart from several advantages, one major drawback which limits the application of alloy anodes in the batteries is excessive variation in their volume after cycling. It can cause material fracture, resulting in the loss of active material as well as an unstable solid-electrolyte interphase.

Different types of nanostructured anodes have been developed to address the challenges of alloy anodes. Chen et al. (2011) have developed Si nanowire anode design to overcome volume expansions during long cycle life. Nanowire anodes were efficiently able to relax the strain compared to solid film and particle electrodes. Empty space between the nanowires accommodates the volume expansion, their robust electrical and mechanical contact provides better 1D electron transport pathways and all the electrode materials become electrochemically active (Figure 2.5). Similarly, Kowalski et al. (2017) electrochemically synthesized 1D core-shell TiO_2/Si nanotubes having well-controlled structure and shape. The anode has good cycle stability, and the expansion of Si without fracture allows the release of stress.

Cu_2Sb has been considered as an important anode material because of its high reversible reaction with lithium, small change in volume and suitable lithium intercalation potential. All the current fabrication methods of Cu_2Sb such as ball milling,

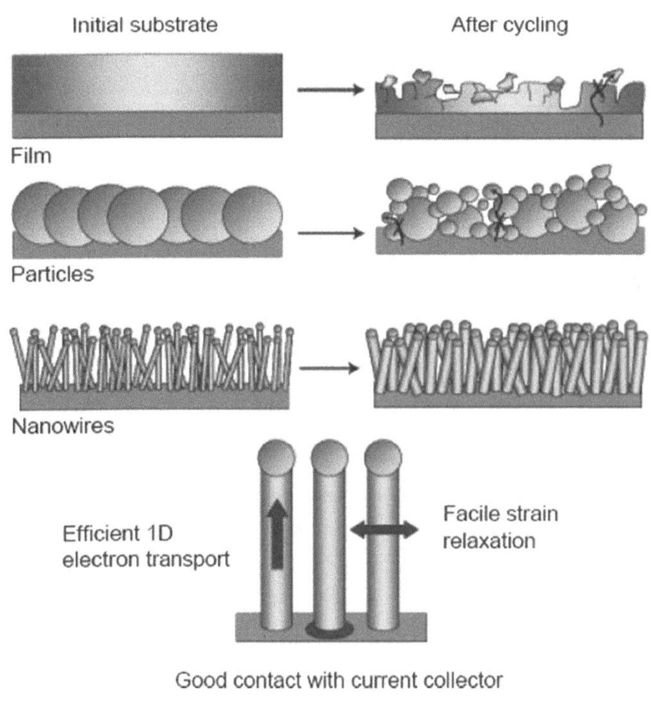

FIGURE 2.5 Advantages of nanowire anodes compared to solid film and particle anodes during electrochemical cycling. Reprinted from Chan et al. (2008) with permission from Nature Publishing Group.

pulse vapor deposition, chemical reduction from solution and crystal growth from melts usually require high reaction temperatures causing the loss of Sb over time (Li et al., 2013). The major challenge during electrochemical synthesis of Cu_2Sb alloy is that the deposition potential of the two metals varies by nearly 130 mV, and copper deposits are preferred because of their lower deposition potential. In addition, salts of Sb do not form neutral aqueous solutions; rather, they precipitate in the form of Sb_2O_3. To electrochemically synthesize crystalline Cu_2Sb alloy for Li-ion batteries, citric acid was added to the reaction solution as a complexing agent. It raised the pH of the solution and prevented the formation of Sb_2O_3. In addition, it improves the solubility of Sb salts and forces the deposition potential of both metals to come closer. As a result of this efficient electrodeposition route, Cu_2Sb alloy has been deposited onto copper substrates at room temperature and pH 6 (Mosby and Prieto, 2008).

Adjustable morphology and hollow structure of ZnSb nanotubes enabled them to accommodate large volume expansions in long cycle life lithium-ion batteries (Saadat et al., 2011). Similarly, a quaternary Fe-Sn-Sb-P composite having a dicranopteris-like structure has been electrochemically synthesized because of its high energy density and long cycle life. Its structure and properties were tailored by controlling deposition parameters, and thus, its dendritic structure allowed accommodation of large volume expansions (Zheng et al., 2012). Electrodeposition of Si on carbon cloth produced Si nanowires having diameters around 20–30 nm with a thin SiO_2 layer of around 1 nm. Thus, they were able to accommodate charge transfer and lithium-ion diffusions as well as volume expansion without compromising Li^+ penetration (Weng and Xiao, 2019). Recently, Liu et al. (2022) have electrodeposited ZnO with an average diameter of 1 µm on carbon cloth to organize a free-standing flexible lithium-ion battery anode material with good electrical contact between the current collector and the active material. Trofimov et al. (2022) have studied the electrochemical properties of electrodeposited silicon fibers used as lithium-ion battery anodes to improve their charging capacity.

2.4.2 Electrochemically Synthesized Nanomaterials for Electrochemical Capacitor Applications

Nanostructured metal oxides such as RuO_2, MnO_2 and $RuO_2.xH_2O$ nanotube composites have been investigated for electrochemical capacitor applications due to pseudocapacitance effect. Hydrous RuO_2 is considered as an excellent example of redox pseudocapacitor having a capacitance of 850 F g^{-1} because of its high conductivity, high reversibility and broad voltage range redox activity (Dmowski et al., 2002). But low porosity, high cost and toxicity of RuO_2 limit its applications and force the researchers to focus on its replacement with other inexpensive nanostructured transition metal nitrides and oxides of Mn, Fe, V, Co, Ni, Mo, etc. Although MnO_2 could be a promising electrode material due to its lower cost, its poor electrical conductivity of around 10^{-6} S cm^{-1} limits its performance. Deposition of MnO_2 on nanoporous gold substrate promoted by ion diffusion and electron transport was suggested by Lang et al. (2011), which reached a specific capacitance of up to 1,145 F g^{-1}, but the synthesis was costly.

However, cost-effective electrode material can also be developed by electrochemical synthesis. Binder-free, highly porous, reduced graphene oxide electrodes for Li-ion capacitors delivering capacitance up to 168 F g^{-1} were electrochemically deposited by Zhan et al. (2020) with well-controlled thickness and weight. MnO_2 has been deposited on carbon aerogel and thus has high surface area because of its mesoporous structure (Li et al., 2007). Its improved electrochemical activity results in the highest specific capacitance for mesoporous MnO_2-carbon aerogel composite electrode (515.5 F g^{-1}) compared to bare carbon aerogel electrode (182 F g^{-1}) and bulk MnO_2-carbon aerogel electrode (327 F g^{-1}). By carefully controlling deposition parameters, multilayered MnO_2 nanosheets (specific capacitance 521.5 F g^{-1}) were synthesized from a reaction solution of 0.01 M $Mn(NO_3)_2$ + 0.01 M thiourea at applied current density of 1.7 mA cm^{-1} (Feng et al., 2009).

As we know, the drawback of MnO_2 for electrochemical capacitor applications was its lower electrical conductivity. Thus, several attempts were made to incorporate conducting materials with MnO_2 to form hybrid electrodes having high specific capacity. For example, Hou et al. (2010) have synthesized MnO_2-CNT conducting polymer electrodes having a specific capacitance of 427 F g^{-1}, while Bao et al. (2011) have used Zn_2SnO_4@MnO_2 nanocable-coated carbon microfibers to develop flexible supercapacitors having a specific capacitance of 642.4 F g^{-1}. However, their synthesis process was complex and non-scalable. A scalable, electrodeposited MnO_2-CNT-sponge hybrid electrode was developed by Chen et al. (2011), and its fabrication process is depicted in Figure 2.6. Galvanostatic electrochemical deposition required a very small current density, provided well-controlled deposited mass by deposition time, and showed excellent adhesion between MnO_2 and CNT sponge. The specific capacitance was found to be as high as 1,230 F g^{-1}.

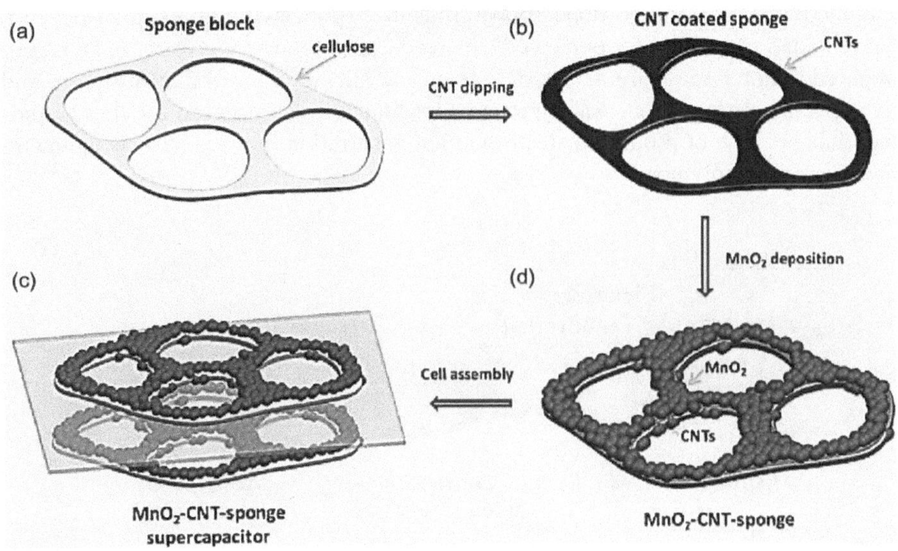

FIGURE 2.6 Fabrication process of MnO_2-CNT-sponge supercapacitors. Reproduced from Chen et al. (2011). Copyright 2011 American Chemical Society.

Ansari et al. (2020) proposed an electrochemical method to synthesize TiN nanoparticles on a Ti foil substrate. These nanoparticles exhibited good reversibility, high charge-discharge rate, high specific capacity, high power density and high energy density along with durable cycle life. An easily controllable, high-yield electrodeposition method was used by Xia et al. (2012) to synthesize a one-dimensional hydroxide and oxide composite Co_3O_4-$Co(OH)_2$ nanomaterials for supercapacitor applications. It can be seen from Figure 2.7 that initially self-supported Co_3O_4 nanowire arrays were hydrothermally synthesized followed by $Co(OH)_2$ nanosheet growth by electrodeposition, and finally, they were thermally annealed to produce porous oxides. The thickness and composition can be easily tailored by varying electrodeposition parameters, and specific capacity of 1,095 F g^{-1} can be attained.

Conducting polymers such as polythiophene, polypyrrole and polyaniline have large π-conjugation length, are suitable for doping/dedoping reactions and can be oxidized or reduced reversibly. Thus, they are used as low-cost, easily processable, high charge density, electroactive pseudocapacitive materials for ECs having improved electrical energy storage and lower self-discharge. Polyaniline is the most widely used conducting polymer for supercapacitor applications. Shaikh et al. (2021) found that the hydrophilic materials having porous structures are useful to improve the supercapacitance of the electrodes. For this purpose, they have electrochemically synthesized hydrophilic nano-nest polyaniline electrodes at room temperature. Reduced graphene oxide hydrogel film was hydrothermally synthesized from a graphene oxide solution by Liu et al. (2021), which is further used for electropolymerization of polyaniline (Figure 2.8). The resulting graphene oxide-polyaniline hydrogel has shown capacitance of up to 853.7 F g^{-1} and retention of above 90% capacitance (741.8 F g^{-1}) after 8,000 cycles. Similarly, Hou et al. (2020) have synthesized sulfuric acid and perchloric acid co-doped polyaniline electrode having good electrical conductivity and roughness for enhanced supercapacitive properties. Wang et al. (2010) prepared aligned nanowire arrays of polyaniline having ordered nanostructure and high specific surface area. An improved capacitance was observed for these nanomaterials because of promoted electrolyte ion penetration into the narrow nanowire diameter of the polymer.

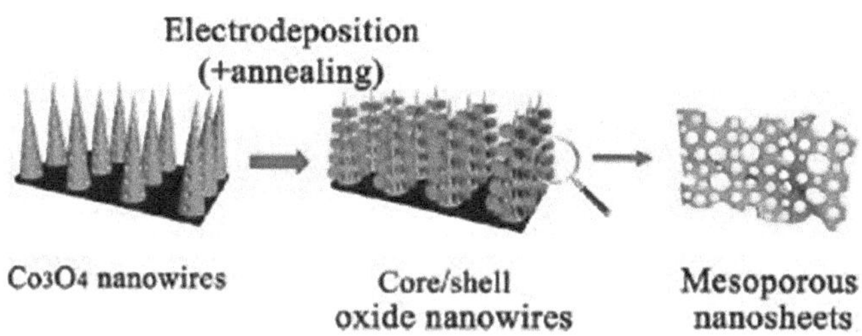

FIGURE 2.7 Schematic representation of the formation of porous hydroxide nanosheets by electrodeposition on Co_3O_4 nanowire arrays followed by thermal annealing to produce porous oxides. Reproduced from Xia et al. (2012). Copyright 2012 American Chemical Society.

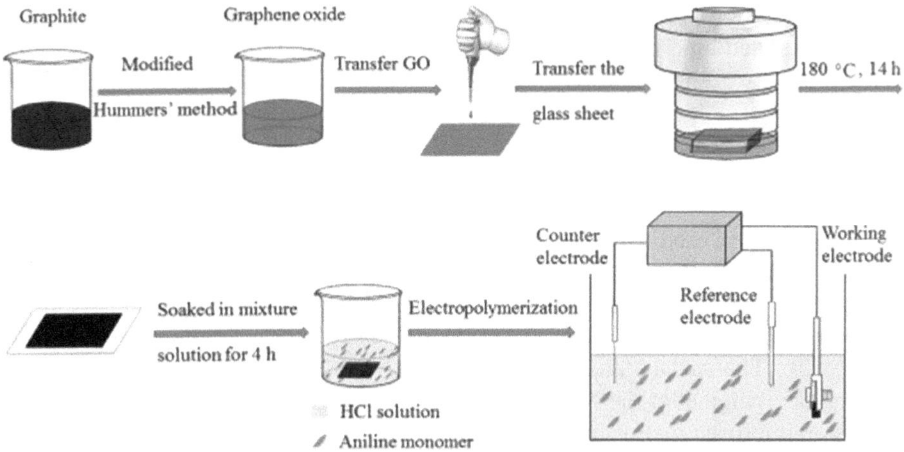

FIGURE 2.8 Schematics of the fabrication process of reduced graphene oxide-polyaniline composite hydrogel film. Reproduced from Liu et al. (2021) with permission from Elsevier Publishers.

2.4.3 Electrochemically Synthesized Nanomaterials for Hydrogen Energy Storage

Hydrogen energy is considered as one of the most clean, greener and promising alternatives to traditional fossil fuels. The gravimetric energy density of hydrogen (33 kWh kg^{-1}) is more than two times higher compared to gasoline (13 kWh kg^{-1}), and thus, hydrogen is one of the excellent energy sources (Gholami and Salavati-Niasari, 2017). It can be used as an energy source for different purposes such as transport, portable electronic and stationary applications. However, storage of hydrogen is a major challenge which significantly limits the utilization of hydrogen energy, particularly for automotive applications. An ideal hydrogen storage material should be easily available and should have low cost, high gravimetric and volumetric hydrogen density, long cycle life, low-temperature hydrogen dissociation, fast kinetics and high degree of reversibility. Conventionally, hydrogen is stored as condensed liquid (–253°C, 5–10 bar) or compressed gas (350–700 bar at room temperature). The use of high-pressure tanks has safety concerns, and even when they are used, the density of hydrogen is only 26.3 g L^{-1} at a high pressure of 700 bar (Felderhoff et al., 2007; Hua et al., 2011). Although condensed liquid hydrogen has a relatively higher volumetric hydrogen density, it needs a huge amount of energy to liquefy hydrogen and suffers from heat loss due to heat transfer (Felderhoff et al., 2007; Eberle et al., 2009).

Several practical hydrogen storage systems such as adsorbents for hydrogen physisorption, metal hydrides, chemical hydrides, complex hydrides and various types of nanomaterials and nanocomposites have been developed in the past few decades. The use of nanomaterials for molecular hydrogen storage provides several advantages over bulk materials. Their high surface area and porous structure can be very useful for encapsulating hydrogen and thus can significantly improve volumetric and gravimetric energy storage densities. In addition, porous structure and higher surface offer some added binding sites in the pores and on the surface, which significantly increase the possibility of storing hydrogen by physisorption. In physisorption, the interacting intermolecular forces are weak, which results in fast kinetics and easy reversibility.

Recently, several electrochemically synthesized nanomaterials have also been used for hydrogen storage application. Wang et al. (2019) have prepared several hydrogen storage Ce-Ni alloys by potentiostatic electrolysis in LiCl-KCl-$CeCl_3$ melts using Ni as cathode. Similarly, Lu et al. (2019) electrodeposited Ni-Mo-Cu coatings from roasted nickel matte precursor in choline chloride-urea deep eutectic solvent using platinum flake as a counter electrode and silver wire as a reference electrode (Figure 2.9). The surface area of Ni-Mo-Cu electrodeposits was found almost 31 times higher than that of the Ni foam substrate. The significantly larger surface area resulted in improved hydrogen evolution performance.

2.5 CONCLUSIONS AND PERSPECTIVES

This chapter demonstrates the highly efficient and facile modern electrochemical synthesis technique for the production of energy storage nanomaterials. It is a sustainable method to produce complex nanostructures and has advantages of high purity and selectivity, low cost, simplicity, low-temperature synthesis, greenness and environmental friendliness. Thickness, stoichiometry, composition and microstructure of the products can be easily controlled by the adjustment of electrodeposition parameters. In addition, they can provide fast electron transport through metal supports to the materials having poor electrical conductivity. This provides proper control for tuning the properties of nanostructures and their applications as energy storage materials.

The role of electrochemically synthesized nanomaterials has been summarized for lithium-ion batteries, electrochemical capacitors and hydrogen energy storage, and it has been found that their energy storage capacity has been significantly enhanced by using nanostructured materials. In brief, nanostructures have enhanced the performance of these devices by (a) providing a large surface area and an open nanostructure which allow facile diffusion of the electroactive ions into the inner regions of the electrode for efficient electrochemical reaction. For hydrogen storage, it boosts the molecular adsorption occurring at the solid-liquid or solid-gas interface. (b) The empty space between the nanostructures provides an elastic buffer to accommodate the strain aroused because of volume variations and improves the mechanical strength, durability and cycle life. (c) By delivering new mechanisms, nanomaterials more efficiently achieve the energy storage such as quantum confinement effect and varying charge-discharge mechanisms.

Looking forward, there are still many aspects which need to be addressed. Electrochemical synthetic routes with more precise control over the shape, size and composition of the nanomaterials to produce more ideal structures and suitable pore sizes will be beneficial. For example, the mechanism of formation of different nanostructures such as nanospheres, nanorods, nanotubes and dendritic and flower-like structures is not very well understood; however, it strongly influences the mechanism of energy storage. Many nanostructured materials have excellent energy

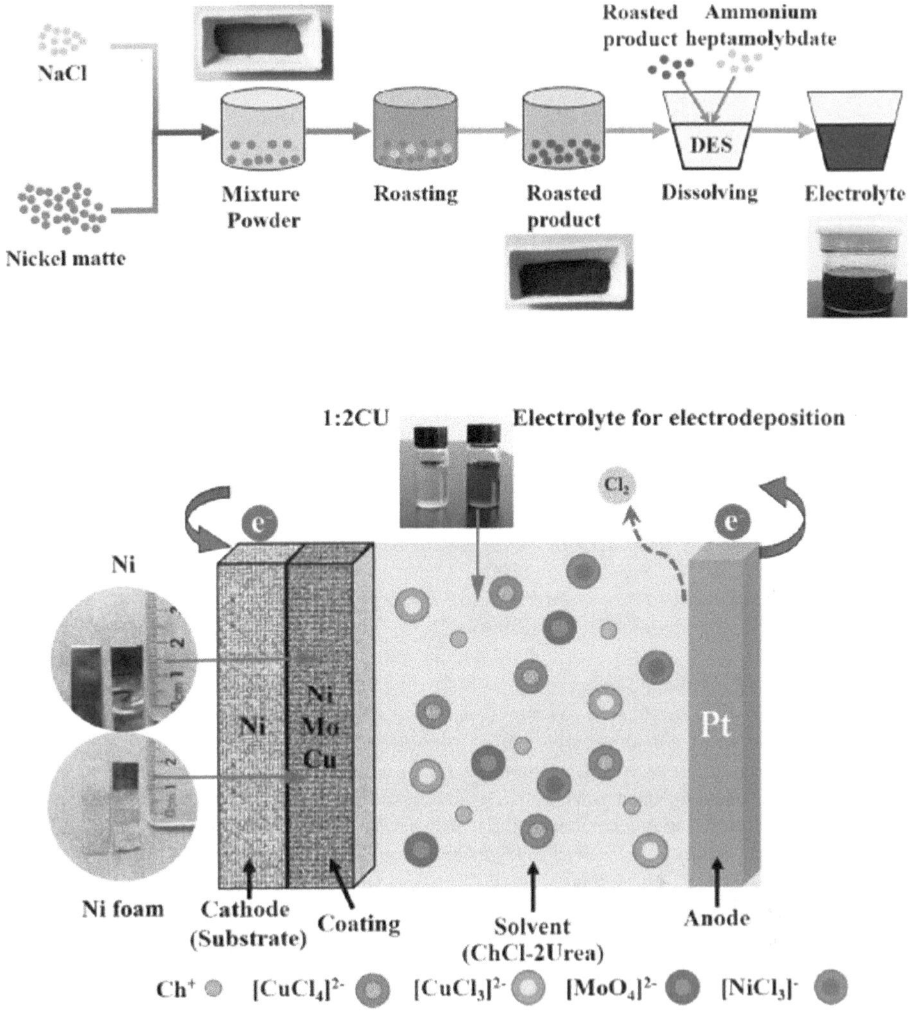

FIGURE 2.9 Schematic representation of electrolyte preparation and electrodeposition processes of Ni-Mo-Cu alloy. Reproduced from Lu et al. (2019) with permission from Elsevier Publishers.

storage properties, but their poor electrical conductivity limits their applicability, which must be improved. Finally, the development of low-cost, high-storage-density nanostructures with new energy storage mechanism is anticipated such as composite and polymer-based materials.

ACKNOWLEDGMENTS

The authors acknowledge the financial support from the University Grant Commission (Grant No: F.30-505/2020-BSR), New Delhi, and Banaras Hindu University, Varanasi (Grant No: AB/7-P-276/39652).

REFERENCES

Ansari, S. A., Khan, N. A., Hasan, Z., Shaikh, A. A., Ferdousi, F. K., Barai, H. S., Lopa, N. S., & Rahman, M. M. (2020). Electrochemical synthesis of titanium nitride nanoparticles onto titanium foil for electrochemical supercapacitors with ultrafast charge/discharge, *Sustainable Energy & Fuels*, 4, 2480–2490.

Bao, L., Zang, J., & Li, X. (2011). Flexible Zn_2SnO_4/MnO_2 core/shell nanocable-carbon microfiber hybrid composites for high-performance supercapacitor electrodes, *Nano Letters*, 11, 1215–1220.

Chan, C. K., Peng, H., Liu, G., McIlwrath, K., Zhang, X. F., Huggins, R. A., & Cui, Y. (2008). High-performance lithium battery anodes using silicon nanowires, *Nature Nanotechnology*, 3(1), 31–35.

Chen, D., Lin, Z., Sartin, M. M., Huang, T.-X., Liu, J., Zhang, Q., Han, L., Li, J. F., Tian, Z.-Q., & Zhan, D. (2020). Photosynergetic electrochemical synthesis of graphene oxide, *Journal of the American Chemical Society*, 142, 6516–6520.

Chen, W., Rakhi, R. B., Hu, L., Xie, X., Cui, Y., & Alshareef, H. N. (2011). High-performance nanostructured supercapacitors on a sponge, *Nano Letters*, 11, 5165–5172.

Dmowski, W., Egami, T., Swider-Lyons, K. E., Love, C. T., & Rolison, D. R. (2002). Local atomic structure and conduction mechanism of nanocrystalline hydrous RuO_2 from X-ray scattering, *The Journal of Physical Chemistry B*, 106, 12677–12683.

Eberle, U., Felderhoff, M., & Schuth, F. (2009). Chemical and physical solutions for hydrogen storage, *Angewandte Chemie International Edition in English*, 48, 6608–6630.

Felderhoff, M., Weidenthaler, C., von Helmolt, R., & Eberle, U. (2007). Hydrogen storage: The remaining scientific and technological challenges, *Physical Chemistry Chemical Physics*, 9, 2643–2653.

Feng, Z.-P., Li, G.-R., Zhong, J.-H., Wang, Z.-L., Ou, Y.-N., & Tong, Y.-X. (2009). MnO_2 multilayer nanosheet clusters evolved from monolayer nanosheets and their predominant electrochemical properties, *Electrochemistry Communications*, 11, 706–710.

Gholami, T., & Salavati-Niasari, M. (2017). Green facile thermal decomposition synthesis, characterization and electrochemical hydrogen storage characteristics of $ZnAl_2O_4$ nanostructure, *International Journal of Hydrogen Energy*, 47(27), 17167–17177.

Hou, L., Zhi, X., Zhang, W., & Zhou, H. (2020). Boosting the electrochemical properties of polyaniline by one-step co-doped electrodeposition for high performance flexible supercapacitor applications, *Journal of Electroanalytical Chemistry*, 863, 114064.

Hou, Y., Cheng, Y., Hobson, T., & Liu, J. (2010). Design and synthesis of hierarchical MnO_2 nanospheres/carbon nanotubes/conducting polymer ternary composite for high performance electrochemical electrodes, *Nano Letters*, 10(7), 2727–2733.

Hua, T. Q., Ahluwalia, R. K., Peng, J. K., Kromer, M., Lasher, S., McKenney, K., Law, K., & Sinha, J. (2011). Technical assessment of compressed hydrogen storage tank systems for automotive applications, *International Journal of Hydrogen Energy*, 36, 3037–3049.

Kowalski, D., Mallet, J., Thomas, S., Nemaga, A. W., Michel, J., Guery, C., Molinari, M., & Morcrette, M. (2017). Electrochemical synthesis of 1D core shell Si/TiO_2 nanotubes for lithium ion batteries, *Journal of Power Sources*, 361, 243–248.

Lang, X. Y., Hirata, A., Fujita, T., & Chen, M. W. (2011). Nanoporous metal/oxide hybrid electrodes for electrochemical supercapacitors, *Nature Nanotechnology*, 6, 232–236.

Li, G.-R., Xu, H., Lu, X.-F., Feng, J.-X., Tong, Y.-X., & Su, C.-Y. (2013). Electrochemical synthesis of nanostructured materials for electrochemical energy conversion and storage, *Nanoscale*, 5, 4056–4069.

Li, Y., Jia, W.-Z., Song, Y.-Y., & Xia, X.-H. (2007). Superhydrophobicity of 3D porous copper films prepared using the hydrogen bubble dynamic template, *Chemistry of Materials*, 19(23), 5758–5764.

Liu, Z., Zhao, Z., Xu, A, Li, W., & Qin, Y. (2021). Facile preparation of graphene/polyaniline composite hydrogel film by electrodeposition for binder-free all-solid-state supercapacitor, *Journal of Alloys and Compounds*, 875, 159931.

Liu, S., Zhu, F., Wang, X., Gao, H., & Meng, Y. (2022). Preparation of ZnO/CC flexible materials for lithium-ion batteries by electrodeposition, *Journal of Materials Science: Materials in Electronics*, 33, 4559–4567.

Lu, Y., Geng, S., Wang, S., Rao, S., Huang, Y., Zou, X., Zhang, Y., Xu, Q., & Lu, X. (2019). Electrodeposition of Ni-Mo-Cu coatings from roasted nickel matte in deep eutectic solvent for hydrogen evolution reaction, *International Journal of Hydrogen Energy*, 44, 5704–5716.

Mockute, D., & Bernotiene, G. (2000). The interaction of additives with the cathode in a mixture of saccharin, 2-butyne-1,4-diol and phthalimide during nickel electrodeposition in a Watts-type electrolyte, *Surface and Coating Technology*, 135, 42–47.

Mosby, J. M., & Prieto, A. L. (2008). Direct electrodeposition of Cu_2Sb for lithium-ion battery anodes, *Journal of American Chemical Society*, 130, 10656–10661.

Saadat, S., Zhu, J., Shahjamali, M. M., Maleksaeedi, S., Tay, Y. Y., Tay, B. Y., Hng, H. H., Ma, J., & Yan, Q. (2011). Template free electrochemical deposition of ZnSb nanotubes for Li ion battery anodes, *Chemical Communications*, 47, 9849–9851.

Schotten, C., Nicholls, T. P., Bourne, R. A., Kapur, N., Nguyen, B. N., & Willans, C. E. (2020). Making electrochemistry easily accessible to the synthetic chemist, *Green Chemistry*, 22, 3358–3375.

Shaikh, S. F., Shaikh, F. F. M., Shaikh, A. V., Ubaidullah, Md., Al-Enizi, A. M., & Pathan, H. M. (2021). Electrodeposited more-hydrophilic nano-nest polyaniline electrodes for supercapacitor application, *Journal of Physics and Chemistry of Solids*, 149, 109774.

Szeptycka, B. (2001). Effect of organic compounds on the electrocrystallization of nickel, *Russian Journal of Electrochemistry*, 37, 684–689.

Trofimov, A. A., Leonova, A. M., Leonova, N. M., & Gevel, T. A. (2022). Electrodeposition of silicon from molten KCl-K_2SiF_6 for lithium-ion batteries, *Journal of Electrochemical Society*, 169, 020537.

Wang, K., Huang, J. Y., & Wei, Z. X. (2010). Conducting polyaniline nanowire arrays for high performance supercapacitors, *Journal of Physical Chemistry C*, 114, 8062–8067.

Wang, S., Li, M., Han, W., Zhang, M., Wang, W., Wang, W., Yang, X., & Sun, Y. (2019). Selective formation of Ce-Ni hydrogen storage alloys by electrodeposition in $LiCl$-KCl-$CeCl_3$ melts using Ni as cathode, *Journal of Alloys and Compounds*, 777, 1211–1221.

Weng, W., & Xiao, W. (2019). Electrodeposited silicon nanowires from silica dissolved in molten salts as a binder-free anode for lithium-ion batteries, *Applied Energy Materials*, 2, 804–813.

Xia, X., Tu, J., Zhang, Y., Chen, J., Wang, X., Gu, C., Guan, C., Luo, J., & Fan, H. J. (2012). Porous hydroxide nanosheets on preformed nanowires by electrodeposition: Branched nanoarrays for electrochemical energy storage, *Chemistry of Materials*, 24(19), 3793–3799.

Zhan, Y., Edison, E., Manalastas, W., Tan, M. R. J., Satish, R., Buffa, A., Madhavi, S., & Mandler, D. (2020). Electrodeposition of highly porous reduced graphene oxide electrodes for Li-ion capacitors, *Electrochimica Acta*, 337, 135861.

Zhang, W. J. (2011). A review of the electrochemical performance of alloy anodes for lithium-ion batteries, *Journal of Power Sources*, 196(1), 13–24.

Zheng, X.-M., Huang, L., Xiao, Y., Su, H., Xu, G.-L., Fu, F., Li, J.-T., & Sun, S.-G. (2012). A dicranopteris-like Fe-Sn-Sb-P alloy as a promising anode for lithium ion batteries, *Chemical Communications*, 48, 6854–6856.

3 Microalgal Biofuels
Recent Developments, Challenges and Future Applications

Saumi Pandey, Nasreen Amin, and Vinod K. Kannaujiya

3.1 INTRODUCTION

Global demand for energy is growing exponentially because of economic development, industrial development, urbanization and overgrowth of population density across the green planet. Recently, it was found that the world population increased continuously by more than 1 billion per generation where developing countries play a major role in the outburst of the population (UN, 2019). Thus, the supply of energy is always limited to fulfill the essential duties in human life. Moreover, an essential and highly accessible form of non-renewable energy (fossil fuels) is depleted continuously because of excessive utilization in transportation and economic sectors. However, fossil fuel combustion is the main contributor of toxic gases like carbon dioxide, sulfur oxide, nitrogen oxide and other greenhouse gases that induce degradation of sustainable life resources and deteriorate the healthy environment of Earth's surface (Shafiee and Topal, 2009). Furthermore, fossil fuels are an exhaustible and limited resource on Earth's surface and incapable to fulfill all the demands of energy of the continuously increasing world population. It has been well documented that non-renewable sources of energy keenly affect our natural climate system by emission of CO_2 and life-threatening gases (Dogan and Seker, 2016; Eckelman et al., 2020; Koroneos et al., 2003; Pata, 2021). In addition, continuous consumption of non-renewable sources causes drastic changes in the natural biosphere and ecological development (Pata, 2021; Shahbaz et al., 2017) which increases life-threatening natural calamities in various geographical areas. Therefore, most of the countries are demanding renewable natural sources of energy for a progressive economy, clean energy and sustainable development of natural biodiversity on Earth's surface (Kannan and Vakeesan, 2016).

Solar radiation is the ultimate source of natural energy on Earth's surface to fulfill all demands of energy of the entire world population (Blaschke et al., 2013). Solar radiation has provided a constant intensity of photonic energy (1,370 W m^{-2}) and spread in variable ranges from 0 to 1,100 W m^{-2} on the ground surface (Ulusoy and Pektas, 2019). It was estimated that about 4 million exajoules of energy reach Earth's

DOI: 10.1201/9781003481157-3

surface annually (Kabir et al., 2018; Van der Hoeven, 2013). However, only a few storage and energy harvesting technologies are available to store solar energy which is incapable to fulfill the demand and supply of energy worldwide (De Vries et al., 2007; Ellabban et al., 2014). Most of the developing countries have spent a lot for searching novel technology to harvest renewable solar energy (Asafu-Adjaye, 2000).

In the current scenario, the major perspective of solar energy research is associated with the reduction of alarming global carbon emissions for a sustainable Earth environment (Kabir et al., 2018). Interestingly, many Asian countries have greater opportunities to utilize solar radiation for longer duration of time due to longer geographical sunshine duration. To reduce greenhouse gas emission, effective solar harvesting is required for the conversion of solar energy to electrical energy which plays a pivotal role in the sustainable development of fuel science (El-Khouly et al., 2017). All photosynthetic organisms manifest dissimilitude in the utilization and conversion of solar energy. In plants, most of the conversion efficiency of solar irradiance ranges from 4.6% to 6%, and in algae, maximum efficiency is about 9% in comparison to plants (Zhu et al., 2008). Microalgae are potent organisms due to short life cycles, high growth rates, high CO_2 fixation rates, stress adaptability and basic nutritional requirements for mass cultivation (Majid et al., 2014). In addition, growth compositions of microalgae are enriched with lipids, carbohydrates and protein which makes them an excellent choice for utilizing as biomass for the production of a variety of biorefinery products in energy sectors (Khoo et al., 2019). However, these microalgae-based biofuels encounter some challenges to manage industrial-scale production. To encounter these problems, metallic and hybrid nanoparticles are used to enhance the growth rate and biomass production (Banerjee et al., 2020). Apart from cultivation, microalgae are also employed for harvesting energy using magnetic nanoparticles due to their scalability, recyclability and cost-effectiveness (Banerjee et al., 2020). Recently, nanomaterial has expanded with various functional group modifications (e.g., amino, gold, hydrophobic and nickel) for the enhancement of biofuel production (Khoo et al., 2020). Microbial fuel cell (MFC) has an advantage over other nanomaterials due to its ecofriendly nature and cost-effective sustainable technology (Pandit and Das, 2015). Photosynthetic microorganisms are commonly used in the development of various products such as fuel cells, biodiesel, bioethanol, biohydrogen, biogas and jet fuels (Saad et al., 2019; Brennan and Owende, 2010). In this study, our main focus is to deliver the potentials of microalgae (algae and blue-green algae) in green energy technology, including their biological characteristics, challenges and recent developments for sustainable energy production.

3.2 PHOTOBIOLOGY OF LIGHT-HARVESTING COMPLEX

Microalgal phototrophs efficiently utilize solar energy for the conversion of carbon dioxide into organic products and the release of life-saving molecular oxygen for sustainable growth of life forms on Earth's surface (Voloshin et al., 2016). This unique biochemical reaction occurs in the specialized part of a cellular system called the thylakoid membrane. There are active membrane vesicles of thylakoid found in lower (blue-green algae) as well as in higher groups of algal photoautotrophs (Vothknecht and Westhoff, 2001). However, structural organizations of evolutionary ancient microalgae (cyanobacteria) are varied from plant or algal phototrophs (Liu, 2016). Evolutionary ancient blue-green algae thylakoid membranes are located in the cytosol embedded with specialized micro domains and proteins such as light-absorbing accessory molecules, pigments, cofactors and electron transport proteins (Nickelsen, 2012; Nickelsen and Zerges, 2013). More specifically, there are several protein complexes such as phycobilisome, photosystem II (PSII), photosystem I (PSI), cytochrome (cyt) b_6f and ATP synthase (ATPase) involved in the electron transport chain for photosynthesis (Yoon et al., 2014). Moreover, thylakoid membranes are arranged in several layers associated with brilliantly colored phycobiliproteins (accessory light-harvesting complex) (Kannaujiya and Sinha, 2017). Most of the phycobiliproteins (PBPs) are made of mainly three sub-proteins such as allophycocyanin (APC, $A^{650-660\,nm}$), phycocyanin (PC, $A^{610-625\,nm}$) and phycoerythrin (PE, $A^{490-570\,nm}$) (Kannaujiya et al., 2019; Kannaujiya et al., 2021; Sidler, 1994). In addition, thylakoid membranes are separated from the thylakoid lumen, which is made up of unique four classes of glycerolipids such as monogalactosyldiacylglycerol (MGDG), digalactosyldiacylglycerol (DGDG), sulfoquinovosyldiacylglycerol (SQDG) and phosphatidylglycerol (PG) (Kobayashi, 2016). PG is the key phospholipid component found in cyanobacteria (blue-green algae) (Wada and Murata, 1998). In addition, some cyanobacterial species have monogalactosyldiacylglycerol (MGlcDG) (Sato, 2015). The lipid bilayer plays an important role in permitting the generation of proton motive force via photosynthetic activities and prevents the diffusion of ions across the membrane (Mullineaux, 2014).

3.3 CELLULAR ELECTRON TRANSPORT MECHANISM

Electron transport chain (ETC) complexes are involved in the process of photosynthesis and respiratory chain reactions which serve as the site for the generation of protons and electrons pool for energy production (Yoon et al., 2014). In a phase of light reaction, solar energy is captured by phycobilisomes and transferred to PSII where the light energy is used for splitting cellular water molecules and releasing oxygen, protons and electrons (Westermark and Steuer, 2016). These electrons help in the reduction of plastoquinone (PQ pool) and are further transported to PSI *via* Cyt b_6f in the Z-scheme pathway. PSI catalyzes the light-driven electron transport, including the oxidation of plastocyanin (PC), Cyt-c and ferredoxin (Fd) (Vermaas, 2001). In linear electron transport, ferredoxin has a strong reductant that helps in the transfer of electrons from ferredoxin to nicotinamide adenine dinucleotide phosphate ($NADP^+$) for the generation of NADPH molecules (Rochaix, 2011). Apart from this linear electron transport pathway, PSI also played a major role in the cyclic electron transport mechanism (Battchikova et al., 2011). In order to balance the ratio of ATP and NADPH in the cell, the cyclic electron transport produces ATP (Larkum et al., 2018). Thus, the electron flow pathway is associated with the electrochemical gradient formed (e^-/H^+) across thylakoid membranes which crucially play an important role in driving ATP synthesis by the ATPase. Most of the ATP and NADPH are utilized for CO_2 fixation in the form of reduced sugars (Nikkanen et al., 2021; Shimakawa et al., 2021). Simultaneously, the cytoplasmic membrane contains key respiratory ETC complexes, including NADPH dehydrogenase (NDH-1, NDH-II), succinate dehydrogenase (SDH) and respiratory terminal oxidase (RTO) which play a major role in generating the proton/electron pool (Liu, 2016). Apart from respiratory electron carrier, cyt b_6f complex and PC (cyt-c) have also been involved in the pathway of the respiratory electron transport system in the thylakoid membrane (Vermaas, 2001). Overall, functional coordination between the two pathways indicates the existence of both respiration and photosynthesis processes occurring in similar membrane compartments (Vermaas, 2001) (Figure 3.1).

3.4 ELECTROCHEMICAL MICROBIAL FUEL CELLS

In the past century, many innovative technologies have been developed for rapid and efficient conversion of solar energy into bioelectricity. Recently, bio-electrochemical technologies have been used in photovoltaic devices for the development of revolutionary artificial photosynthesis systems for the generation of bioelectricity. In addition, MFCs and photosynthetic microbial fuel cells (PMFCs) have played a pivotal role in the transformation of chemical energy (organic substrates) directly into electrical energy through vital phototrophic electron transport mechanism (Greenman et al., 2019; Ndayisenga et al., 2018). An MFC is a unique bio-electrochemical system for generation of bioelectricity based on the uses of biochemical reactions inside cells (Allen and Bennetto, 1993; Rosenbaum et al., 2010). The integrated approach of photosynthetic organisms in MFCs leads to the establishment of PMFCs. In substrate oxidation, electrons are generated and transferred to the anode surface via mediators (Pandit and Das, 2015) whereas protons (H^+) pass through the proton-exchange membrane and reach the cathode through an electric circuit for generation of electricity (Montpart et al., 2014; Shukla and Kumar, 2018; Sevda et al., 2019). Extracellular electrons are transferred to the anode directly through cytochromes of the cell membrane (Schröder, 2007; Sund et al., 2007). It has been well documented that model cyanobacterium *Synechocystis* sp. PCC 6803 deposited on the cathode exhibited a positive response with anodic *Shewanella* sp. under illumination conditions which plays an important role as a biocatalyst and oxygenator (Yoon et al., 2014). Recently, PMFC and MFC have been used in single as well as dual-chambered cells (Rabaey et al., 2003; Logan et al., 2005) such as anode and cathode by proton-exchange membrane (Liu and Logan, 2004; Liu et al., 2005) for rapid generation of bioelectricity. The role of microalgae in bioelectricity production through fuel cell technology is well established (Table 3.1).

The use of nanomaterial in the fabrication of MFC and PMFC components is used particularly for modification of electrodes (anode and cathode) and improvement in the performance and efficacy of electron conductivity, power density, corrosion-resistant, thermal stability and cost-effectiveness (Zou et al., 2017). In addition, certain metal nanoparticles (i.e., silver, copper, gold), 2-D nanomaterials (graphene), carbon nanotubes and nanocomposites are well known for enhancement of bioenergy production through MFC/PMFC nanotechnology (Khoo et al., 2020; Slate et al., 2019). Certain MFCs have been reported to produce a maximum power of 5.85 W m^{-2} (Rosenbaum et al., 2006). Interestingly, PMFCs also help in the reduction of global carbon dioxide for a healthy environment (Cao et al., 2009). Therefore, utilization of the thylakoid membrane of plants, microalgae and phototrophic bacteria is the

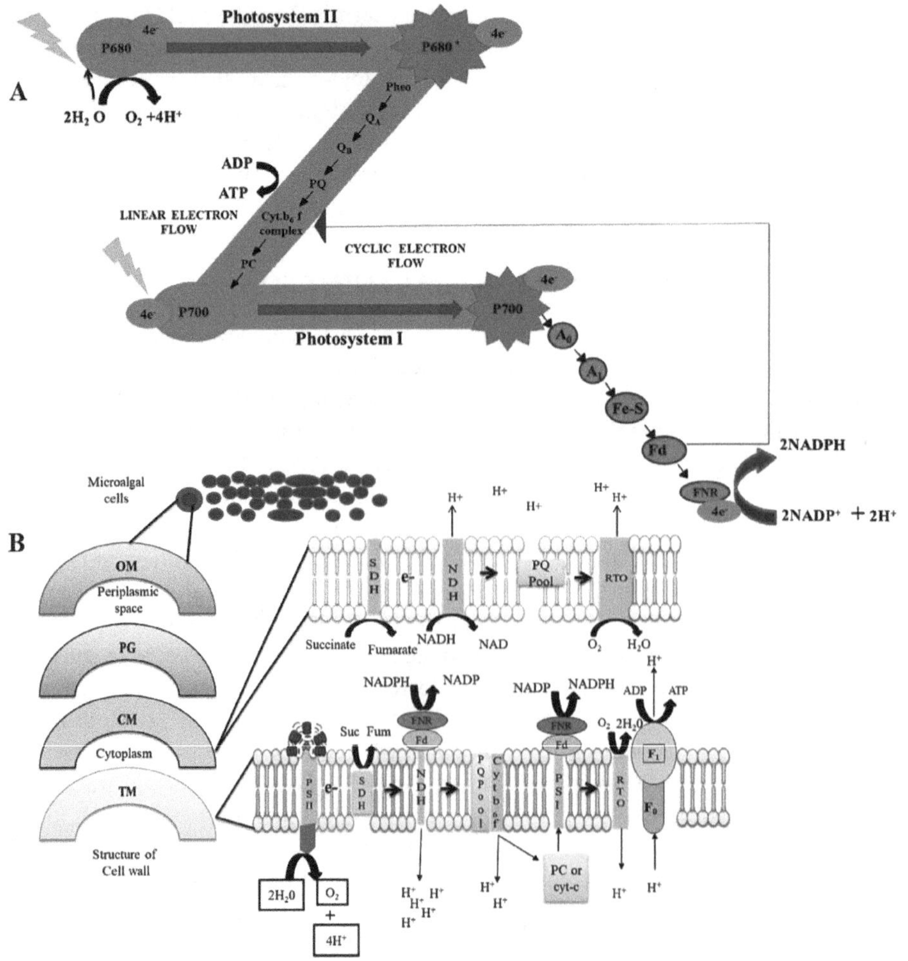

FIGURE 3.1 Diagrammatic representation of Z-scheme: (a) electron transport chain, (b) in microalgae (blue-green algae/cyanobacteria). The key regulatory protein complexes in both cytoplasmic and thylakoid membranes play a vital role in photosynthesis and respiration. Abbreviations: OM, outer membrane; PG, peptidoglycan layer; CM, cytoplasmic membrane; TM, thylakoid membrane; Fd, ferredoxin; FNR, ferredoxin nicotinamide adenine dinucleotide phosphate ($NADP^+$) oxidoreductase; PC, plastocyanin; PQ, plastoquinone; PSI, photosystem I; PSII, photosystem II, NDH, NADPH dehydrogenase; SDH, succinate dehydrogenase; RTO, respiratory terminal oxidase; P680, reaction center (PSII); Pheo, pheophytin; PQ, plastoquinone molecule;; Cyt.b_6f, cytochrome complex consists of cytochrome b_6 and cytochrome f; PC, plastocyanin; P700, reaction center chlorophyll (Chl) in PSI; Ao, primary electron acceptor of PSI; $A_{1,}$ phylloquinone; FeS, iron-sulfur protein.

ultimate resource for the development of MFC- and PMFC-based electricity generation (Rosenbaum et al., 2010). Apart from the photosynthetic thylakoid membrane, the anode and cathode also played distinguished roles in the regulation of generation of bioelectricity (Figure 3.2).

TABLE 3.1
Power Generation Capacity and Advantages o10.882 ptf Microalgae in Single- and Dual-Chambered Photosynthetic Fuel Cells for Energy Production

S.No.	Microalgae	Type of PEM	MFC/PMFC	Power Density (W/m^2)	Advantages	References
	Arthrospira maxima	Ion exchange	Dual chamber	–	Proton conductivity, thermostable	Inglesby and Fisher, (2013)
	Chlamydomonas reinhardtii	Nafion 211	Dual chamber	0.01295	Good mechanical and chemical stability, high cation conductivity	Lan et al. (2013) and Peron et al. (2010)
	Chlorella vulgaris	Sterion	Dual chamber	0.0135	Resistance and durable toward chemicals	del Campo et al. (2013)
	Chlorella vulgaris	Nafion membrane	Dual chamber	0.187	Thermotolerant, role in proton conduction	Liu et al. (2015)
	Chlorella vulgaris	Ultrex proton-exchange membrane (CMI-7000)	Dual chamber	0.0053	Cost-effective, high cation selectivity	Lakaniemi et al. (2012)
	Chlorella vulgaris	Nafion 112	Dual chamber	–	Proton transportation without the aid of solvent and water	Mitra and Hill, (2012) and Peng et al. (2009)
	Chlorella vulgaris	Glass fiber filters	Single chamber	–	Cost-effective, high power density, limits oxygen permeability, biomass growth resistant	Velasquez-Orta et al. (2009)
	Chlorella pyrenoidosa	Nafion 112	Dual chamber	0.0302	Proton transportation without the aid of solvent and water	Xu et al. (2015) and Peng et al. (2009)
	Chlorella pyrenoidosa	Nafion 112	Dual chamber	0.0025	Proton transportation without the aid of solvent and water	Xu et al. (2015) and Peng et al. (2009)
	Chlamydomonas reinhardtii	–	Single chamber	0.075	–	Nishio et al. (2013)
	Cyanobacteria	Microfiltration membrane	Single chamber	0.072	Role in wastewater treatment at the initial stage	Zhao et al. (2012) and Anis et al. (2019)

(Continued)

TABLE 3.1 (*Continued*)
Power Generation Capacity and Advantages o10.882 ptf Microalgae in Single- and Dual-Chambered Photosynthetic Fuel Cells for Energy Production

S.No.	Microalgae	Type of PEM	MFC/PMFC	Power Density (W/m^2)	Advantages	References
	Cyanobacteria	–	Single chamber	0.114	–	Yuan et al. (2011)
	Desmodesmus sp. AS	Cation exchange membrane	Dual chamber	0.0642	Proton conductivity, thermostable	Wu et al. (2014)
	Microcystis aeruginosa	CMI-7000	Dual chamber	–	Cost-effective, high cation selectivity	Wang et al. (2012)
	Microcystis aeruginosa IPP	Nafion 117	Dual chamber	–	Ion exchange membrane, role in fabrication of fuel cells	Cai et al. (2013) and Yang et al. (2004)
	Mixed algae	Nafion-117	Single chamber	3.55×10^{-6}	Ion exchange membrane, role in fabrication of fuel cells	Subhash et al. (2013) and Yang et al. (2004)
	Mixed algae	Nafion-117	Single chamber	0.076	Ion exchange membrane, role in fabrication of fuel cells	Chandra et al. (2012) and Yang et al. (2004)
	Mixed algae	Salt bridge	Dual chamber	0.207	Inexpensive, lower oxygen permeability, easy handling	Singhvi and Chhabra (2013)
	Nostoc sp. ATCC 27893	–	–	0.1		Sekar et al. (2014)
	Synechococcus leopoliensis	Ceramic membrane	Stack of 9MFCS	–	Inexpensive, scale-up MFC	Walter et al. (2015) and Winfield et al. (2016)
	Scenedesmus obliquus	Nafion 117	Dual chamber	0.102	Ion exchange membrane, role in fabrication of fuel cells	Kondaveeti et al. (2014) and Yang et al. (2004)
	Synechocystis PCC-6803	Proton-exchange membrane	Dual chamber	0.0723	High cation selectivity, less expensive than nafion	Ma et al. (2012)
	Spirulina platensis	–	Single chamber	0.0065	–	Fu et al. (2010b)

Note: For details see references Daud et al. (2015), Shukla and Kumar (2018) and Cao et al. (2019).

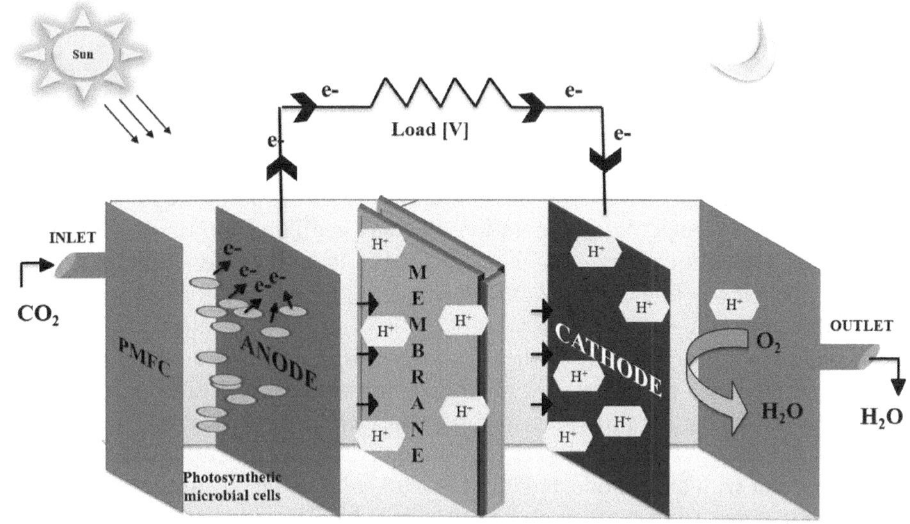

FIGURE 3.2 Schematic representation of photosynthetic microalgal fuel cells showing electron transfer mechanism from ETC system to anode and cathode for generation of bioelectricity.

3.4.1 Anodic Bioelectricity Mechanism

Microalgal cells residing in wastewater serves as an easy inoculum for generation of fuel cell-based bioelectricity through an anode (Rashid et al., 2013; Martin et al., 2010; Shukla and Kumar, 2018). Interestingly, maximum power output generation was recorded by using natural substrates such as algal biomass, activated sludge and lipids (Rashid et al., 2013). In response to power generation efficiency, an experiment was conducted on nine marine microalgae, namely *Isochrysis* sp., *Dunaliella* sp., *Nanochloropsis* sp., *Synechocystis* sp., *Tetraselmis gracilis*, *Chaetoceros calcitrans*, *Chlorella salina*, *Pavlova* sp. and *Dicarteria* sp., where *Tetraselmis gracilis* exhibited maximum power output (0.792 V) in the process of bioelectricity generation (Ramanathan et al., 2011). Interestingly, *Chlorella pyrenoidosa* was used as an electron donor at the anode which exhibited maximum power output (6.03 W m^{-2}) at 2,500 lux (Xu et al., 2015). A comparative study between microalgae-based MFC and acetate-based MFC reveals that microalgae-based MFC performed with maximum power (1.92 W m^{-2}) as compared to chemical-based MFC (Cui et al., 2014). Thus, microalgae have a promising role at the anode for the efficient generation of electricity.

3.4.2 Cathodic Bioelectricity Mechanism

Biocathode is an inventive approach where phototrophic microorganisms function as biocatalysts to initiate electrochemical reduction on the cathode surface (Huang et al., 2011; Sun et al., 2012). Microalgae in MFC cathode chamber perform photosynthesis and convert carbon dioxide to produce biomass and oxygen (Wu et al., 2013). The continuous supply of microalgae generates oxygen at the cathode, accelerating the reaction in a more favorable direction as compared to the addition of a catalyst or salt at the cathodic side (Ren et al., 2013; Lay et al., 2015). Interestingly, biofilm thickness of biomass has also played a major role in the production of bioelectricity (Gajda et al., 2015; Behera et al., 2010). A similar study was done by using *Scenedesmus obliquus* as a biocathode (Kakarla and Min, 2014). Light is one of the major parameters that induce bioelectricity generation whereas negligible power generation is found during the dark conditions (Xiao et al., 2012). In addition to variation in light and intensities, it was found that *Chlamydomonas reinhardtii* performed well in red light with a maximum power density of 12.95 mW m^{-2} during electricity generation (Lan et al., 2013). The carbon dioxide generated at the anodic compartment of MFCs is utilized by a cathode called microbial carbon capture (MCC) that is useful for microalgae during the process of photosynthesis (Cui et al., 2014). This mechanism has been used for the development of an integrated approach for the setup of a photobioreactor with biocathode and utilization of subtraction of nitrogen (96%) and phosphorous (55%) from municipal wastewater in bioelectricity generation (Sevda et al., 2019). The above result gives a new way for the utilization of algal biocathodes for municipal wastewater treatment without an additional requirement of water and nutrients (ElMekawy et al., 2014).

3.5 APPLICATIONS OF MICROALGAE IN THE ENERGY SECTOR

Microalgae have been tested as a viable biofuel feedstock in the contemporary period due to their rapid growth rate, high energy content, low-cost culture technology and remarkable capacity for CO_2 fixation and O_2 evolution for a healthy environment. Now, microalgae-based biofuel technologies have been implemented for large-scale production of biofuels (Medipally et al., 2015; Hossain et al., 2019; Mobin et al., 2019; Saad et al., 2019) (Figure 3.3). Recently, a "Biorefinery" concept has been introduced which encompasses methods and equipment to produce bioelectricity, biofuels and high-value-added chemicals (Siddiki et al., 2022). This concept is designed for sustainable processing of microalgal biomass into a variety of commercial products and clean energy (Markou and Nerantzis, 2013).

3.5.1 Bio-photovoltaic Cells

Nowadays, photosynthetic microbes are more frequently used in PMFCs or bio-photovoltaic cells for generation of bioelectricity. MFCs are unique and highly efficient for harvesting photonic energy (Lee et al., 2015). The usage of photosynthetic microorganisms in the field of photovoltaics has played a major role in the area of clean energy technology (Logan et al., 2006). Moreover, this could also be used to remove nitrogen elements from water (Xiao et al., 2012) and sequestration of carbon dioxide from the atmosphere (Wang et al., 2010). The bio-photovoltaics technique has much more advantages over present photovoltaic cells due to cost-effective efficient utilization of solar energy by photosynthetic microbes (Rosenbaum and Schröder, 2010). In the past century, most of the studies were more focused on whole-cell technology with external

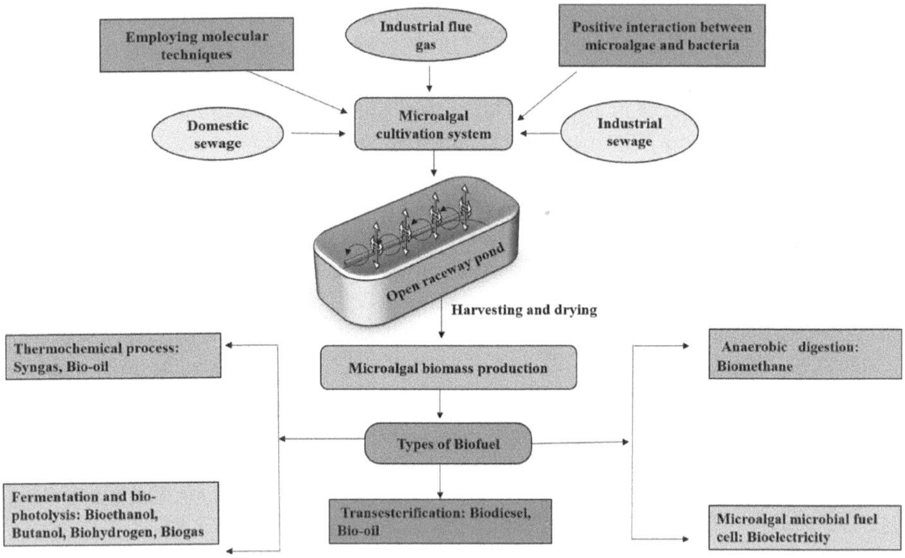

FIGURE 3.3 A pictorial overview of different sources and strategies for the production of biofuel from microalgae.

electron mediators for energy generation, but handling and commercialization of electron mediators are very critical to maintain in energy devices (McCormick et al., 2011). Nowadays, photosynthetic organisms are directly cultured on the electrode surface for generation of photocurrent (Bond and Lovely, 2003; Reguera et al., 2006; Franks et al., 2010). Recently, it was found that algal biofilm coated over anode surface produces high power density as compared to algal suspensions used at anode position (Periasamy et al., 2022).

3.5.2 Biodiesel

Microalgae are smaller in size, mostly unicellular and have wide adaptability in various environmental conditions (Leite et al., 2013). The biomass of microalgae is an indispensable source of various types of biofuels. Biodiesel (clean fuel) is one of the promising solutions to counteract with increasing demands of fuel without posing any threat to our green environment (Elrayies, 2018). Previously, it was obtained from the vegetable oil of plants such as jatropha, oil palm, canola, soy, waste cooking oil, animal fats and lipids (Edwards, 2008; Elrayies, 2018). Nowadays, microalgae are the richest source with many precursor elements required for the synthesis of biofuel. Recently, diatoms have been considered as third-generation microalgae biofuel (Mourya et al., 2022). Microalgal biodiesel production has sequential steps from site selection to level of production (Mata et al., 2010; Ahmad et al., 2011). For efficient production of biodiesel from microalgae, whole processes are designed into two main units, the first unit is used for algal cultivation and the second unit is concerned with biodiesel production from biomass (Faried et al., 2017). Sedimentation and solvent extraction processes are made handier for the isolation of crude oil from algal biomass (Tercero et al., 2014; Faried et al., 2017). Interestingly, the process of pyrolysis and hydrothermal liquefaction are the main steps involved in the production of bio-oil (Jena and Das, 2011; Li et al., 2019). The bio-oil obtained from algae yields 23 times more than palm oil and 300 times more than soyabean oil (Ray et al., 2022). In terms of quality and yield, temperature and lipid content are major parameters for biofuel production (Yoo et al., 2015). Moreover, nitrogen and oxygen contents are a kind of undesirable products in direct combustion for preparation of biocrude oil (Biller et al., 2015). Oils and lipids are extracted *via* organic solvent, cold pressing, Soxhlet and supercritical fluid methods (Bahadar and Khan, 2013). In addition, several fractionation processes have also been employed for obtaining desired products of bio-oil (Biller et al., 2015; Chen et al., 2015; Elliott et al., 2015). Interestingly, algal storage biomass of lipids and starch are also utilized for the formation of biohydrogen, bioethanol and biodiesel (Hallenbeck et al., 2016). The quality of microalgal biodiesel for application in automobiles has been regulated by the American Society for Testing and Materials (ASTM) (Ashokkumar et al., 2014; Sazzad et al., 2016). There are many ways for the production of biodiesel in an industrial scale (Figure 3.4).

In the past few years, magnetic nanoparticles (MNPs) were used in the process of hydrolysis of microalgal cells by immobilizing cellulase (Duraiarasan et al., 2016). In the process of hydrolysis, the maximum yield was recorded at about 93.56% under optimal conditions (Khoo et al., 2020). Recently, certain other nanomaterials like CaO and MgO have been introduced as catalysts or catalyst support for lipid accumulation, extraction and transesterification process of microalgal biomass for high production of biodiesel (Zhang et al., 2013). However, there are still a number of gaps for feasible application of nanomaterials for large-scale production of biodiesel.

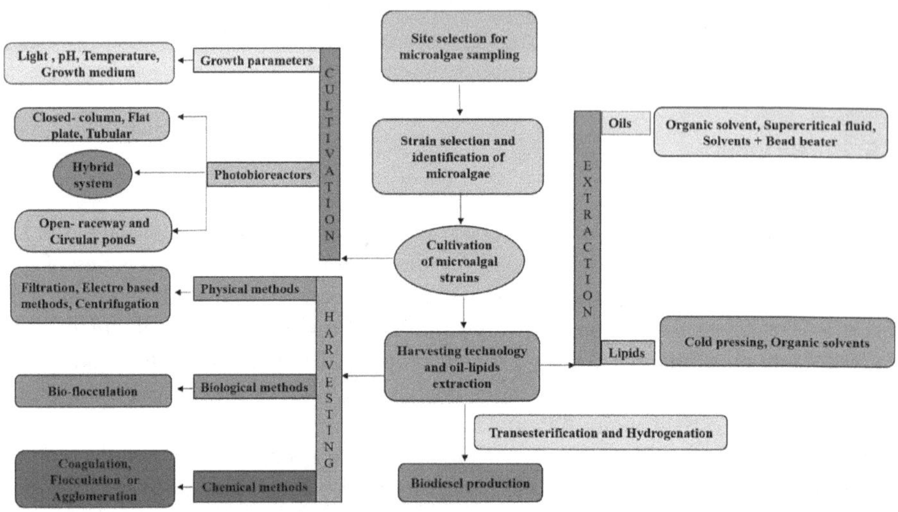

FIGURE 3.4 A simple illustration of biodiesel production from microalgae. For details see references Mata et al. (2010), Ahmad et al. (2011), Bahadar and Khan (2013) and Faried et al. (2017).

3.5.3 Bioethanol

Bioethanol is ethyl alcohol (Et-OH or C_2H_5OH) used as an alternative transportation fuel in the past century (Balat and Balat, 2009). The scientific community has shown keen interest in the production of bioethanol to substitute gasoline (Lam and Lee, 2015). Previously, ethanol was produced from sugar materials (wheat, sugar beet, sugarcane and corn) and blended with gasoline from 5% to 100% in an SI engine for obtaining higher octane rating, density, decreasing heating value and latent heat of vaporization (Harun et al., 2010; Masum et al., 2013; Lam and Lee, 2015). Ethanol blend in lower concentration has resulted in emission of higher amounts of nitrogen-containing gases (NO_x) which increase pollution in the environment. However, NO_x emission was reduced in a higher proportion (85%) of blending of pure ethanol (Masum et al., 2013). The second generation of bioethanol production was based on lignocellulosic (*Miscanthus* crop) biomass which does not involve any food commodity (Guo et al., 2013). The major drawback of the second generation of fuel is that it requires an additional step of lignin purification followed by hydrolysis and fermentation to isolate bioethanol (Cardona and S'anchez, 2007). Third generation of bioethanol was based on aquatic microorganisms, that is, mostly microalgal biomass (Jambo et al., 2016). Microalgal species have great potential to accumulate a major amount of carbohydrate which can be further used as a substrate for fermentation in bioethanol production (Radakovits et al., 2010; Harun et al., 2010; Lam and Lee, 2015). Most of the microalgae have 21%–64% carbohydrate content in dry-weight biomass (Ho et al., 2012; Schwenzfeier et al., 2011; Lam and Lee, 2015; Bondioli et al., 2012). Certain common examples that are being used in bioethanol production are *Scenedesmus abundans*, *Scenedesmus obliquus* and *Tetraselmis suecica* (Lam and Lee, 2015). Industrially, ethanol is produced through acidic (e.g., HCl, H_2SO_4 and HNO_3) and basic (KOH, NaOH and Na_2CO_3) chemical pre-treatment to break down lignocellulosic biomass (Sun and Cheng, 2002; Bensah and Mensah, 2013). However, this is not a highly feasible technology of production due to requirements of high pressure and temperature for carbohydrate degradation (Chen et al., 2013; Lam and Lee, 2015). In alkali treatment, NaOH of concentration 0.75 w/v%, 120°C for 30 min was used to obtain the optimum yield of bioethanol from *Chlorococcum infusionum* (Harun et al., 2011; Lam and Lee, 2015). In the process of carbohydrate degradation, two major enzymes namely α-amylase (endoamylase, EC 3.2.1.1) and amyloglucosidase (exoamylase, EC 3.2.1.3) were utilized in bioethanol production (Van Der Maarel et al., 2002; Richardson et al., 2000; Choi et al., 2010). Recently, various nanotechnological approaches have also been involved for enhancing the process of extraction and yield of carbohydrates for bioethanol production (Sirajunnisa and Surendhiran, 2016). In the enzymatic approach, nanofibrous membrane (polyacrylonitrile) was used to immobilize the cellulase enzyme of *Chlorella* sp. for rapid lyases of cell wall and releasing of sugar molecules for industrial scale of bioethanol production (Fu et al., 2010a).

3.5.4 Biohydrogen

The inherent potential of microalgae for biohydrogen production has attracted the industry at the global scale for development as alternative resources of fossil fuel. Certain genera of microalgae are commonly used in biohydrogen production, including

Anabaena, *Botryococcus*, *Scenedesmus*, *Chlorella*, *Tetraspora*, *Chlorococcum* and *Nostoc* (Eroglu and Melis, 2011). The presence of hydrogenase enzyme in microalgae makes it the most suitable organism for biohydrogen production (Ghirardi et al., 2000; Eroglu and Melis, 2011). Biohydrogen production is a very clean technology, economically sustainable without any detrimental impact on green environment.

Biohydrogen transforms the conventional use of hydrocarbon fuels due to high energy efficiency (Hosseini and Wahid, 2016; Aziz, 2016). Bio-photolysis is the key step for biohydrogen production that is derived from protons and electrons during the splitting of water at photosystem II under anaerobic conditions (Nagarajan et al., 2017) by degradation of carbon compounds (Mathews and Wang, 2009). Interestingly, a unit weight of H_2 (Hydrogen) gas has the capacity to generate 141.65 MJ kg^{-1} of heating value (Khetkorn et al., 2017). Hydrogen energy can be directly employed in the vehicle and industrial sectors for generation of electricity.

3.5.5 Biogas

The production of biogas from anaerobic digestion of microalgal biomass is one of the promising approaches in the biofuel sector (Kendir and Ugurlu, 2018). The anaerobic digestion is an obligate microbiological process for the decomposition of organic matter under oxygen-free conditions (Srirangan et al., 2012). The biogas is made up of methane (55%–70%), carbon dioxide (30%–45%), a trace amount of H_2S (50–2,000 ppm), nitrogen, ammonia water, oxygen and hydrocarbon (Zabed et al., 2020; Kendir and Ugurlu, 2018). In the process of biomass combustion, gasification reaction occurs in the presence of high temperature and micro-oxygen environment with a catalyst to decompose biomolecules into carbon dioxide and hydrogen (Chow et al., 2013). The process of catalytic hydrothermal gasification and hydrothermal liquefaction function at a supercritical state under average temperature and pressure (Jiao et al., 2017; Biller et al., 2015). The advantages associated with microalgal biogas production are minimum resource requirements, maximum yield, economically cost-effective and minimal sludge production (Saratale et al., 2018; Zabed et al., 2020). In 1957, Golueke and co-workers had reported an effective process of anaerobic digestion by using *Scenedesmus* sp. and *Chlorella* sp. (Golueke et al., 1957). The yield of biogas depends on three basic parameters – protein, lipid and carbohydrate content of microalgae, type of species and environment of culture conditions (Sialve et al., 2009; Kendir and Ugurlu, 2018).

3.5.6 Environmental Biosensor

An effluent from industries, hospitals, agricultural farms and houses causes detrimental effects on organisms residing in water bodies. In recent scenarios, the overgrowth of population leads to a massive increase in pollution in water, air and land that causes heavy accumulation of toxic elements in a global environment. Biosensors could be of a great utility for the early detection of polluting substances for reducing hazardous effects on the environment. Nowadays, algal biosensors have emerged as a sensing tool for the detection of metals, volatile organic compounds, herbicides (Naessens et al., 2000) and chemical warfare agents (Altamirano et al., 2004). Microalgal biosensor measures the metabolic activity of organisms (Brayner et al., 2011).

3.6 CONCLUSION AND FUTURE PERSPECTIVE

Overgrowth of the global population causes detrimental impacts on human survivability by limiting space and resource availability. Energy is the basic and daily need for the existence of life on Earth's surface. Therefore, demand and utilization of energy have continuously increased by many folds over the past century. Overdependence on limited non-renewable energy sources has tremendously induced environmental pollution, causing life-threatening diseases and instability of the Earth's ecosystem. To overcome the challenges of energy, most of the innovation targeted renewable sources of energy like solar radiation, wind energy, hydro energy, biomass energy and geothermal energy. Solar radiation is the ultimate source of energy for long-term utilization in various sectors. Microalgae are the smallest group of photosynthetic organisms that provide a promising solution for harvesting solar energy and the development of various kinds of biofuels. Over the last decades, many experimental researches have been formulated worldwide for unmasking the potentials of microalgae in the bioenergy field. The synergy of photosynthetic systems in microalgae with nanotechnology has been put forth with the hope of meeting the demands of an exponentially growing population. The prime concerns related to production for commercialization depend on the selection of high-performance microalgal strains, yield efficiency and cost-effectiveness in large-scale industries. There is a need for the development of an approach to maximize the utilization of the photosynthetic efficiency of microalgae in nanotechnology.

ACKNOWLEDGMENTS

This work was supported by IoE-Seed Grant, BHU, Varanasi (scheme no. 6031) and Core Research Grant, DST-SERB, New Delhi (CRG/2020/001323) to V. K. Kannaujiya. Saumi Pandey and Nasreen Amin would like to thank UGC, BHU for granting fellowship.

REFERENCES

Ahmad, A. L., Yasin, N. M., Derek, C. J. C., & Lim, J. K. (2011). Microalgae as a sustainable energy source for biodiesel production: A review. *Renewable and Sustainable Energy Reviews*, 15(1), 584–593.

Allen, R. M., & Bennetto, H. P. (1993). Microbial fuel-cells. *Applied Biochemistry and Biotechnology*, 39(1), 27–40.

Altamirano, M., Garcıa-Villada, L., Agrelo, M., Sánchez-Martın, L., Martın-Otero, L., Flores-Moya, A.,... & Costas, E. (2004). A novel approach to improve specificity of algal biosensors using wild-type and resistant mutants: An application to detect TNT. *Biosensors and Bioelectronics*, 19(10), 1319–1323.

Anis, S. F., Hashaikeh, R., & Hilal, N. (2019). Microfiltration membrane processes: A review of research trends over the past decade. *Journal of Water Process Engineering*, 32, 100941.

Asafu-Adjaye, J. (2000). The relationship between energy consumption, energy prices and economic growth: Time series evidence from Asian developing countries. *Energy Economics*, 22(6), 615–625.

Ashokkumar, V., Agila, E., Sivakumar, P., Salam, Z., Rengasamy, R., & Ani, F. N. (2014). Optimization and characterization of biodiesel production from microalgae *Botryococcus* grown at semi-continuous system. *Energy Conversion and Management*, 88, 936–946.

Aziz, M. (2016). Integrated hydrogen production and power generation from microalgae. *International Journal of Hydrogen Energy*, 41(1), 104–112.

Bahadar, A., & Khan, M. B. (2013). Progress in energy from microalgae: A review. *Renewable and Sustainable Energy Reviews*, 27, 128–148.

Balat, M., & Balat, H. (2009). Recent trends in global production and utilization of bio-ethanol fuel. *Applied Energy*, 86(11), 2273–2282.

Banerjee, I., Dutta, S., Pohrmen, C. B., Verma, R., & Singh, D. (2020). Microalgae-based carbon sequestration to mitigate climate change and application of nanomaterials in algal biorefinery. *Octa Journal of Biosciences*, 8, 129–136.

Battchikova, N., Eisenhut, M., & Aro, E. M. (2011). Cyanobacterial NDH-1 complexes: Novel insights and remaining puzzles. *Biochimica et Biophysica Acta (BBA)-Bioenergetics*, 1807(8), 935–944.

Behera, M., Jana, P. S., & Ghangrekar, M. M. (2010). Performance evaluation of low-cost microbial fuel cell fabricated using earthen pot with biotic and abiotic cathode. *Bioresource Technology*, 101(4), 1183–1189.

Bensah, E. C., & Mensah, M. (2013). Chemical pretreatment methods for the production of cellulosic ethanol: Technologies and innovations. *International Journal of Chemical Engineering*. doi:10.1155/2013/719607.

Biller, P., Sharma, B. K., Kunwar, B., & Ross, A. B. (2015). Hydroprocessing of bio-crude from continuous hydrothermal liquefaction of microalgae. *Fuel*, 159, 197–205.

Blaschke T, Biberacher M, Gadocha S, Schardinger I (2013). 'Energy landscapes': Meeting energy demands and human aspirations. *Biomass and Bioenergy*, 55, 3–16.

Bond, D. R., & Lovley, D. R. (2003). Electricity production by *Geobacter sulfurreducens* attached to electrodes. *Applied and Environmental Microbiology*, 69(3), 1548–1555.

Bondioli, P., Della Bella, L., Rivolta, G., Zittelli, G. C., Bassi, N., Rodolfi, L.,... & Tredici, M. R. (2012). Oil production by the marine microalgae *Nanochloropsis* sp. F&M-M24 and *Tetraselmis suecica* F&M-M33. *Bioresource Technology*, 114, 567–572.

Brayner, R., Couté, A., Livage, J., Perrette, C., & Sicard, C. (2011). Micro-algal biosensors. *Analytical and Bioanalytical Chemistry*, 401(2), 581–597.

Brennan, L., & Owende, P. (2010). Biofuels from microalgae-a review of technologies for production, processing, and extractions of biofuels and co-products. *Renewable and Sustainable Energy Reviews*, 14(2), 557–577.

Cai, P. J., Xiao, X., He, Y. R., Li, W. W., Zang, G. L., Sheng, G. P.,... & Yu, H. Q. (2013). Reactive oxygen species (ROS) generated by cyanobacteria act as an electron acceptor in the biocathode of a bio-electrochemical system. *Biosensors and Bioelectronics*, 39(1), 306–310.

Cao, X., Huang, X., Liang, P., Boon, N., Fan, M., Zhang, L., & Zhang, X. (2009). A completely anoxic microbial fuel cell using a photo-biocathode for cathodic carbon dioxide reduction. *Energy & Environmental Science*, 2(5), 498–501.

Cao, Y., Mu, H., Liu, W., Zhang, R., Guo, J., Xian, M., & Liu, H. (2019). Electricigens in the anode of microbial fuel cells: Pure cultures versus mixed communities. *Microbial Cell Factories*, 18(1), 1–14.

Cardona, C. A., & Sánchez, Ó. J. (2007). Fuel ethanol production: Process design trends and integration opportunities. *Bioresource Technology*, 98(12), 2415–2457.

Chandra, R., Subhash, G. V., & Mohan, S. V. (2012). Mixotrophic operation of photo-bioelectrocatalytic fuel cell under anoxygenic microenvironment enhances the light dependent bioelectrogenic activity. *Bioresource Technology*, 109, 46–56.

Chen, C. Y., Zhao, X. Q., Yen, H. W., Ho, S. H., Cheng, C. L., Lee, D. J.,... & Chang, J. S. (2013). Microalgae-based carbohydrates for biofuel production. *Biochemical Engineering Journal*, 78, 1–10.

Chen, W. H., Lin, B. J., Huang, M. Y., & Chang, J. S. (2015). Thermochemical conversion of microalgal biomass into biofuels: A review. *Bioresource Technology*, 184, 314–327.

Choi, S. P., Nguyen, M. T., & Sim, S. J. (2010). Enzymatic pretreatment of *Chlamydomonas reinhardtii* biomass for ethanol production. *Bioresource Technology*, 101(14), 5330–5336.

Chow, M. C., Jackson, W. R., Chaffee, A. L., & Marshall, M. (2013). Thermal treatment of algae for production of biofuel. *Energy & Fuels*, 27(4), 1926–1950.

Cui, Y., Rashid, N., Hu, N., Rehman, M. S. U., & Han, J. I. (2014). Electricity generation and microalgae cultivation in microbial fuel cell using microalgae-enriched anode and bio-cathode. *Energy Conversion and Management*, 79, 674–680.

Daud, S. M., Kim, B. H., Ghasemi, M., & Daud, W. R. W. (2015). Separators used in microbial electrochemical technologies: Current status and future prospects. *Bioresource Technology*, 195, 170–179.

De Vries BJ, Van Vuuren DP, Hoogwijk MM (2007). Renewable energy sources: Their global potential for the first-half of the 21st century at a global level: An integrated approach. *Energy Policy*, 35(4), 2590–2610.

del Campo, A. G., Cañizares, P., Rodrigo, M. A., Fernández, F. J., & Lobato, J. (2013). Microbial fuel cell with an algae-assisted cathode: A preliminary assessment. *Journal of Power Sources*, 242, 638–645.

Dogan, E., & Seker, F. (2016). Determinants of CO_2 emissions in the European Union: The role of renewable and non-renewable energy. *Renewable Energy*, 94, 429–439.

Duraiarasan, S., Razack, S. A., Manickam, A., Munusamy, A., Syed, M. B., Ali, M. Y.,... & Mohiuddin, M. S. (2016). Direct conversion of lipids from marine microalga *C. salina* to biodiesel with immobilised enzymes using magnetic nanoparticle. *Journal of Environmental Chemical Engineering*, 4(1), 1393–1398.

Eckelman, M. J., Huang, K., Lagasse, R., Senay, E., Dubrow, R., & Sherman, J. D. (2020). Health care pollution and public health damage in the United States: An update: Study examines health care pollution and public health damage in the United States. *Health Affairs*, 39(12), 2071–2079.

Edwards, M. R. (2008). *Green Algae Strategy: End Biowar I and Engineer Sustainable Food and Biofuels*. LuLu Press.

El-Khouly, M. E., El-Mohsnawy, E., & Fukuzumi, S. (2017). Solar energy conversion: From natural to artificial photosynthesis. *Journal of Photochemistry and Photobiology C: Photochemistry Reviews*, 31, 36–83.

Ellabban O, Abu-Rub H, Blaabjerg F (2014). Renewable energy resources: Current status, future prospects and their enabling technology. *Renewable and Sustainable Energy Reviews*, 39, 748–764.

Elliott, D. C., Biller, P., Ross, A. B., Schmidt, A. J., & Jones, S. B. (2015). Hydrothermal liquefaction of biomass: Developments from batch to continuous process. *Bioresource Technology*, 178, 147–156.

ElMekawy, A., Hegab, H. M., Vanbroekhoven, K., & Pant, D. (2014). Techno-productive potential of photosynthetic microbial fuel cells through different configurations. *Renewable and Sustainable Energy Reviews*, 39, 617–627.

Elrayies, G. M. (2018). Microalgae: Prospects for greener future buildings. *Renewable and Sustainable Energy Reviews*, 81, 1175–1191.

Eroglu, E., & Melis, A. (2011). Photobiological hydrogen production: Recent advances and state of the art. *Bioresource Technology*, 102(18), 8403–8413.

Faried, M., Samer, M., Abdelsalam, E., Yousef, R. S., Attia, Y. A., & Ali, A. S. (2017). Biodiesel production from microalgae: Processes, technologies and recent advancements. *Renewable and Sustainable Energy Reviews*, 79, 893–913.

Franks, A. E., Malvankar, N., & Nevin, K. P. (2010). Bacterial biofilms: The powerhouse of a microbial fuel cell. *Biofuels*, 1(4), 589–604.

Fu, C. C., Hung, T. C., Chen, J. Y., Su, C. H., & Wu, W. T. (2010a). Hydrolysis of microalgae cell walls for production of reducing sugar and lipid extraction. *Bioresource Technology*, 101(22), 8750–8754.

Fu, C. C., Hung, T. C., Wu, W. T., Wen, T. C., & Su, C. H. (2010b). Current and voltage responses in instant photosynthetic microbial cells with *Spirulina platensis*. *Biochemical Engineering Journal*, 52(2–3), 175–180.

Gajda, I., Greenman, J., Melhuish, C., & Ieropoulos, I. (2015). Self-sustainable electricity production from algae grown in a microbial fuel cell system. *Biomass and Bioenergy*, 82, 87–93.

Ghirardi, M. L., Zhang, L., Lee, J. W., Flynn, T., Seibert, M., Greenbaum, E., & Melis, A. (2000). Microalgae: A green source of renewable H_2. *Trends in Biotechnology*, 18(12), 506–511.

Golueke, C. G., Oswald, W. J., & Gotaas, H. B. (1957). Anaerobic digestion of algae. *Applied Microbiology*, 5(1), 47–55.

Greenman, J., Gajda, I., & Ieropoulos, I. (2019). Microbial fuel cells (MFC) and microalgae; photo microbial fuel cell (PMFC) as complete recycling machines. *Sustainable Energy & Fuels*, 3(10), 2546–2560.

Guo, H., Daroch, M., Liu, L., Qiu, G., Geng, S., & Wang, G. (2013). Biochemical features and bioethanol production of microalgae from coastal waters of Pearl River Delta. *Bioresource Technology*, 127, 422–428.

Hallenbeck, P. C., Grogger, M., Mraz, M., & Veverka, D. (2016). Solar biofuels production with microalgae. *Applied Energy*, 179, 136–145.

Harun, R., Danquah, M. K., & Forde, G. M. (2010). Microalgal biomass as a fermentation feedstock for bioethanol production. *Journal of Chemical Technology & Biotechnology*, 85(2), 199–203.

Harun, R., Jason, W. S. Y., Cherrington, T., & Danquah, M. K. (2011). Exploring alkaline pre-treatment of microalgal biomass for bioethanol production. *Applied Energy*, 88(10), 3464–3467.

Ho, S. H., Chen, C. Y., & Chang, J. S. (2012). Effect of light intensity and nitrogen starvation on CO_2 fixation and lipid/carbohydrate production of an indigenous microalga *Scenedesmus obliquus* CNW-N. *Bioresource Technology*, 113, 244–252.

Hossain, N., Mahlia, T. M. I., & Saidur, R. (2019). Latest development in microalgae-biofuel production with nano-additives. *Biotechnology for Biofuels*, 12(1), 1–16.

Hosseini, S. E., & Wahid, M. A. (2016). Hydrogen production from renewable and sustainable energy resources: Promising green energy carrier for clean development. *Renewable and Sustainable Energy Reviews*, 57, 850–866.

Huang, L., Regan, J. M., & Quan, X. (2011). Electron transfer mechanisms, new applications, and performance of biocathode microbial fuel cells. *Bioresource Technology*, 102(1), 316–323.

Inglesby, A. E., & Fisher, A. C. (2013). Downstream application of a microbial fuel cell for energy recovery from an *Arthrospira maxima* fed anaerobic digester effluent. *RSC Advances*, 3(38), 17387–17394.

Jambo, S. A., Abdulla, R., Azhar, S. H. M., Marbawi, H., Gansau, J. A., & Ravindra, P. (2016). A review on third generation bioethanol feedstock. *Renewable and Sustainable Energy Reviews*, 65, 756–769.

Jena, U., & Das, K. C. (2011). Comparative evaluation of thermochemical liquefaction and pyrolysis for bio-oil production from microalgae. *Energy & Fuels*, 25(11), 5472–5482.

Jiao, J. L., Wang, F., Duan, P. G., Xu, Y. P., & Yan, W. H. (2017). Catalytic hydrothermal gasification of microalgae for producing hydrogen and methane-rich gas. *Energy Sources, Part A: Recovery, Utilization, and Environmental Effects*, 39(9), 851–860.

Kabir, E., Kumar, P., Kumar, S., Adelodun, A. A., & Kim, K.-H. (2018). Solar energy: Potential and future prospects. *Renewable and Sustainable Energy Reviews*, 82, 894–900.

Kakarla, R., & Min, B. (2014). Photoautotrophic microalgae *Scenedesmus obliquus* attached on a cathode as oxygen producers for microbial fuel cell (MFC) operation. *International Journal of Hydrogen Energy*, 39(19), 10275–10283.

Kannan, N., & Vakeesan, D. (2016). Solar energy for future world: A review. *Renewable and Sustainable Energy Reviews*, 62, 1092–1105.

Kannaujiya, V. K., & Sinha, R. P. (2017). Impacts of diurnal variation of ultraviolet-B and photosynthetically active radiation on phycobiliproteins of the hot-spring cyanobacterium *Nostoc* sp. strain HKAR-2. *Protoplasma*, 254(1), 423–433.

Kannaujiya, V. K., Kumar, D., Pathak, J., & Sinha, R. P. (2019). Phycobiliproteins and their commercial significance. In: Mishra, A.K., Tiwari, D.N., Rai, A.N. (eds) *Cyanobacteria from Basic Science to Applications* (pp. 207–216). Academic Press.

Kannaujiya, V. K., Kumar, D., Singh, V., & Sinha, R. P. (2021). Advances in phycobiliproteins research: Innovations and commercialization. In: Sinha, R. P., Häder, Donat-P. (eds) *Natural Bioactive Compounds* (pp. 57–81). Academic Press.

Kendir, E., & Ugurlu, A. (2018). A comprehensive review on pretreatment of microalgae for biogas production. *International Journal of Energy Research*, 42(12), 3711–3731.

Khetkorn, W., Rastogi, R. P., Incharoensakdi, A., Lindblad, P., Madamwar, D., Pandey, A., & Larroche, C. (2017). Microalgal hydrogen production-A review. *Bioresource Technology*, 243, 1194–1206.

Khoo, K. S., Chia, W. Y., Tang, D. Y. Y., Show, P. L., Chew, K. W., & Chen, W. H. (2020). Nanomaterials utilization in biomass for biofuel and bioenergy production. *Energies*, 13(4), 892.

Khoo, K. S., Lee, S. Y., Ooi, C. W., Fu, X., Miao, X., Ling, T. C., & Show, P. L. (2019). Recent advances in biorefinery of astaxanthin from *Haematococcus pluvialis*. *Bioresource Technology*, 288, 121606.

Kobayashi, K. (2016). Role of membrane glycerolipids in photosynthesis, thylakoid biogenesis and chloroplast development. *Journal of Plant Research*, 129(4), 565–580.

Kondaveeti, S., Choi, K. S., Kakarla, R., & Min, B. (2014). Microalgae *Scenedesmus obliquus* as renewable biomass feedstock for electricity generation in microbial fuel cells (MFCs). *Frontiers of Environmental Science & Engineering*, 8(5), 784–791.

Koroneos, C., Spachos, T., & Moussiopoulos, N. (2003). Exergy analysis of renewable energy sources. *Renewable Energy*, 28(2), 295–310.

Lakaniemi, A. M., Tuovinen, O. H., & Puhakka, J. A. (2012). Production of electricity and butanol from microalgal biomass in microbial fuel cells. *BioEnergy Research*, 5(2), 481–491.

Lam, M. K., & Lee, K. T. (2015). Bioethanol production from microalgae. In: Kim, S.K. (ed) *Handbook of Marine Microalgae* (pp. 197–208) Academic Press, Elsevier.

Lan, J. C. W., Raman, K., Huang, C. M., & Chang, C. M. (2013). The impact of monochromatic blue and red LED light upon performance of photo microbial fuel cells (PMFCs) using *Chlamydomonas reinhardtii* transformation F5 as biocatalyst. *Biochemical Engineering Journal*, 78, 39–43.

Larkum, A. W. D., Szabo, M., Fitzpatrick, D., & Raven, J. A. (2018). Cyclic electron flow in cyanobacteria and eukaryotic algae. In: Barber, J., Ruban. A. V. (eds) *Photosynthesis and Bioenergetics* (pp. 305–343) World Scientific Publishing.

Lay, C. H., Kokko, M. E., & Puhakka, J. A. (2015). Power generation in fed-batch and continuous up-flow microbial fuel cell from synthetic wastewater. *Energy*, 91, 235–241.

Lee, D. J., Chang, J. S., & Lai, J. Y. (2015). Microalgae-microbial fuel cell: A mini review. *Bioresource Technology*, 198, 891–895.

Leite, G. B., Abdelaziz, A. E., & Hallenbeck, P. C. (2013). Algal biofuels: Challenges and opportunities. *Bioresource Technology*, 145, 134–141.

Li, F., Srivatsa, S. C., & Bhattacharya, S. (2019). A review on catalytic pyrolysis of microalgae to high-quality bio-oil with low oxygeneous and nitrogenous compounds. *Renewable and Sustainable Energy Reviews*, 108, 481–497.

Liu, H., Cheng, S., & Logan, B. E. (2005). Production of electricity from acetate or butyrate using a single-chamber microbial fuel cell. *Environmental Science & Technology*, 39(2), 658–662.

Liu, H., & Logan, B. E. (2004). Electricity generation using an air-cathode single chamber microbial fuel cell in the presence and absence of a proton exchange membrane. *Environmental Science & Technology*, 38(14), 4040–4046.

Liu, L. N. (2016). Distribution and dynamics of electron transport complexes in cyanobacterial thylakoid membranes. *Biochimica et Biophysica Acta (BBA) - Bioenergetics*, 1857(3), 256–265.

Liu, T., Rao, L., Yuan, Y., & Zhuang, L. (2015). Bioelectricity generation in a microbial fuel cell with a self-sustainable photocathode. *The Scientific World Journal*. doi:10.1155/2015/864568.

Logan, B. E., Hamelers, B., Rozendal, R., Schröder, U., Keller, J., Freguia, S.,... & Rabaey, K. (2006). Microbial fuel cells: Methodology and technology. *Environmental Science & Technology*, 40(17), 5181–5192.

Logan, B. E., Murano, C., Scott, K., Gray, N. D., & Head, I. M. (2005). Electricity generation from cysteine in a microbial fuel cell. *Water Research*, 39(5), 942–952.

Ma, M., Cao, L., Ying, X., & Deng, Z. (2012). Study on the performance of photosynthetic microbial fuel cells powered by *Synechocystis* PCC-6803. *Kezaisheng Nengyuan/Renewable Energy Resources*, 30(5), 42–46.

Majid, M., Shafqat, S., Inam, H., Hashmi, U., & Kazi, A. G. (2014). Production of algal biomass. In: Hakeem, K.R., Jawaid, M., Rashid, U. (eds) *Biomass and Bioenergy* (pp.207–224). Springer.

Markou, G., & Nerantzis, E. (2013). Microalgae for high-value compounds and biofuels production: A review with focus on cultivation under stress conditions. *Biotechnology Advances*, 31(8), 1532–1542.

Martin, E., Savadogo, O., Guiot, S. R., & Tartakovsky, B. (2010). The influence of operational conditions on the performance of a microbial fuel cell seeded with mesophilic anaerobic sludge. *Biochemical Engineering Journal*, 51(3), 132–139.

Masum, B. M., Masjuki, H. H., Kalam, M. A., Fattah, I. R., Palash, S. M., & Abedin, M. J. (2013). Effect of ethanol-gasoline blend on NOx emission in SI engine. *Renewable and Sustainable Energy Reviews*, 24, 209–222.

Mata, T. M., Martins, A. A., & Caetano, N. S. (2010). Microalgae for biodiesel production and other applications: A review. *Renewable and Sustainable Energy Reviews*, 14(1), 217–232.

Mathews, J., & Wang, G. (2009). Metabolic pathway engineering for enhanced biohydrogen production. *International Journal of Hydrogen Energy*, 34(17), 7404–7416.

McCormick, A. J., Bombelli, P., Scott, A. M., Philips, A. J., Smith, A. G., Fisher, A. C., & Howe, C. J. (2011). Photosynthetic biofilms in pure culture harness solar energy in a mediatorless bio-photovoltaic cell (BPV) system. *Energy & Environmental Science*, 4(11), 4699–4709 [photosystem II at the alga *Scenedesmus obliquus*, *Int. J. Res. Rev. Appl.* pp. 162–168, 2008].

Medipally, S. R., Yusoff, F. M., Banerjee, S., & Shariff, M. (2015). Microalgae as sustainable renewable energy feedstock for biofuel production. *BioMed Research International*, doi:10.1155/2015/519513.

Mitra, P., & Hill, G. A. (2012). Continuous microbial fuel cell using a photoautotrophic cathode and a fermentative anode. *The Canadian Journal of Chemical Engineering*, 90(4), 1006–1010.

Mobin, S. M., Chowdhury, H., & Alam, F. (2019). Commercially important bioproducts from microalgae and their current applications: A review. *Energy Procedia*, 160, 752–760.

Montpart, N., Ribot-Llobet, E., Garlapati, V. K., Rago, L., Baeza, J. A., & Guisasola, A. (2014). Methanol opportunities for electricity and hydrogen production in bioelectrochemical systems. *International Journal of Hydrogen Energy*, 39(2), 770–777.

Mourya, M., Khan, M. J., Ahirwar, A., Schoefs, B., Marchand, J., Rai, A.,... & Vinayak, V. (2022). Latest trends and developments in microalgae as potential source for biofuels: The case of diatoms. *Fuel*, 314, 122738.

Mullineaux, C. W. (2014). Electron transport and light-harvesting switches in cyanobacteria. *Frontiers in Plant Science*, 5, 7.

Naessens, M., Leclerc, J. C., & Tran-Minh, C. (2000). Fiber optic biosensor using Chlorella vulgaris for determination of toxic compounds. *Ecotoxicology and Environmental Safety*, 46(2), 181–185.

Nagarajan, D., Lee, D. J., Kondo, A., & Chang, J. S. (2017). Recent insights into biohydrogen production by microalgae-From biophotolysis to dark fermentation. *Bioresource Technology*, 227, 373–387.

Ndayisenga, F., Yu, Z., Yu, Y., Lay, C. H., & Zhou, D. (2018). Bioelectricity generation using microalgal biomass as electron donor in a bio-anode microbial fuel cell. *Bioresource Technology*, 270, 286–293.

Nickelsen, J., & Zerges, W. (2013). Thylakoid biogenesis has joined the new era of bacterial cell biology. *Frontiers in Plant Science*, 4(458), 1–4.

Nickelsen, K. (2012). From the Red Drop to the Z-scheme of photosynthesis. *Annalen der Physik*, 524(11), A157–A160.

Nikkanen, L., Solymosi, D., Jokel, M., & Allahverdiyeva, Y. (2021). Regulatory electron transport pathways of photosynthesis in cyanobacteria and microalgae: Recent advances and biotechnological prospects. *Physiologia Plantarum*. doi:10.1093/jxb/erad118

Nishio, K., Hashimoto, K., & Watanabe, K. (2013). Digestion of algal biomass for electricity generation in microbial fuel cells. *Bioscience, Biotechnology, and Biochemistry*, 120833. doi:10.1271/bbb.120982.

Pandit, S., & Das, D. (2015). Role of Microalgae in Microbial Fuel Cell. In: Das, D. (ed) *Algal Biorefinery: An Integrated Approach*. Springer, Cham.

Pata, U. K. (2021). Renewable and non-renewable energy consumption, economic complexity, CO 2 emissions, and ecological footprint in the USA: Testing the EKC hypothesis with a structural break. *Environmental Science and Pollution Research*, 28(1), 846–861.

Peng, C., Yang, Y., Wang, L., Huang, M., & Shi, X. F. (2009). A novel kind of proton exchange membrane: Characters and proton transport mechanism. *Chinese Journal of Polymer Science*, 27(5), 755–760.

Periasamy, V., Jaafar, M. M., Chandrasekaran, K., Talebi, S., Ng, F. L., Phang, S. M.,... & Iwamoto, M. (2022). Langmuir-Blodgett graphene-based films for algal biophotovoltaic fuel cells. *Nanomaterials*, 12(5), 840.

Peron, J., Mani, A., Zhao, X., Edwards, D., Adachi, M., Soboleva, T.,... & Holdcroft, S. (2010). Properties of Nafion(r) NR-211 membranes for PEMFCs. *Journal of Membrane Science*, 356(1–2), 44–51.

Rabaey, K., Lissens, G., Siciliano, S. D., & Verstraete, W. (2003). A microbial fuel cell capable of converting glucose to electricity at high rate and efficiency. *Biotechnology Letters*, 25(18), 1531–1535.

Radakovits, R., Jinkerson, R. E., Darzins, A., & Posewitz, M. C. (2010). Genetic engineering of algae for enhanced biofuel production. *Eukaryotic Cell*, 9(4), 486–501.

Ramanathan, G., Birthous, R. S., Abirami, D., & Highcourt, D. (2011). Efficacy of marine microalgae as exoelectrogen in microbial fuel cell system for bioelectricity generation. *World Journal of Fish and Marine Sciences*, 3(1), 79–87.

Rashid, N., Cui, Y. F., Rehman, M. S. U., & Han, J. I. (2013). Enhanced electricity generation by using algae biomass and activated sludge in microbial fuel cell. *Science of the Total Environment*, 456, 91–94.

Ray, A., Nayak, M., & Ghosh, A. (2022). A review on co-culturing of microalgae: A greener strategy towards sustainable biofuels production. *Science of the Total Environment*, 802, 149765.

Reguera, G., Nevin, K. P., Nicoll, J. S., Covalla, S. F., Woodard, T. L., & Lovley, D. R. (2006). Biofilm and nanowire production leads to increased current in *Geobacter sulfurreducens* fuel cells. *Applied and Environmental Microbiology*, 72(11), 7345–7348.

Ren, Y., Pan, D., Li, X., Fu, F., Zhao, Y., & Wang, X. (2013). Effect of polyaniline-graphene nanosheets modified cathode on the performance of sediment microbial fuel cell. *Journal of Chemical Technology & Biotechnology*, 88(10), 1946–1950.

Richardson, S., Nilsson, G. S., Bergquist, K. E., Gorton, L., & Mischnick, P. (2000). Characterisation of the substituent distribution in hydroxypropylated potato amylopectin starch. *Carbohydrate Research*, 328(3), 365–373.

Rochaix, J. D. (2011). Reprint of: Regulation of photosynthetic electron transport. *Biochimica et Biophysica Acta (BBA) - Bioenergetics*, 1807(8), 878–886.

Rosenbaum, M., & Schröder, U. (2010). Photomicrobial solar and fuel cells. *Electroanalysis: An International Journal Devoted to Fundamental and Practical Aspects of Electroanalysis*, 22(7–8), 844–855.

Rosenbaum, M., He, Z., & Angenent, L. T. (2010). Light energy to bioelectricity: Photosynthetic microbial fuel cells. *Current Opinion in Biotechnology*, 21(3), 259–264.

Rosenbaum, M., Zhao, F., Schröder, U., & Scholz, F. (2006). Interfacing electrocatalysis and biocatalysis with tungsten carbide: A high-performance, noble-metal-free microbial fuel cell. *Angewandte Chemie International Edition*, 45(40), 6658–6661.

Saad, M. G., Dosoky, N. S., Zoromba, M. S., & Shafik, H. M. (2019). Algal biofuels: Current status and key challenges. *Energies*, 12(10), 1920.

Saratale, R. G., Kumar, G., Banu, R., Xia, A., Periyasamy, S., & Saratale, G. D. (2018). A critical review on anaerobic digestion of microalgae and macroalgae and co-digestion of biomass for enhanced methane generation. *Bioresource Technology*, 262, 319–332.

Sato, N. (2015). Is monoglucosyldiacylglycerol a precursor to monogalactosyldiacylglycerol in all cyanobacteria? *Plant and Cell Physiology*, 56(10), 1890–1899.

Sazzad, B. S., Fazal, M. A., Haseeb, A. S. M. A., & Masjuki, H. H. (2016). Retardation of oxidation and material degradation in biodiesel: A review. *RSC Advances*, 6(65), 60244–60263.

Schröder, U. (2007). Anodic electron transfer mechanisms in microbial fuel cells and their energy efficiency. *Physical Chemistry Chemical Physics*, 9(21), 2619–2629.

Schwenzfeier, A., Wierenga, P. A., & Gruppen, H. (2011). Isolation and characterization of soluble protein from the green microalgae *Tetraselmis* sp. *Bioresource Technology*, 102(19), 9121–9127.

Sekar, N., Umasankar, Y., & Ramasamy, R. P. (2014). Photocurrent generation by immobilized cyanobacteria via direct electron transport in photo-bioelectrochemical cells. *Physical Chemistry Chemical Physics*, 16(17), 7862–7871.

Sevda, S., Garlapati, V. K., Sharma, S., Bhattacharya, S., Mishra, S., Sreekrishnan, T. R., & Pant, D. (2019). Microalgae at niches of bioelectrochemical systems: A new platform for sustainable energy production coupled industrial effluent treatment. *Bioresource Technology Reports*, 7, 100290.

Shafiee S, Topal E (2009). When will fossil fuel reserves be diminished? *Energy Policy*, 37(1), 181–189.

Shahbaz, M., Solarin, S. A., Hammoudeh, S., & Shahzad, S. J. H. (2017). Bounds testing approach to analyzing the environment Kuznets curve hypothesis with structural beaks: The role of biomass energy consumption in the United States. *Energy Economics*, 68, 548–565.

Shimakawa, G., Kohara, A., & Miyake, C. (2021). Characterization of light-enhanced respiration in cyanobacteria. *International Journal of Molecular Sciences*, 22(1), 342.

Shukla, M., & Kumar, S. (2018). Algal growth in photosynthetic algal microbial fuel cell and its subsequent utilization for biofuels. *Renewable and Sustainable Energy Reviews*, 82, 402–414.

Sialve, B., Bernet, N., & Bernard, O. (2009). Anaerobic digestion of microalgae as a necessary step to make microalgal biodiesel sustainable. *Biotechnology Advances*, 27(4), 409–416.

Siddiki, S. Y. A., Mofijur, M., Kumar, P. S., Ahmed, S. F., Inayat, A., Kusumo, F.,... & Mahlia, T. M. I. (2022). Microalgae biomass as a sustainable source for biofuel, biochemical and biobased value-added products: An integrated biorefinery concept. *Fuel*, 307, 121782.

Sidler, W.A. (1994). Phycobilisome and Phycobiliprotein Structures. In: Bryant, D.A. (ed) *The Molecular Biology of Cyanobacteria*. Advances in Photosynthesis, vol 1. Springer, Dordrecht.

Singhvi, P., & Chhabra, M. (2013). Simultaneous chromium removal and power generation using algal biomass in a dual chambered salt bridge microbial fuel cell. *Journal of Bioremediation and Biodegradation*, 4(5), 190.

Sirajunnisa, A. R., & Surendhiran, D. (2016). Algae-A quintessential and positive resource of bioethanol production: A comprehensive review. *Renewable and Sustainable Energy Reviews*, 66, 248–267.

Slate, A. J., Whitehead, K. A., Brownson, D. A., & Banks, C. E. (2019). Microbial fuel cells: An overview of current technology. *Renewable and Sustainable Energy Reviews*, 101, 60–81.

Srirangan, K., Akawi, L., Moo-Young, M., & Chou, C. P. (2012). Towards sustainable production of clean energy carriers from biomass resources. *Applied Energy*, 100, 172–186.

Subhash, G. V., Chandra, R., & Mohan, S. V. (2013). Microalgae mediated bio-electrocatalytic fuel cell facilitates bioelectricity generation through oxygenic photomixotrophic mechanism. *Bioresource Technology*, 136, 644–653.

Sun, Y., & Cheng, J. (2002). Hydrolysis of lignocellulosic materials for ethanol production: A review. *Bioresource Technology*, 83(1), 1–11.

Sun, Y., Wei, J., Liang, P., & Huang, X. (2012). Microbial community analysis in biocathode microbial fuel cells packed with different materials. *AMB Express*, 2(1), 1–8.

Sund, C. J., McMasters, S., Crittenden, S. R., Harrell, L. E., & Sumner, J. J. (2007). Effect of electron mediators on current generation and fermentation in a microbial fuel cell. *Applied Microbiology and Biotechnology*, 76(3), 561–568.

Tercero, E. A. R., Domenicali, G., & Bertucco, A. (2014). Autotrophic production of biodiesel from microalgae: An updated process and economic analysis. *Energy*, 76, 807–815.

Ulusoy ÖF, Pektaş E (2019). Recent trends and issues in energy conservation technologies. *Heritage and Sustainable Development*, 1 (1), 33–40.

United Nations. (2019). *World Population Prospects 2019: Data Booklet*. Statistical Papers United Nations, Population and vital Statistics Report.

Van der Hoeven M (2013). *World Energy Outlook 2012*. International Energy Agency.

Van Der Maarel, M. J., Van der Veen, B., Uitdehaag, J. C., Leemhuis, H., & Dijkhuizen, L. (2002). Properties and applications of starch-converting enzymes of the α-amylase family. *Journal of Biotechnology*, 94(2), 137–155.

Velasquez-Orta, S. B., Curtis, T. P., & Logan, B. E. (2009). Energy from algae using microbial fuel cells. *Biotechnology and Bioengineering*, 103(6), 1068–1076.

Vermaas, W. F. (2001). Photosynthesis and respiration in cyanobacteria. doi:10.1038/npg.els.0001670.

Voloshin, R. A., Rodionova, M. V., Zharmukhamedov, S. K., Hou, H., Shen, J. R., & Allakhverdiev, S. I. (2016). Components of natural photosynthetic apparatus in solar cells. In: Najafpour, M. M. (ed) *Applied Photosynthesis e New Progress* (pp. 161–88). InTech.

Vothknecht, U. C., & Westhoff, P. (2001). Biogenesis and origin of thylakoid membranes. *Biochimica et Biophysica Acta (BBA)-Molecular Cell Research*, 1541(1–2), 91–101.

Wada, H., & Murata, N. (1998). Membrane Lipids in Cyanobacteria. In: Paul-André, S., Norio, M. (eds) *Lipids in Photosynthesis: Structure, Function and Genetics. Advances in Photosynthesis and Respiration*, vol 6. Springer, Dordrecht.

Walter, X. A., Greenman, J., Taylor, B., & Ieropoulos, I. A. (2015). Microbial fuel cells continuously fuelled by untreated fresh algal biomass. *Algal Research*, 11, 103–107.

Wang, H., Liu, D., Lu, L., Zhao, Z., Xu, Y., & Cui, F. (2012). Degradation of algal organic matter using microbial fuel cells and its association with trihalomethane precursor removal. *Bioresource Technology*, 116, 80–85.

Wang, X., Feng, Y., Liu, J., Lee, H., Li, C., Li, N., & Ren, N. (2010). Sequestration of CO_2 discharged from anode by algal cathode in microbial carbon capture cells (MCCs). *Biosensors and Bioelectronics*, 25(12), 2639–2643.

Westermark, S., & Steuer, R. (2016). Toward multiscale models of cyanobacterial growth: A modular approach. *Frontiers in Bioengineering and Biotechnology*, 4, 95.

Winfield, J., Gajda, I., Greenman, J., & Ieropoulos, I. (2016). A review into the use of ceramics in microbial fuel cells. *Bioresource Technology*, 215, 296–303.

Wu, X. Y., Song, T. S., Zhu, X. J., Wei, P., & Zhou, C. C. (2013). Construction and operation of microbial fuel cell with *Chlorella vulgaris* biocathode for electricity generation. *Applied Biochemistry and Biotechnology*, 171(8), 2082–2092.

Wu, Y. C., Wang, Z. J., Zheng, Y., Xiao, Y., Yang, Z. H., & Zhao, F. (2014). Light intensity affects the performance of photo microbial fuel cells with *Desmodesmus* sp. A8 as cathodic microorganism. *Applied Energy*, 116, 86–90.

Xiao, L., Young, E. B., Berges, J. A., & He, Z. (2012). Integrated photo-bioelectrochemical system for contaminants removal and bioenergy production. *Environmental Science & Technology*, 46(20), 11459–11466.

Xu, C., Poon, K., Choi, M. M., & Wang, R. (2015). Using live algae at the anode of a microbial fuel cell to generate electricity. *Environmental Science and Pollution Research*, 22(20), 15621–15635.

Yang, C., Srinivasan, S., Bocarsly, A. B., Tulyani, S., & Benziger, J. B. (2004). A comparison of physical properties and fuel cell performance of Nafion and zirconium phosphate/Nafion composite membranes. *Journal of Membrane Science*, 237(1–2), 145–161.

Yoo, G., Park, M. S., Yang, J. W., & Choi, M. (2015). Lipid content in microalgae determines the quality of biocrude and energy return on investment of hydrothermal liquefaction. *Applied Energy*, 156, 354–361.

Yoon, S., Lee, H., Fraiwan, A., Dai, C., & Choi, S. (2014). A microsized microbial solar cell: A demonstration of photosynthetic bacterial electrogenic capabilities. *IEEE Nanotechnology Magazine*, 8(1), 24–29.

Yuan, Y., Chen, Q., Zhou, S., Zhuang, L., & Hu, P. (2011). Bioelectricity generation and microcystins removal in a blue-green algae powered microbial fuel cell. *Journal of Hazardous Materials*, 187(1–3), 591–595.

Zabed, H. M., Akter, S., Yun, J., Zhang, G., Zhang, Y., & Qi, X. (2020). Biogas from microalgae: Technologies, challenges and opportunities. *Renewable and Sustainable Energy Reviews*, 117, 109503.

Zhang, X. L., Yan, S., Tyagi, R. D., & Surampalli, R. Y. (2013). Biodiesel production from heterotrophic microalgae through transesterification and nanotechnology application in the production. *Renewable and Sustainable Energy Reviews*, 26, 216–223.

Zhao, J., Li, X. F., Ren, Y. P., Wang, X. H., & Jian, C. (2012). Electricity generation from Taihu Lake cyanobacteria by sediment microbial fuel cells. *Journal of Chemical Technology & Biotechnology*, 87(11), 1567–1573.

Zhu, X. G., Long, S. P., & Ort, D. R. (2008). What is the maximum efficiency with which photosynthesis can convert solar energy into biomass? *Current Opinion in Biotechnology*, 19(2), 153–159.

Zou, L., Qiao, Y., Zhong, C., & Li, C. M. (2017). Enabling fast electron transfer through both bacterial outer-membrane redox centers and endogenous electron mediators by polyaniline hybridized large-mesoporous carbon anode for high-performance microbial fuel cells. *Electrochimica Acta*, 229, 31–38.

4 Photoelectrochemical (PEC) Solar Cells

Shreya Rastogi, Shalini Singh, and Pankaj Srivastava

4.1 INTRODUCTION

An ever-growing population of the world demands more energy for all kinds of work. The increasing global need for energy production, conservation and management has intensified interest in developing more effective means of power generation. Currently, a majority of energy is obtained from non-renewable sources, such as coal, natural gas and petroleum, that are present in our earth's crust in a limited amount and also require costly explorations and potentially dangerous mining and drilling, thus making energy production expensive. Moreover, fossil fuel combustion causes unbearable harm to our environment (Figure 4.1), for example CO_2 emitted from the combustion of fossil fuel contributes to global warming to a large extent.

Alternative renewable energy sources, which are reliable, plentiful and will potentially be very cheap, may be considered once the technology and infrastructure for their exploitation is improved. It includes solar, heat, wind, geothermal, hydropower and tidal energy and biofuels that are grown and harvested without fossil fuels. Among the various options for renewable energy sources that are available to us, sunlight seems to be the most promising source of energy for several reasons.

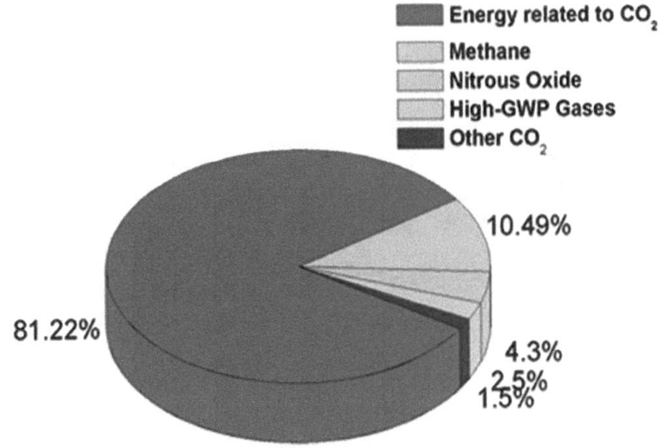

FIGURE 4.1 Diagram showing CO_2 emission caused by the combustion of fossil fuels.

DOI: 10.1201/9781003481157-4

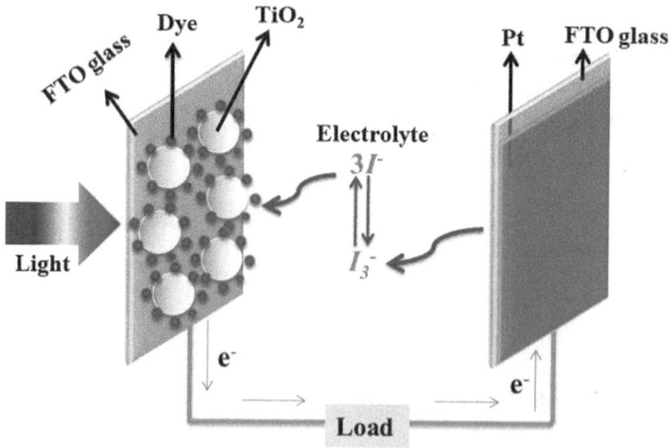

FIGURE 4.2 First, second and third generations of solar cells.

i. It is ecologically pure and the solar radiation received continuously on the earth far outweighs our energy requirement.
ii. Its availability is assured until the survival of the human civilization on this planet.
iii. It can be converted to a usable form of energy through pollution-free devices.

So, with the enormous energy available in the form of sunlight, the only challenging task before all concerned is how to harvest it conveniently, efficiently and above all economically. Research revealed that covering 0.1% of the Earth's surface with solar cells having an efficiency of 10% would gratify the present global requirements [1].

A solar cell is a solid electrical device that converts solar energy directly to electricity. There are two fundamental functions of solar cells: photogeneration of charge carriers (electrons and holes) in a light-absorbing material and separation of the charge carriers to a conductive port to transmit electricity. There are three generations of solar cells as shown in Figure 4.2 [2].

The first generation of solar cells is fully commercialized and most developed among the three generations. They are made of single- or multi-crystalline Si. Second-generation solar cells are developing slowly. The third generation of solar cells is in early phase and many modifications are in progress for its development. Dye-sensitized solar cells (DSSCs) can be considered to be a part of the third-generation cells [3]. Adding to the wide range of solar cells, hybrid solar cells based on inorganic and organic compounds are very promising with regard to efficiency of such cells.

4.1.1 Solar Cell

The systems to harvest incident sunlight and convert it into electrical energy are photovoltaic devices, called solar cells. In these cells, the photocurrent is generated through charge separation and collection of free electrons and holes under solar irradiation. Solar cells based on silicon that come in wafer-like monocrystalline [4], polycrystalline and amorphous forms and categorized as the first generation are usually doped with group V and group III elements in a *p–n* junction to achieve high-efficient charge separation of electron–hole pairs. These solar cells demonstrate a good performance with more than 20% of power conversion efficiency (PCE) as well as high stability. Multi-junction cells, also named tandem cells, cooperate with concentrated solar power systems, and the highest recorded conversion efficiency is 40%. They are rigid and usually lose efficiency under higher temperature or imperfect illumination angle, thus restraining their application. The second-generation solar cells, thin film solar cells, are composed of amorphous silicon, cadmium telluride (CdTe) and copper indium gallium diselenide (CIGS) [5] and have outstanding performance, as high as 20% PCE. They are made from layers of semiconductive materials with only a few micrometre-thick films that can make devices flexible and reduce the production cost. However, this technique is still limited by vacuum processes and high-temperature treatments in manufacturing and also restricted by resource scarcity. The emerging solar cells employing organic dye, quantum dots, conductive polymers and perovskite materials with new photovoltaic mechanisms are assigned to the third-generation solar cells, which are currently under laboratory investigations. Among this generation, DSSCs have many advantages over traditional silicon-based counterparts, such as low cost, mechanical robustness and ability to operate under imperfect irradiation conditions. More importantly, the hybrid structure of DSSCs separates the charge transport from charge separation which reduces the efficiency loss by electron recombination. Recently, perovskite solar cells evolving from DSSCs became competitive with an unprecedented growth in PCE from 3.8% to more than 20% in less than five years. The perovskite thin film can also be used as a top coating in tandem cells to improve the performance of the original solar cells at a much lower extra cost. Overall, the third-generation solar cells are promising for commercialization because of their low-cost, readily available source materials and low energy expenditure in fabrication [6]. However, the reproducibility and the long-term stability of these solar cells are still the major concerns.

4.1.2 Photoelectrochemical (PEC) Solar Cell

As a result of consistent efforts, several methods have been developed for the conversion of sunlight into some convenient forms such as thermal, chemical or electrical energy. Among all the possible methods, solar energy conversion through photoelectrochemical (PEC) cells, based primarily on (semiconductor/electrolyte) junction, is extremely attractive for the following reasons:

1. Solid-state *p–n* junction-based photovoltaic cells are quite expensive as they are made with a single crystal of semiconductors and *p–n* junctions have to be made by a complicated diffusion technology. On the other hand, in PEC cells, cheaper polycrystalline semiconducting materials can be used in the form of pellets or thin films (prepared by spray pyrolysis, sol–gel, sputtering, chemical vapour deposition, etc.) without much loss of cell efficiency, and a junction is spontaneously formed by mere immersion of the

semiconductor electrode in the electrolyte; no special technique is required for this purpose. So, PEC cells can be made so cheaply so as to become competitive in terms of cost.

2. PEC cells provide a means for storing solar energy in the form of high-energy chemical fuels (e.g. H_2, O_2, C_2H_5OH and NH_3) which can easily be transported and utilized in different ways (as heat via combustion or electricity via fuel cells), when sunlight is not available.
3. By suitable coupling, it is possible to construct a rechargeable PEC cell that can store, in situ, electrical energy produced from light energy. One such cell is $CdSe/S^{2-}$, $/Ag_2S/Pt$.

Functioning of PEC cell: A PEC cell usually consists of a semiconductor photoelectrode and a metal counter electrode, both immersed in an appropriate electrolyte solution containing some supporting electrolyte (like tetrabutyl ammonium perchlorate) and a redox couple (such as I_2/I^-, $Fe(CN)_6{}^{4-}/Fe(CN)_6{}^{3-}$, and V^{3+}/V^{2+}).

An isolated semiconductor is characterized by its band-gap (E_g), the energy difference between the upper edge (E_v) of the valence band (constituted by highest occupied energy level) and the lower edge (E_c) of the conduction band (constituted by lowest partially occupied or empty energy level). When such semiconductor is immersed in an electrolyte solution containing a redox couple having formal redox potential $E^o{}_{redox}$, the Fermi level (E_f) of the semiconductor, which represents the electron transfer energy level in the semiconductor, shifts in such a way so as to match with the redox potential of the redox couple. To maintain the initial separation between E_c & E_f and E_c & E_v unchanged, the E_c as well as E_v also shift downwards in *n*-type semiconductor and upwards in *p*-type semiconductor. As a result, a band bending occurs. This creates a depletion region inside and near the surface of the semiconductor electrode that is in contact with the electrolyte solution. When this semiconductor electrode, immersed in the electrolyte solution, is illuminated with light of appropriate energy, $h\nu > E_g$, electrons are excited from the valence band to conduction band, leaving behind a hole (positive centre). These photogenerated charge carriers move in the opposite direction, electron towards the bulk and hole towards the surface of the semiconductor, due to the potential drop in the depletion region. This results in the shifting of the Fermi level towards its initial position and hence band bending is reduced. The extent of the reduction in band bending is the measure of the photovoltage (V_{photo}). Photogenerated electrons go to metal electrode through bulk of the semiconductor and external circuit (load) and take part in the following reaction:

$Ox. + e^- \rightarrow Red.$ at metal electrode

However, the holes migrate to the surface of the semiconductor electrode and take part in the following reaction:

$Red. + p^+(hole) \rightarrow Ox.$ at semiconductor electrode

So, there is no net chemical reaction in the cell and hence such cells are called "regenerative cells". In this way, a PEC cell converts light to electrical energy.

When redox couple is not used in the cell and the medium of the electrolyte is water, it gets oxidized at the semiconductor electrode producing O_2 and reduced at the metal electrode producing H_2. When a cell is used in this mode, it is called "photoassisted electrolysis (PAE)" cell. Using a PEC cell in this mode, we can accomplish other up-hill reactions ($\Delta G > 0$) and produce high-energy chemical fuels (such as NH_3 and C_2H_5OH).

4.1.3 Dye-Sensitized Solar Cells (DSSCs)

In DSSCs, the sensitizers play a very crucial role as they harvest solar light and transfer the energy via electron transfer to a suitable material (wide band-gap semiconductor) to produce electricity. Several natural compounds, organic dyes and metal complexes have been employed as sensitizers in DSSCs. The integral architecture of DSSCs was first proposed by Grätzel and O'Regan in 1991 [7]. Since then, improvement in device designs along with the surge in new materials for light-absorbing sensitizers (i.e. dye molecules) and redox electrolytes has further improved their performance. Based on the National Renewable Energy Laboratory (NREL) statistics, the highest lab-recorded efficiency of DSSCs to date is about 12%. Even though their efficiencies are still lower than the first- and second-generation cells, the fabrication of DSSCs is cost-effective by utilizing inexpensive and abundant wide band-gap semiconductive materials as the photoanodes, such as titania (TiO_2) and zinc oxide (ZnO). Some emerging solar cells have adopted a hybrid structure of DSSCs, such as quantum dots-sensitized solar cells and perovskite-sensitized solar cells. Quantum dots, perovskite crystals or the photosynthetic pigment-protein complexes are employed as novel light-absorbing sensitizers to replace organic dyes, and their fundamental properties involve photon capture, energy transfer and charge separation processes.

DSSCs mimic natural photosynthesis to convert sunlight into electricity [8]. The operation principle of DSSCs differs from the conventional *p–n* junction-based solar cells. Briefly, the photosynthesis processes take place in the chloroplast where solar energy is absorbed by the pigments (such as chlorophylls) in the photosystems and converted into electrons in the reaction centres to trigger a series of chemical reactions. In DSSCs, the dye molecules similar to the chlorophylls in green leaves are excited by absorbing visible light. In contrast to conventional *p–n* junction solar cells, the charge separation occurs at the sensitizer/TiO_2 interface followed by electron diffusion in the TiO_2 network, finally flowing to the cathode through the external circuit as photocurrent. DSSCs utilize a separate media for charge generation (occurs within the dye) and charge transport (occurs in the TiO_2 matrix), which greatly reduces the possibility of charge recombination. Concurrently, the oxidized dye is reduced to its ground state by the oxidation of I^- into I_3^-; subsequently, I_3^- will be reduced at the cathode by accepting electrons to complete the whole regeneration process. Overall, this system converts solar energy into electricity without any net consumption of chemicals, thus the DSSCs can continuously supply power.

4.1.4 Components and Functioning of DSSCs

DSSCs mainly comprise four components: (a) photoanode (semiconducting metal oxide deposited on the surface of transparent conducting oxide (TCO) substrate, typically fluorine-doped tin oxide (FTO) glass); (b) dye/sensitizer adsorbed onto the surface of the semiconductor thin film; (c) electrolyte containing redox couple; and (d) counter electrode (typically Pt-coated FTO glass) (Figure 4.3).

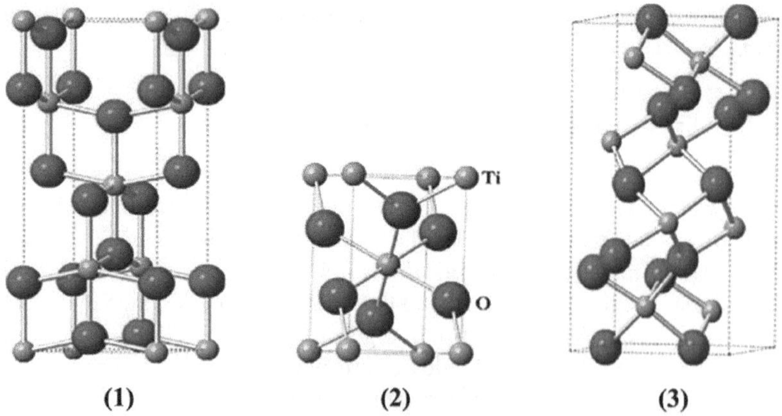

FIGURE 4.3 Schematic diagram of DSSC assembly.

4.1.4.1 Semiconductor Electrode (Photoanode)

The photoanode is also known as working electrode in DSSCs. The photoanode usually consists of semiconducting wide band-gap metal oxide deposited on the surface of TCO substrate, typically FTO glass. The functions of TCO substrate in DSSCs are to support the semiconductor layer and to collect the current. An ideal TCO substrate should possess high optical transparency and low electrical resistivity. High transparency is essential to allow the good transmittance of sunlight through the substrate without redundant absorption, while low resistivity is crucial in facilitating the charge transfer process and reducing the energy loss [9]. Several studies reported that the FTO substrate is superior to the indium tin oxide (ITO) substrate as photoelectrode in DSSCs due to its stable resistivity at elevated temperature [10,11]. In DSSC fabrication, the preparation of photoanode usually involves the deposition and sintering of TiO_2 paste on TCO substrate at high temperature (~450°C) in order to improve the electrical contact.

In the last few years, numerous methods have been chosen for the synthesis of anatase TiO_2 as nanoparticles, nanowires, nanorods, nanotubes, nanosheets, nanobowls and mesoporous materials such as aerogels, inverse opals and photonic materials. For the synthesis of TiO_2 structures, different methods have been adopted such as sol–gel, hydrothermal, micelle and inverse micelle, solvothermal, sonochemical, chemical vapour deposition, direct oxidation, physical vapour deposition, microwave deposition techniques and electrodeposition [12–18].

TiO_2: Titanium dioxide is a non-toxic semiconductor with a wide band-gap, which is commonly used as a white pigment in toothpaste, paints, paper, etc. The material has unique properties because of its excellent chemical stability. The electron injection of TiO_2 is greater and faster due to electronic coupling and density states of TiO_2. It has a high dielectric constant that provides efficient electrostatic shielding of the injected electron from the oxidized dye, thus avoiding electron recombination before reduction of the dye by the redox electrolyte. Moreover, high refractive index of TiO_2 results in efficient diffuse scattering of light inside the porous photoelectrode that

enhances light adsorption. High porosity and high surface area for good dye adsorption hence broaden the absorption spectrum. High light-scattering effect increases the photoresponse in TiO_2 film because when the size of a particle is comparable with the wavelength of light, the travelling distance of light increases, so that it undergoes multiple reflection, and eventually, more photons are absorbed within the film.

TiO_2 normally occurs in three crystal modification structures, namely rutile, anatase and brookite. Rutile is the thermodynamically stable phase and anatase is preferred for DSSCs due to its larger band-gap ($E_g = 3.2$ eV for anatase compared to $E_g = 3.0$ eV for rutile, corresponding to an absorption edge of λ_g 390 nm and 410 nm, respectively) [19]. The high refractive index of TiO_2 ($n = 2.5$ for anatase) results in efficient diffuse scattering of light inside the porous photoelectrode, which significantly enhances the light absorption [20] (Figure 4.4). As far as the structure is concerned, in the case of anatase, the next octahedra share a common edge while in brookite, points of attachment are both the corner and the edge. Moreover, rutile is built of parallel chains of octahedrons, a little deformed, but each of them is connected to the ten neighbouring same units whereas octahedrons in anatase have a warpage of the prism, and each octahedron is connected to eight others. In anatase, Ti–Ti distance is longer and Ti–O is shorter than it is in the rutile form.

Further, TiO_2 can be used as a photocatalyst in the splitting of water induced by UV radiation. The hole potential in the valence band is low enough to oxidize most of the organic compounds, which results in a wide range of applications of titanium dioxide on self-cleaning surfaces. In comparison to other semiconductors with a similar band-gap energy, TiO_2 is not subject to photodegradation under the influence of excitation. The advantages of titanium dioxide thin films are a large value of refractive index (over 2.3) and very good transparency (over 90%) over a wide spectral range (from about 320 to about 6000 nm). In addition to the good optical properties of TiO_2, it has also many other desirable properties such as high mechanical resistance and long-term stability. Therefore, titanium oxide thin films are widely used as anti-reflective coatings in optics and photovoltaic cells. In DSSCs, anatase titanium dioxide form is most often used.

Doping has a major effect on the band structure and trap states of TiO_2, which in turn affect important properties such as the conduction band energy, charge

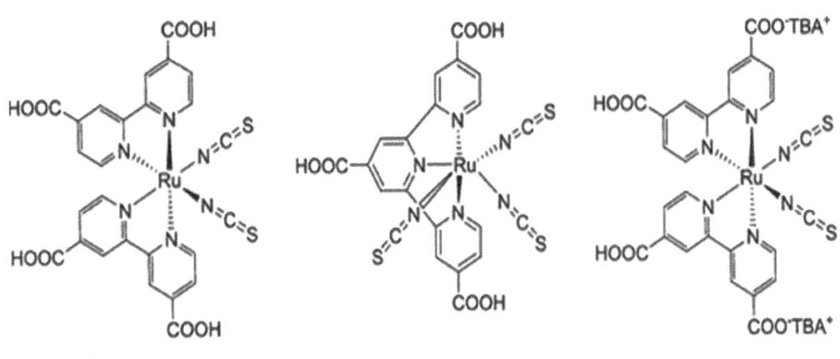

FIGURE 4.4 Crystal structure of TiO_2: (1) anatase, (2) rutile and (3) brookite.

transport, recombination and collection. Numerous efforts have been made towards the synthesis of high surface area anatase from different precursors viz. $Ti(O^nBu)_4$, $Ti(O^iPr)_4$, $TiCl_4$, $TiCl_3$ and $Ti(SO_4)_2$ and by different synthesis routes such as hydrothermal, thermo-hydrolysis and sol–gel method. The defect states of TiO_2 are highly dependent on the synthesis method. The chemical and physical properties of TiO_2 can be modified by the doping of metallic ions, and various dopants have successfully been used to increase device performance successfully [21–27].

4.1.4.2 Dye Sensitizer

In DSSCs, a dye is one of the key components for high power conversion efficiencies as it is responsible for light harvesting and generation of photoexcited electrons. In recent years, considerable developments have been made in the engineering of novel dye structures in order to enhance the performance of the system. In DSSCs, the dye/sensitizer is usually anchored on the surface of metal oxide. An ideal dye should possess several characteristics viz. (a) high molar extinction coefficient in the visible and near-infrared regions and (b) appropriate lowest unoccupied molecular orbital (LUMO) and highest occupied molecular orbital levels for both efficient charge injection into the conduction band (CB) of semiconducting metal oxide and dye regeneration from electrolyte, respectively, good solubility and photostability [28]. The dye can influence the cell efficiency in three different ways, namely the efficiency in absorbing the incident photons, efficiency in converting the incident photon to electron–hole pairs and lastly, the efficiency in the charge transfer process [29,30]. Ruthenium-based organometallic dyes (e.g. N3, N719 and black dyes) are the most efficient dyes due to their superior light absorption, durability and most importantly, the metal–ligand charge transfer transition that allows the photogenerated charges to be injected into TiO_2 efficiently [31] (Figure 4.5). Both the charge transfer process and photon-to-electron conversion are very efficient in ruthenium dyes [32].

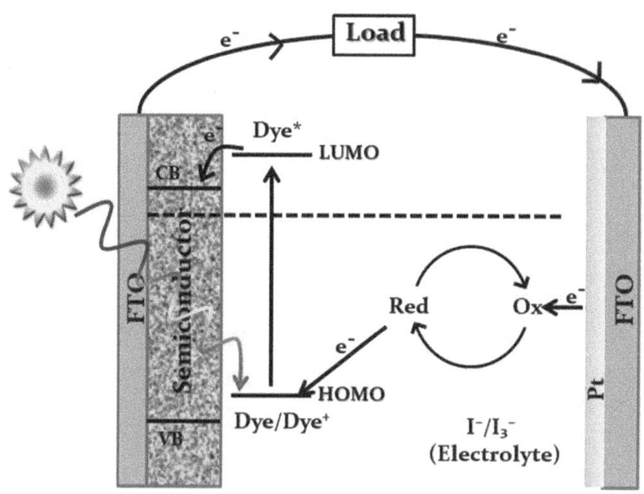

FIGURE 4.5 Commercial ruthenium dyes used in DSSCs.

Extensive research focussed on enhancing the light-harvesting property of ruthenium dyes has been made. In particular, amphiphilic homologues of the pioneering ruthenium-based N3 dye have been developed. These amphiphilic dyes display several advantages compared to the N3 dye such as:

1. A higher ground state pKa of the binding moiety, thus increasing electrostatic binding onto the TiO_2 surface at lower pH values.
2. The decreased charge on the dye attenuates the electrostatic repulsion in between adsorbed dye units and thereby increasing the dye loading.
3. The oxidation potential of these complexes is cathodically shifted compared to that of the N3 sensitizer, which increases the reversibility of the ruthenium III/II couple, leading to enhanced stability.

The use of ruthenium complexes as sensitizer in DSSCs has significantly expanded new possibilities because these complexes have intense charge transfer absorption in almost the entire visible range, greater excitation lifetime and good (electro) chemical stability. The best photovoltaic performance both in terms of conversion yield and long-term stability has so far been achieved with polypyridyl complexes of ruthenium. Sensitizers having the general structure $ML_2(X)_2$ where L stands for 2,2′-bipyridyl-4,4′-dicarboxylic acid; M is Ru and X presents a halide, cyanide, thiocyanate, acetyl acetonate, thiocarbamate or water substituent. Many of these Ru complexes used in DSSCs have reached more than 10% solar cell efficiency under standard conditions. Generally, these complexes consist of Ru metal ion with ancillary ligands having at least one anchoring group. Light absorption is due to the metal-to-ligand charge transfer (MLCT) process. Ancillary ligands, typically bipyridines or terpyridines, can be tuned by different substituents (alkyl, aryl, heterocycle, etc.) to change the photophysical and electrochemical properties of the complexes which improve the photovoltaic performance. Anchoring groups are employed to link the dye with the semiconductor and facilitate the injection of the excited electron into the CB of the semiconductor. N3, N719, Z709 and N749 dyes are considered as the reference dyes for DSSCs and used as a base for designing other ruthenium photosensitizers by changing ancillary ligands.

Reports that focus on the engineering of Ru(II) dyes with different ligands in order to improve overall device efficiency and strategies viz. incorporating potential electron-donor functionalities (e.g. triazolylpyridine and butyloxy-substituted benzene ring), replacing the thiocyanate ligands with other chelating anions (e.g. cyclometalates and pyridyl azolate) and modifying anchoring ligands are other focussed research areas. Besides the organometallic dyes, organic dyes such as porphyrin have also been utilized in DSSC applications. Natural sensitizers or chlorophyll dye extracted from leaves such as Pandannusamaryllifolius and papaya have been investigated as well [33,34].

4.1.4.3 Electrolyte

In DSSCs, the electrolyte functions as an electrically conducting medium that transports electronic charge between the working electrode and counter electrode, as well as allows the regeneration of oxidized dye. The most popularly used electrolyte in

DSSCs is iodide/triiodide (I/I_3^-) redox couple in an organic solvent, normally acetonitrile [35]. The excellent performance of (I/I_3^-) liquid-based electrolyte is attributed to its several interesting properties, namely low recombination loss, extremely fast dye regeneration and slow penetration into semiconducting metal oxide film [36]. However, some undesirable intrinsic properties also exist in a liquid electrolyte, which affect the long-term stability of DSSCs. The main concern is the evaporation of volatile iodide ions that will decrease the charge carrier concentration, resulting in cell degradation. Besides, the leakage of toxic organic solvent will also lead to environmental pollution [9]. In order to overcome the disadvantages of liquid electrolytes, other types of electrolytes such as quasi-solid-state, solid-state and room temperature ionic liquid electrolytes have been investigated for DSSC applications.

4.1.4.4 Counter Electrode

Electro-catalytic property of the counter electrode is important in governing the Photo Voltaic (PV) performance of DSSCs. Without the catalytic layer, the TCO substrate has a very high charge transfer resistance (>106 $\Omega = cm^2$) in iodide/triiodide electrolyte, which makes it a very poor counter electrode [37]. Platinum is the standard catalyst deposited on the counter electrode due to its high catalytic activity, ability to reduce the over-potential for redox reaction and high resistance to corrosion against electrolytes [38,39]. Deposition of platinum on the TCO substrate can be implemented by using a wide range of methods, namely spray pyrolysis, sputtering and doctor blading technique [40]. Although platinum is the most efficient catalyst for counter electrode to date, its expensiveness makes it unsuitable for the low-cost approach of DSSCs [41]. Other alternatives such as carbon black [42], carbon nanotubes (CNTs) [43] and conducting polymers such as poly(3,4-ethylenedioxythiophene) doped with toluene sulphonate anions [20,44,45] were employed as a catalyst for the realization of a platinum-free counter electrode.

4.1.4.5 The Basic Operating Principle of DSSCs

The working principle of DSSCs is based on the photogeneration of an electron, as observed in photosynthesis. At the heart of a DSSC is a mesoporous layer composed of nanometre-sized particles of a wide band-gap semiconductor oxide, such as TiO_2, ZnO or SnO_2, that have been sintered together to allow electronic conduction to take place. Photoexcitation of the sensitizer dye results in the injection of an electron into the CB of the oxide, generating the oxidized form of the dye. After electron injection, the ground state of the dye is subsequently restored by electron donation from the electrolyte reductant, which in turn is regenerated by the reduction of the electrolyte oxidant at the counter electrode (Figure 4.6). The difference between the Fermi level of the oxide and the redox potential of the electrolyte determines the voltage generated by the cell under illumination. The overall operation of the cell involves excitation process, injection process, generation of energy, regeneration of dye and electron recapture.

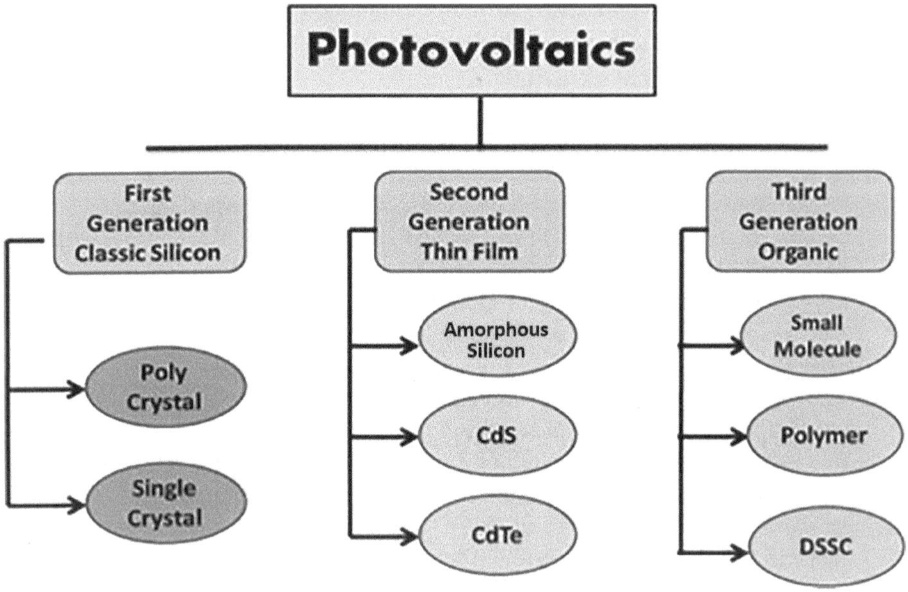

FIGURE 4.6 Schematic diagram of DSSCs' operating mechanism.

4.1.4.6 Advantages and Disadvantages of DSSCs

Advantages

i. The fundamental advantage of the DSSC system is the spatial separation of the electron- and hole-transporting components. This design hugely suppresses the charge recombination and allows efficient charge collection through microns-thick material.
ii. DSSCs have very low fabrication cost as well as the compatibility with flexible substrates, and this feature facilitates the variety of appearances to commercial market. DSSCs normally allow radiating away heat much easier and therefore work at lower internal temperatures.
iii. The efficiency of DSSCs has dramatically improved (11%) with the addition of various additives to the hole-transporting material. Modifying TiO_2 film surface structure and slowing the photochemical degradation of dye could improve the stability of DSSCs.
iv. The rate that the dye regains the electrons from the electrolyte, the recombination rate is quite slow and rate for electron transfer between counter electrodes to the species in the electrolyte is extremely fast.
v. DSSCs can work even in dim conditions. For example, DSSCs are able to work under cloudy skies indoor light and non-direct sunlight, whereas traditional solar cells would suffer a "cutout" at some lower limit of illumination.
vi. A practical advantage is that the mechanical robustness of DSSCs does not directly lead to a higher PCE.

vii. Traditional silicon cells are very fragile and hence require protective cautions; DSSCs normally allow radiating away heat much easier and therefore work at lower internal temperatures.

Disadvantages

i. DSSCs have low efficiency compared to traditional semiconductors and long-term stability issues; these are two serious concerns. The dye molecules are highly efficient at converting the absorbed photons into electrons in the semiconducting material (TiO_2) which ultimately produce the current. The photon absorption rate depends upon the solar flux spectrum and upon the absorption spectrum range of photoelectrodes.
ii. The major disadvantage to the DSSC design is the liquid electrolyte, which has temperature stability and sealing problems. The liquid electrolyte can freeze at low temperatures, which further leads to power production and potentially ending physical damage. The liquid might expand at higher temperatures, and then result in a series sealing problem. Therefore, replacing the liquid electrolyte with a solid-state material attracts a lot of attention.
iii. A third major drawback is volatile organic compounds in the electrolyte solution, which must be sealed carefully. Otherwise, these hazardous elements will harm the environment and human health. In flexible DSSCs, the solvents might permeate plastic substrates. This will preclude large-scale outdoor application and further integration into a flexible surface or structure.

4.2 CONCLUSIONS

Exploration of a potential energy source is one of the major challenges of this century as the demand for energy keeps increasing with the growth of population, industrial developments and advancement of technology. Solar power is supposed to play a pivotal role as a renewable energy source to meet the growing demand of energy worldwide. Solar cells, which convert an abundant amount of received solar energy into electricity, are potentially attractive candidates because of the easy fabrication and working of these devices.

ACKNOWLEDGEMENTS

The authors are thankful to Banaras Hindu University and CSIR for providing institutional and financial support.

REFERENCES

[1] Grätzel, M. (2001). Photoelectrochemical cells, *Nature*, 414, 338–344.
[2] Zhang, Q., & Cao, G. (2011). Nanostructured photoelectrodes for dye-sensitized solar cells, *Nano Today*, 6, 91–109.

[3] Safriani, L., Primawati, W. P., Mulyana, C., Susilawati, T., & Aprilia, A. (2017). Fabrication of semi-quasi solid DSSC using spiro material as hole transport material, *Materials Science and Engineering*, 196, 012014.

[4] Goodrich, A., Hacke, P., Wang, Q., Sopori, B., Margolis, R., James, T. L., & Woodhouse, M. (2013). A wafer-based monocrystalline silicon photovoltaics road map: Utilizing known technology improvement opportunities for further reductions in manufacturing costs, *Solar Energy Materials and Solar Cells*, 114, 110–135.

[5] Raturi, A., & Fepuleai, Y. (2010). Photosynthesis in a test tube-dye sensitized solar cells as a teaching tool, *Renewable Energy*, 35, 1010–1013.

[6] Garnett, E., & Yang, P. (2010). Light trapping in silicon nanowire solar cells, *Nano Letters*, 10, 1082–1087.

[7] O'regan, B., Grätzel, M. (1991). A low-cost, high-efficiency solar cell based on dye-sensitized, *Nature*, 353, 737–740.

[8] Gong, J., Liang, J., & Sumathy, K. (2012). Review on dye-sensitized solar cells (DSSCs): Fundamental concepts and novel materials, *Renewable and Sustainable Energy Reviews*, 16, 5848–5860.

[9] Sima, C., Grigoriu, C., & Antohe, S. (2010). Comparison of the dye-sensitized solar cells performances based on transparent conductive ITO and FTO, *Thin Solid Films*, 519, 595–597.

[10] Kwak, D. J., Moon, B. H., Lee, D. K., Park, C. S., & Sung, Y. M. (2011). Comparison of transparent conductive indium tin oxide, titanium-doped indium oxide, and fluorine-doped tin oxide films for dye-sensitized solar cell application, *Journal of Electrical Engineering and Technology*, 6, 684–687.

[11] Ye, M., Wen, X., Wang, M., Iocozzia, J., Zhang, N., Lin, C., & Lin, Z. (2015). Recent advances in dye-sensitized solar cells: From photoanodes, sensitizers and electrolytes to counter electrodes, *Materials Today*, 18, 155–162.

[12] Liao, J. Y., He, J. W., Xu, H., Kuang, D. B., & Su, C. Y. (2012). Effect of TiO_2 morphology on photovoltaic performance of dye-sensitized solar cells: Nanoparticles, nanofibers, hierarchical spheres and ellipsoid spheres, *Journal of Materials Chemistry*, 22, 7910–7918.

[13] Park, H., Kim, W. R., Jeong, H. T., Lee, J. J., Kim, H. G., & Choi, W. Y. (2011). Fabrication of dye-sensitized solar cells by transplanting highly ordered TiO_2 nanotube arrays, *Solar Energy Materials and Solar Cells*, 95, 184–189.

[14] Song, L., Jiang, Q., Du, P., Yang, Y., Xiong, J., & Cui, C. (2014). Novel structure of TiO_2-ZnO core shell rice grain for photoanode of dye-sensitized solar cells, *Journal of Power Sources*, 261, 1–6.

[15] Song, K., Jang, I., Song, D., Kang, Y. S., & Oh, S. G. (2014). Echinoid-like particles with high surface area for dye-sensitized solar cells, *Solar Energy*, 105, 218–224.

[16] Wu, W. Q., Rao, H. S., Feng, H. L., Guo, X. D., Su, C. Y., & Kuang, D. B. (2014). Morphology-controlled cactus-like branched anatase TiO_2 arrays with high light-harvesting efficiency for dye-sensitized solar cells, *Journal of Power Sources*, 260, 6–11.

[17] Suhaimi, S., Shahimin, M. M., Alahmed, Z. A., Chyský, J., & Reshak, A. H. (2015). Materials for enhanced dye-sensitized solar cell performance: Electrochemical application, *International Journal of Electrochemical Science*, 10, 2859–2871.

[18] Chen, X., & Mao, S. S. (2007). Titanium dioxide nanomaterials: Synthesis, properties, modifications and applications, *Chemical Reviews*, 107, 2891–2959.

[19] Lee, J. J., Rahman, M. M., Sarker, S., Nath, N. D., Ahammad, A. S., & Lee, J. K. (2011). Metal oxides and their composites for the photoelectrode of dye sensitized solar cells, *Advances in Composite Materials for Medicine and Nanotechnology*, 1, 181–210.

[20] Che´rubin, N. S. (2009). Dye-Sensitized Solar Cells Based on Perylene Derivatives. Ph.D. Thesis, University of Kassel, Germany.

[21] Kushwaha, R., Chauhan, R., Srivastava, P., & Bahadur, L. (2015). Synthesis and characterization of nitrogen-doped TiO_2 samples and their application as thin film electrodes in dye-sensitized solar cells, *Journal of Solid State Electrochemistry*, 19, 507–517.

[22] Neetu, Maurya, I. C., Gupta, A. K., Srivastava, P., & Bahadur, L. (2017). Extensive enhancement in power conversion efficiency of dye-sensitized solar cell by using Al-doped TiO_2 photoanode, *Journal of Solid State Electrochemistry*, 21, 1229–1241.

[23] Kaewsaenee, J., Visal-athaphand, P., Supaphol, P., & Pavarajarn, V. (2011). Effects of magnesium and zirconium dopants on characteristics of titanium (IV) oxide fibers prepared by combined sol-gel and electrospinning techniques, *Industrial & Engineering Chemistry Research*, 50, 8042–8049.

[24] Gupta, A. K., Srivastava, P., & Bahadur, L. (2016). Improved performance of Ag-doped TiO_2 synthesized by modified sol-gel method as photoanode of dye-sensitized solar cell, *Applied Physics A*, 122, 724.

[25] Raguram, T., & Rajni, K. S. (2022). Synthesis and characterisation of Cu-Doped TiO_2 nanoparticles for DSSC and photocatalytic applications, *International Journal of Hydrogen Energy*, 47, 4674–4689.

[26] Moradzaman, M., Mohammadi, M. R., & Nourizadeh, H. (2015). Efficient dye-sensitized solar cells based on CNTs and Zr-doped TiO_2 nanoparticles, *Materials Science in Semiconductor Processing*, 40, 383–390.

[27] Mahmoud, Z. H., AL- Bayati, R. A. & Khadom, A. A. (2022). Electron transport in dye-sanitized solar cell with tin-doped titanium dioxide as photoanode materials, *Journal of Materials Science: Materials in Electronics*, 33, 5009–5023.

[28] Grätzel, M. (1994). Highly efficient nanocrystalline photovoltaic devices, *Platinum Metals Review*, 38, 151–159.

[29] Zhang, S., Yang, X., Numata, Y., & Han, L. (2013). Highly efficient dye-sensitized solar cells: Progress and future challenges, *Energy & Environmental Science*, 6, 1443–1464.

[30] Ryan, M. (2009). PGM highlights: Progress in ruthenium complexes for dye sensitised solar cells, *Platinum Metals Review*, 53, 216–218.

[31] Nazeeruddin, M. K., Kay, A., Rodicio, I., Humphry-Baker, R., Müller, E., Liska, P., & Grätzel, M. (1993). Conversion of light to electricity by cis-X2bis (2,2′-bipyridyl-4,4′-dicarboxylate) ruthenium (II) charge-transfer sensitizers (X= Cl-, Br-, I-, CN-, and SCN-) on nanocrystalline titanium dioxide electrodes, *Journal of the American Chemical Society*, 115, 6382–6390.

[32] Wang, M., Grätzel, C., Zakeeruddin, S. M., & Grätzel, M. (2012). Recent developments in redox electrolytes for dye-sensitized solar cells, *Energy & Environmental Science*, 5, 9394–9405.

[33] Al-Alwani, M. A. M., Mohamad, A. B., Kadhum, A. A. H., Ludin, N. A., Safie, N. E., Razali, M. Z., Ismail, M., & Sopian, K. (2017). Natural dye extracted from Pandannusamaryllifolius leaves as sensitizer in fabrication of dye-sensitized solar cells, *International Journal of Electrochemical Science*, 12, 747–761.

[34] Arifin, Z., Soeparman. S., Widhiyanuriyawan, D., & Sutanto, B. (2017). Performance enhancement of dye-sensitized solar cells (DSSCs) using a natural sensitizer, *International Journal of Photoenergy*, 1788, 1–5.

[35] Sengupta, D., Das, P., Mondal, B., & Mukherjee, K. (2016). Effects of doping, morphology and film-thickness of photo-anode materials for dye sensitized solar cell application: A review, *Renewable and Sustainable Energy Reviews*, 60, 356–376.

[36] Hauch, A., & Georg, A. (2001). Diffusion in the electrolyte and charge-transfer reaction at the platinum electrode in dye-sensitized solar cells, *Electrochimica Acta*, 46, 3457–3466.

[37] Lan, J. L., Wang, Y. Y., Wan, C. C., Wei, T. C., Feng, H. P., Peng, C., & Hsu, W. C. (2010). The simple and easy way to manufacture counter electrode for dye-sensitized solar cells, *Current Applied Physics*, 10, S168–S171.

[38] Kim, S. H., & Park, C. W. (2013). Novel application of platinum ink for counter electrode preparation in dye sensitized solar cells, *Bulletin of the Korean Chemical Society*, 34, 831–836.
[39] Hagfeldt, A., Boschloo, G., Sun, L., Kloo, L., & Pettersson, H. (2010). Dye-sensitized solar cells, *Chemical Reviews*, 110, 6595–6663.
[40] Murakami, T. N., Ito, S., Wang, Q., Nazeeruddin, M. K., Bessho, T., Cesar, I., & Grätzel, M. (2006). Highly efficient dye-sensitized solar cells based on carbon black counter electrodes, *Journal of the Electrochemical Society*, 153, A2255–A2261.
[41] Gemeiner, P., Pavličková, M., Hatala, M., Hvojnik, M., Homola, T., & Mikula, M. (2022). The effect of secondary dopants on screen-printed PEDOT: PSS counter-electrodes for dye-sensitized solar cells, *Journal of Applied Polymer Science*, 139, 51929.
[42] Kay, A., & Grätzel, M. (1996). Low cost photovoltaic modules based on dye sensitized nanocrystalline titanium dioxide and carbon powder, *Solar Energy Materials and Solar Cells*, 44, 99–117.
[43] Shahzad, N., Perveen, T., Pugliese, D., Haq, S., Fatima, N., Salman, S. M., Tagliaferro, A., & Shahzad, M. I. (2022). Counter electrode materials based on carbon nanotubes for dye-sensitized solar cells, *Renewable and Sustainable Energy Reviews*, 159, 112196.
[44] Lee, J. J., Rahman, M. M., Sarker, S., Nath, N. D., Ahammad, A. S., Lee, J. K. (2011). Metal oxides and their composites for the photoelectrode of dye sensitized solar cells, *Advances in Composite Materials for Medicine and Nanotechnology*, 1, 181–210.
[45] Folli, A., Bloh, J. Z., Lecaplain, A., Walker, R., & Macphee, D. E. (2015). Properties and photochemistry of valence-induced-Ti^{3+} enriched (Nb, N)-codoped anatase TiO_2 semiconductors, *Physical Chemistry Chemical Physics*, 17, 4849–4853.

5 Halide Perovskite for Photovoltaic Applications

S. K. S. Patel and Sajith Kurian

5.1 INTRODUCTION

The growing population and industrialization have caused an increase in energy demand. As a result, clean, inexpensive and renewable energy sources are developed (Snaith, 2013). Silicon-based solar cells generate more than 21% of the world's photovoltaic energy (Chapin et al., 1954; Morgano et al., 2011). Its high cost and serious environmental impact encourage scientists and industrialists worldwide to look for alternatives that can address the issue while providing a reasonable level of performance. As a result, several new energy technologies such as dye-sensitized solar cells (DSSCs) (O'Regan and Gratzel, 1991; Kakiage et al., 2015), organic photovoltaic cells (OPVs) (Azzopardi et al., 2011), organic solar cells (OSCs) based on bulk heterojunction (BHJ) (Rafique et al., 2017) and quantum dot solar cells (QDSCs) (Kamat, 2008; Chen et al., 2021) have been developed as low-cost alternatives over silicon-based solar cells in the last two decades. It is still not possible to achieve the conversion efficiencies necessary to compete in the energy market at this time, but advancements are being made. In this scenario, the organic–inorganic hybrid halide perovskites, like methylammonium lead iodide ($CH_3NH_3PbI_3$ or $MAPbI_3$), which were initially employed in DSSCs as light absorbers by Kojima et al. (2009), have been identified to have the potential to meet these conditions. These hybrid materials possess impressive photoelectric properties, large dielectric constants, high optical absorption coefficients and low exciton binding energies because of their perovskite structure (Jeon et al., 2015). Over the past decade, perovskite solar cells (PSCs) have been the subject of extensive research, with their overall efficiency now comparable to or better than that of silicon-based solar cells.

As a result of their tunable band gaps, high charge mobility, high absorption coefficients, long carrier lifetime and long carrier diffusion length, hybrid halide PSCs have emerged as a very hot research area among the photovoltaic community (Niezgoda et al., 2017). The research efforts into the design of device architectures, optimization of interfacial characteristics and careful control of the morphology of each functional layer (Gu et al., 2016) have enhanced the efficiency of PSCs. In 2009, the compound $MAPbX_3$ ($X = Br$ and I) was first used in liquid electrolyte-based DSSCs (Kojima et al., 2009), and the power conversion efficiencies (PCEs) were 3.13% and 3.81% for $X = Br$ and $X = I$, respectively. In 2011, a quantum dot solar cell with $MAPbI_3$ nanocrystal was prepared, and a PCE of 6.54% was reported (Parida et al., 2011). In just a span of two years, Snaith (2013) reported a planar heterojunction

DOI: 10.1201/9781003481157-5

perovskite solar cell with a PCE of 15%. Since then, this research area has seen a tremendous increase in the number of publications, each reporting either a higher efficiency or stability, a new material, a layered material, QDs, hybrid perovskites or all-inorganic perovskites. The maximum photoelectric PCEs of the PSCs have been found to be over 26.0% according to the NREL Report, 2021 [16].

This chapter discusses the crystal structure of 3D and 2D perovskites and other related structures, solar cell device fabrication, the specific function of each layer in the solar cell, the operating mechanisms of the PSCs and the advances in the deposition methods of perovskite light-absorbing layer. This chapter concludes with an outline of the future research directions in this field.

5.2 CRYSTAL STRUCTURE OF PEROVSKITE

Calcium titanate ($CaTiO_3$) is a well-known mineral and, more than that, it forms the basis of a class of compounds known as "perovskites" with a basic formulation of ABX_3. $CaTiO_3$ was discovered by Gustav Rose in 1839, and he gave the name "perovskites" in honour of the mineralogist Lev Alekseyevich von Perovski. There are many inorganic perovskites having the formula ABX_3, like $BaTiO_3$, $SrTiO_3$ and $PbTiO_3$. The crystal structure of cubic ABX_3 perovskite consists of corner-sharing BX_6 octahedra with an "A" cation occupying the 12-fold coordination site formed in the middle of the cube of eight such octahedra, as illustrated in Figure 5.1. In the formula of perovskite, ABX_3, X is often oxygen, but ions like F^- and Cl^- can also be accommodated.

The hybrid organic–inorganic perovskites also have the same general formula, ABX_3, like that of the inorganic perovskites with an exception that the "A" cation site is occupied by an organic cation instead of a metal cation. The structure of the hybrid perovskite can be considered as follows: it consists of an organic cation of larger size (*A*), an inorganic cation of smaller size (*B*) and a halide anion (*X*), as shown in Figure 5.1.

However, the formation of perovskite is dependent on the ionic radii of the cations A (r_A) and B (r_B) and the halide anion X (r_X) (Niezgoda et al., 2017). The thermal stability of the BX_6 octahedra can be determined by the octahedral factor μ. μ is defined as the ratio of r_B and the r_X. The stable metal halide perovskite has been found to have a value of μ between 0.442 and 0.895 (Yi et al., 2019).

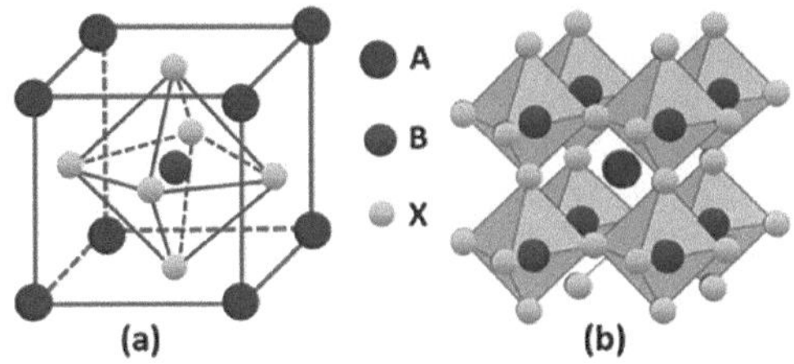

FIGURE 5.1 The perovskite structure ($A=RNH_3$, $B=M$ (metal) and $C=X$ (halogen)): (a) and (b) two ways of representation. Reprinted with permission from Yi et al. (2019).

$$\mu = \frac{r_B}{r_X}$$

The Goldschmidt tolerance factor t needs to be in the range $0.8 \leq t \leq 1.0$ for a perovskite to form, where t is defined as follows (Yi et al., 2019):

$$t = \frac{r_A + r_X}{\sqrt{2}(r_B + r_X)}$$

The tolerance factor is crucial to determining the shape of the structure. When the tolerance factor (t) equals one, the compound has a perfect cubic structure as is common for all perovskites. A perovskite structure may appear distorted in many cases. A distorted perovskite structure with an orthorhombic, tetragonal or rhombohedral crystal structure, for example, is formed when the t-values are $0.80 < t < 0.89$ (Figure 5.2). The very small size of the A cation results in a trigonal structure if $t < 0.8$. Furthermore, for a $t > 1$, the A cation forms a hexagonal structure instead of a perovskite structure.

So, the size of the ions has an important role in deciding the perovskite structure. For example, if we take B = Pb and X = I and consider $t = 1$, we can find that the organic cation A should be a small one consisting of two or three atoms in order to form a three-dimensional (3D) perovskite compound. A good example is $CH_3NH_3PbI_3$ (Knop et al., 1990).

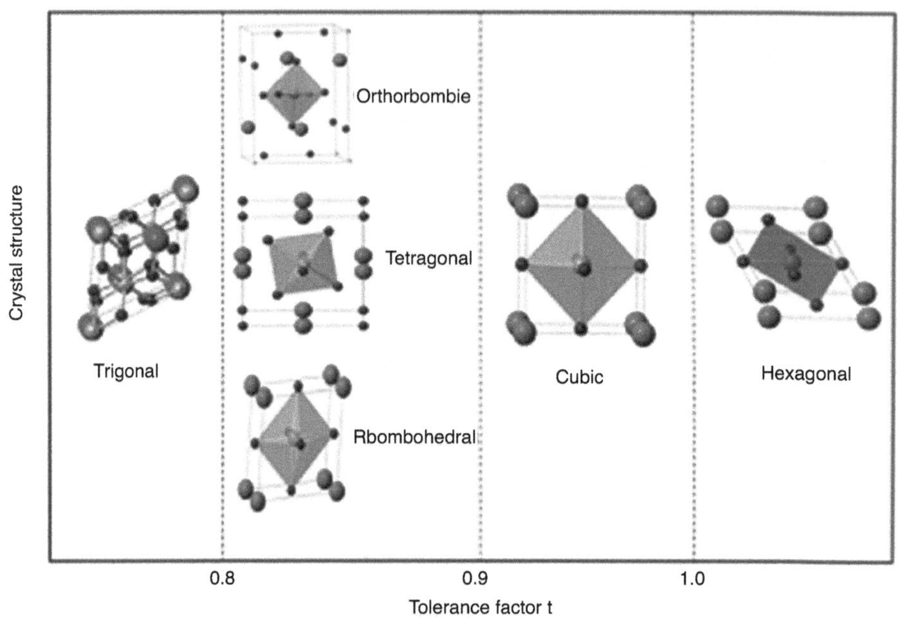

FIGURE 5.2 Correlation between the crystal structure of perovskite and the value of tolerance factor (t). Reprinted with permission from Yi et al. (2019).

As per the size constraints in the 3D perovskites with general formula ABX_3, A site is occupied by monovalent organic cation, and methylammonium ($CH_3NH_3^+$, MA^+) or formamidinium ($NH_2CHNH_2^+$, FA^+) is the one with a suitable band gap for visible light absorption and shows maximum efficiency. B is a divalent metal cation (Pb^{2+}, Sn^{2+}, Ge^{2+}, Mg^{2+} and Ca^{2+}) and halide anion X (Cl^-, Br^- or I^-) occupying the core and apex of the octahedra, respectively. The metal–halogen octahedra are joined together to form a 3D network structure, as shown in Figure 5.3.

- A = an organic cation—methylammonium ($CH_3NH_3^+$) or formamidinium ($NH_2CHNH_2^+$)
- B = a big inorganic cation—usually lead(II) (Pb^{2+})
- X_3 = a slightly smaller halogen anion—usually chloride (Cl^-) or iodide (I^-)

So, the options are very limited, if we stick to the three-dimensional structure having $t = 1$. As the size of the organic cation increases, the tolerance factor becomes much greater than 1 ($t > 1$), and in such cases, a two-dimensional (2D) layered structure forms by keeping the organic group away from the inorganic sheets by spacer groups (Mitzi, 2007).

The layered perovskites consist of two different typical groups of compounds: $(R\text{-}NH_3)_2MX_4$ and $(NH_3\text{-}R\text{-}NH_3)MX_4$, where R is an aliphatic or aromatic compound. In the first case, the inorganic layers are separated by bilayers of organic ammonium cations, where the organic group interacts through π-π interaction (R—aromatic) or through van der Waals force (R–alkyl) (Figure 5.4a) and the inorganic layers connect to the organic layers by hydrogen bonds provided by N-H----X interactions (Figure 5.5). In the second class of compounds, the inorganic layers connect to the organic layers by hydrogen bonds provided by N-H----X interactions at both ends of the organic cation (Figure 5.4b). The structure of a typical $(R\text{-}NH_3)_2MX_4$ system is shown in Figure 5.5.

The possibility of accommodating a variety of organic cations is more in 2D layered perovskite than in 3D perovskite where only groups with a very limited number of atoms can occupy organic cations. Due to this, the properties of the 2D layered

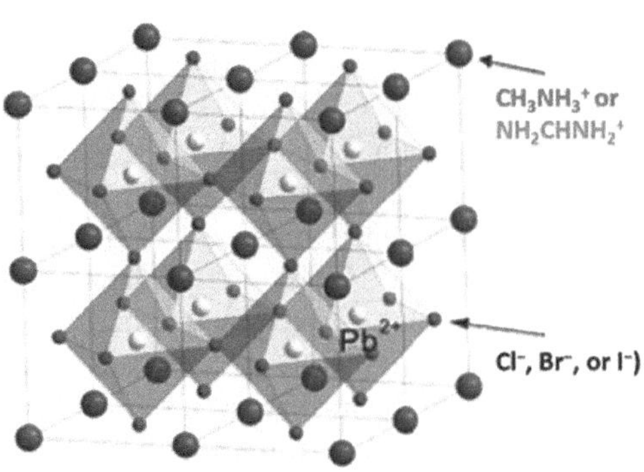

FIGURE 5.3 Typical perovskite cubic lattice structure. Reprinted with permission from Chen et al. (2014).

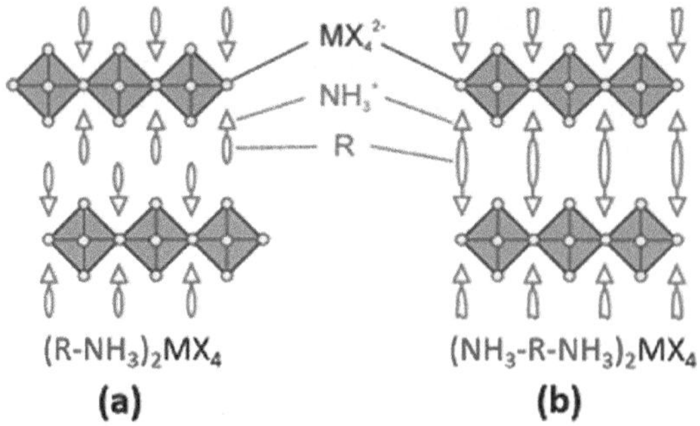

FIGURE 5.4 Schematic representation of 2D layered perovskite: (a) with $(R\text{-}NH_3)_2$-type organic group and (b) $(NH_3\text{-}R\text{-}NH_3)$-type organic group. Reprinted with permission from Mitzi (2001).

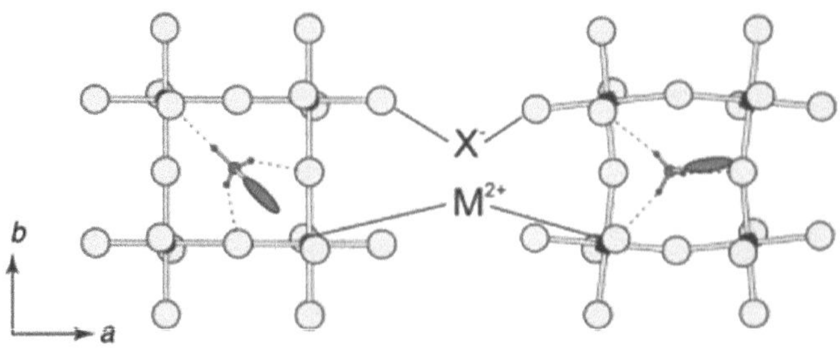

FIGURE 5.5 Schematic representation of bridging between an organic layer and inorganic layer of 2D layered perovskite with $(R\text{-}NH_3)_2$-type organic group. Reprinted with permission from Mitzi (2001).

perovskites can be tuned by varying the organic cation and thus have generated an intensive research interest in this particular group of compounds.

For all-inorganic perovskite materials, the A site is occupied by an inorganic metal cation (Cs^+ and Rb^+), while the B and X sites are occupied by a divalent metal cation (Pb^{2+}, Sn^{2+}, Ge^{2+}, Mg^{2+} and Ca^{2+}) and halide anion X (Cl^-, Br^- or I^-), respectively, which is similar to that in the hybrid perovskite.

In addition to the general representative formula, ABX_3, perovskites with $A_3B_2X_9$, $A_2B\,B'X_6$, A_2BX_6, A_4BX_6 and so on are also reported in the literature. They are termed as double perovskites, layered double perovskites and so on. In general, perovskites can be classified into 3D, 2D, 1D and 0D perovskites just like the case of the classification of nanomaterials. As explained above, a smaller organic cation and inorganic cations at A site lead to the 3D structure and the occupation of a bigger organic cation at A site leads to the formation of the 2D layered structure. In the case of halide perovskites, 1D and 2D structures are uncommon (Zeng et al., 2020).

Why is this material so important in the electronic industry? As it is demonstrated above, the hybrid organic–inorganic perovskite material is a layered material with one inorganic layer followed by one organic layer. This configuration imparts

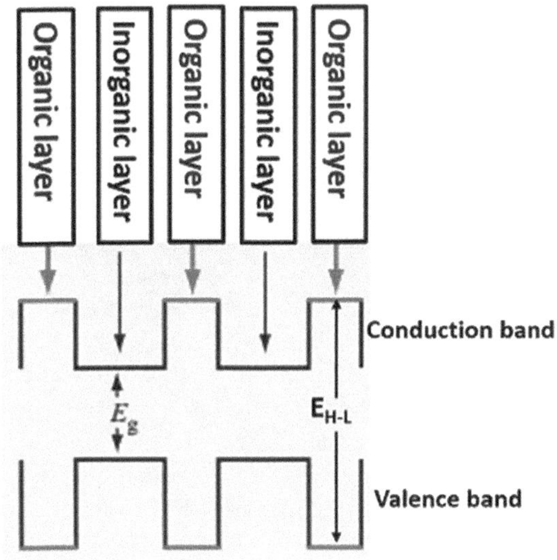

FIGURE 5.6 Schematic structure of HOMO-LUMO gap of 2D layered hybrid perovskites.

interesting and potentially useful electronic properties for the material. It was found that the HOMO-LUMO gap for organic layers is large compared to the inorganic layers and the dielectric constant of organic material is lower than the dielectric constant of inorganic materials (Tanaka et al., 2002). So, the HOMO-LUMO picture can be visualized in such a way that the organic layer forms a barrier and the inorganic layer forms a well for the carriers, and as the organic and inorganic layers repeat alternately, the HOMO and LUMO form a self-organized multiple quantum well structure, as shown in Figure 5.6.

The beauty of this multiple quantum well structure is that the width and height of the quantum well (both the barrier and the well) can be tuned by modifying the organic, inorganic and halogen contents. As a result, the absorption and thus the emission wavelength can be tuned from 320 nm to 800 nm, the entire visible range by substituting the organic, inorganic and halogen contents in the perovskite. This is a very interesting and extremely important/useful property of the perovskites, which can be used for solar cell applications to convert the solar energy into electrons and to function as luminescent materials exhibiting a wide range of colours.

5.3 PEROVSKITE SOLAR CELL DEVICE STRUCTURE

A perovskite solar cell typically consists of a transparent conductive oxide (TCO) electrode, charge transport layers, such as electron transport layer (ETL) and hole transport layer (HTL), the light-harvesting organic–inorganic hybrid halide or all-inorganic halide perovskite layers and the metal contact material known as the counter electrode. The light-harvesting perovskite layer is sandwiched between the ETL and the HTL. PSCs can be classified into two categories: one with the conventional (n-i-p) structure and the other with inverted (p-i-n) structures depending

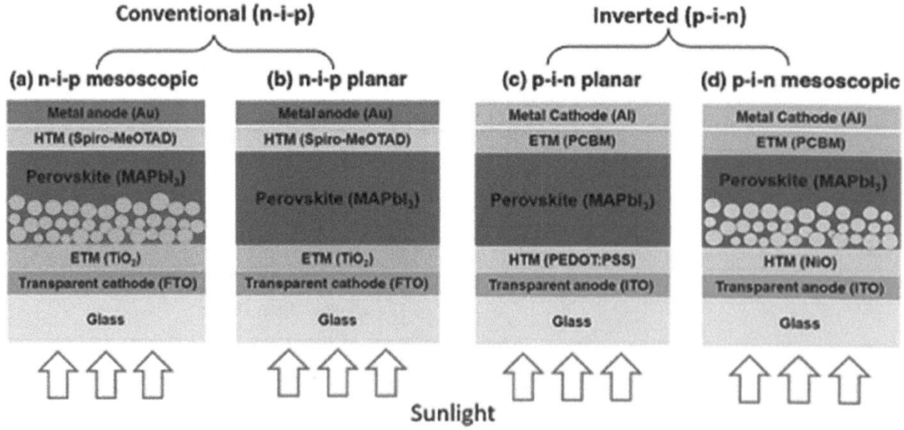

FIGURE 5.7 Schematic diagram showing the layered structure of four typical perovskite solar cells: (a) n-i-p mesoscopic, (b) n-i-p planar, (c) p-i-n planar and (d) p-i-n mesoscopic. Reprinted with permission from Hussain et al. (2018).

on which transport (electron/hole) layer is encountered by the incident light first (Hussain et al., 2018). A perovskite layer is placed on top of the ETL, followed by the HTL and finally an electrode for collecting holes. A PSC structure with inverted perovskite is constructed with a thin perovskite layer above the HTL. In contrast, OSCs use an inverted architecture, where the hole-collecting electrode is above the active layer, as opposed to this convention in OSCs. There are two types of conventional and inverted structures: mesoscopic and planar. Mesoscopic structures have mesoporous layers, whereas planar structures contain only planar layers (Hussain et al., 2018). The conventional and inverted structures of PSCs are schematically shown in Figure 5.7.

5.4 MATERIALS FOR PEROVSKITE SOLAR CELL

Our discussion in this section will focus on the TCO, electron-transporting materials, and hole-transporting materials that comprise each layer of a perovskite solar cell. A detailed discussion on the light-harvesting perovskite layer is followed in section 5.6.

5.4.1 Transparent Conductive Oxide (TCO) Layer

Bottom electrode of a PSC is n-type metal oxide typically on the TCO, either fluorine-doped tin oxide (FTO) for conventional (n-i-p) devices with TiO_2 ETL processed at higher temperatures or indium tin oxide (ITO) for inverted (p-i-n) devices processed at lower temperatures. TCO helps to achieve high open-circuit voltage (V_{OC}) and overall power conversion efficiency (Hussain et al., 2018).

5.4.2 Electron Transport Layer (ETL)

ETL in PSCs can be organic or inorganic. The ETL's primary function is to collect and transfer charges after electrons are injected from the light-absorbing perovskite layer and to prevent holes from migrating to the photoelectrode. As a result, it achieves effective charge separation and reduces charge carrier recombination (Mahmood et al., 2017). In conventional (n-i-p) PSCs, TiO_2, which has a wide band gap, has been widely studied as an effective electron transport material. ZnO and other n-type semiconductor materials are employed in flexible PSCs and ETLs (Mahmood et al., 2017).

In inverted (p-i-n) PSCs, an organic ETL of PCBM (one of the fullerene derivatives, [6,6]-phenyl-C61-butyric acid methyl ester) and PFN (poly[9,9-dioctylfluorene-9,9-bis(N,N-dimethylpropyl)fluorene]) is used to collect electrons, which is being found to be better than that of TiO_2 (Hussain et al., 2018).

5.4.3 Hole Transport Layer (HTL)

The HTL is used to receive and transport holes generated in the perovskite light-absorbing layer to the metal electrode at the surface. The application of doped 2,2′,7,7′-tetrakis(N,N-di-p-methoxyphenylamine)-9,9′-spirobifluorene (spiro-OMeTAD) and poly(3,4-ethylenedioxythiophene) polystyrene sulphonate (*PEDOT:PSS*) as an HTL witnessed a rapid increase in the PCEs of typical PSCs. PEDOT:PSS remains the most commonly used HTL in devices with organic charge transport layers, and spiro-OMeTAD is most commonly used in devices with oxide ETL (Hussain et al., 2018). Poly[bis(4-phenyl)(2,4,6-trimethylphenyl)amine (PTAA) is also used as HTL material due to its electron-rich components. The use of PTAA can substantially improve the open-circuit voltage (V_{OC}) and fill factor (FF) of the cells. The solar cells fabricated making use of the mixed composition $(FAPbI_3)_{0.85}(MAPbBr_3)_{0.15}$ perovskite and the novel anthradithiopene (ADT)-based HTLs show PCEs up to 17.6% under 1 sun illumination compared to the 18.1% observed when using the benchmark compound spiro-OMeTAD (Santosh et al., 2021).

5.4.4 Mesoporous Layer (ML)

A mesoporous oxide layer is often used to increase the area of light harvesting perovskite layer. Introduction of TiO_2 and/or ZrO_2 nanoparticles coated with porous carbon is found to enhance the electronic conductivity. The configuration, which includes the mesoporous TiO_2 layer, exhibits the highest efficiency under 1 sun illumination (Saliba et al., 2016). With a double layer of mesoporous TiO_2 and ZrO_2 covered by porous carbon, a remarkable 12.8% efficiency has been achieved in a PSC (Yin et al., 2015). Some representative devices and their architectures and performances are shown in Table 5.1.

5.5 WORKING PRINCIPLE OF PEROVSKITE SOLAR CELLS

A schematic representation of the energy band alignment of PSCs is illustrated in Figure 5.8. The perovskite solar cell system has three main functioning steps: (1) absorption of photons, (2) charge transport and (3) charge extraction

TABLE 5.1
Some Representative Devices, Architectures and their Parameters and Efficiencies

Device Architecture	HTM	Jsc ($mAcm^{-2}$)	V_{oc}/V	FF	η%	References
$FTO/TiO_2/MAPbI_3/HTM/Au$	Spiro-OMeTAD	21.52	1.04	0.672	15.08	Deng et al. (2017)
$FTO/TiO_2/MAPbI_3/HTM/Au$	Spiro-OMeTAD	19.8	0.924	0.663	12.1	Chen et al. (2016)
$FTO/HTM/MAPbI_3/PCBM/Ag$	Cu:CrOx	16.02	0.98	0.7	10.99	Ping et al. (2016)
$TO/TiO_2/MAPbI_3/HTM/Au$	Cu_2ZnSnS_4	20.54	1.06	0.587	12.75	Wu et al. (2015)
$FTO/TiO_2/MAPbI_3/Au$	None	17.8	0.905	0.65	10.49	Shi et al. (2014)
$ITO/Y\text{-}TiO_2/MAPbI_{3-x}Cl_x/HTM/Au$	Spiro-OMeTAD	21.45	1.08	0.74	17.14	Zhang et al. (2017)
$ITO/Y\text{-}TiO_2/MAPbI_3-xClx/HTM/Au$	Spiro-OMeTAD	22.75	1.13	0.75	19.3	Zhou et al. (2014)
$FTO/c\text{-}TiO_2/MAPbI_3-xClx/HTM/Ag$	Spiro-OMeTAD	21.5	1.07	0.67	15.4	Liu et al. (2014)
$FTO/c\text{-}TiO_2/mp\text{-}TiO_2/MAPbI_3-xBrx/HTM/Au$	PTAA	19.5	1.09	0.76	16.2	Nam et al. (2014)
$FTO/c\text{-}TiO_2/FAPbI_3/MAPbI_3/HTM/Au$	Spiro-OMeTAD	20.97	1.032	0.74	16.01	Lee et al. (2014)
$FTO/TiO_2/(FAPbI_3)0.85(MAPbBr_3)0.15/HTM/Carbon$	PEDOT	23.5	1.05	0.69	17.0	Jiang et al. (2017)
$FTO/TiO_2/FA0.8MA0.1Cs0.1/HTM/Au$	Spiro-OMeTAD	21.5	1.155	0.73	18.1	Matsui et al. (2017)
$FTO/c\text{-}TiO_2/CsPbI_3$ QDs/HTM/MoOx/Al	Spiro-OMeTAD	13.47	1.23	0.65	10.77	Swarnkar et al. (2016)
$ITO/c\text{-}TiO_2/CsPbI_2Br/HTM/Ag$	Spiro-OMeTAD	13.99	1.10	0.67	10.34	Niezgoda et al. (2017)
$FTO/c\text{-}TiO2/Cs_{0.925}K_{0.075}PbI_2Br/HTM/Au$	Spiro-OMeTAD	11.6	1.18	0.73	10.0	Nam et al. (2017)
$ITO/SnO_2/ZnO/CsPbI_2Br/HTM/MoO_3/Ag$	Spiro-OMeTAD	15.0	1.23	78.8	14.6	Yan et al. (2018)
$FTO/TiO_2/SmBr_3/CsPbIBr_2/HTM/Au$	Spiro-OMeTAD	12.75	1.17	73.0	10.88	Subhani et al. (2019)
$FTO/c\text{-}TiO_2/CsPbBr_3/CsPb_3Br_5/HTM/Ag$	Spiro-OMeTAD	8.48	1.296	75.9	8.34	Li et al. (2018)

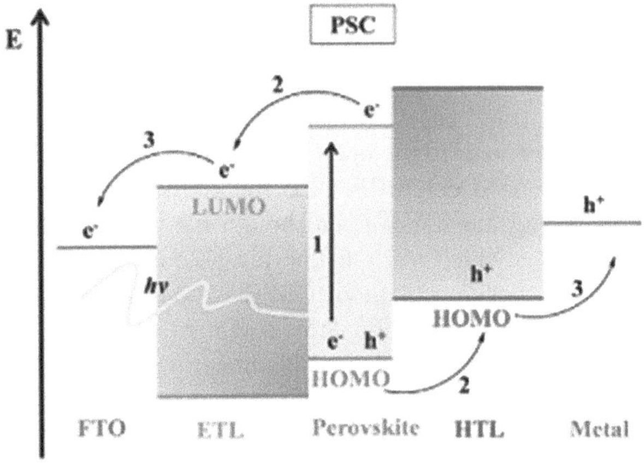

FIGURE 5.8 Band diagram and operation principle in PSC: (1) absorption of photon and free charge generation, (2) charge transport and (3) charge extraction. Reprinted with permission from Hussain et al. (2018).

(Marinova et al., 2017). As the perovskite layer absorbs photons, it produces excitons (electron-hole pairs), which are dissociated to produce charge carriers (free electrons and holes). A dissociation of excitons occurs between the perovskite and charge transport layers; ETL and HTL, respectively, collect these free electrons and holes. As the free electrons from the ETL migrate to the anode, which is FTO, simultaneously the holes from the HTL migrate to the metal electrode (cathode) (Stoumpos et al., 2013). Commonly used metal electrodes are noble metals, such as Au, Ag and Pt. Finally, the FTO and metal electrode are connected and the photocurrent is generated in the external circuit.

5.6 DEPOSITION METHODS OF THE PEROVSKITE LAYER

PSCs are highly dependent on the ETL and HTL structures as well as the device structure. In addition, the method of synthesis of the light-absorbing perovskite layer has a major influence on its properties. Seven kinds of perovskite thin film deposition methods were reported till now, which include one-step precursor deposition (OSPD), sequential deposition process (SDP), two-step spin-coating deposition (TSSD), dual-source vacuum deposition (DSVD), sequential vapour deposition (SVD), vapour-assisted solution process (VASP) and spray-coating deposition (SCD). We can classify these deposition methods into two categories: solution-based deposition methods and vapour-based deposition methods. Table 5.2 shows the devices fabricated using the above methods for the deposition of the perovskite absorber layer and the PCE achieved in each method.

5.6.1 Solution-Based Deposition Methods

Miyasaka and coworkers (2009) were the first to report a OSPD method for the perovskite absorber layer (Kojima et al., 2009). The perovskite precursor solution

(~40 wt%) containing MAX and PbX_2 in appropriate proportions in suitable solvents like γ-butyrolactone (GBL) and *N,N*-dimethylformamide (DMF) was stirred overnight at a certain temperature (e.g. 80°C) before spin coating on the substrate. Thermal annealing at ~100°C ensured the removal of the solvent and the complete transformation of the precursor to the crystalline perovskite. The precursor concentration and/or spinning speed determines the thickness of the perovskite film and can be tuned by changing the same. Even though the uncontrolled crystallization process of the perovskite, which leads to the pin-hole areas and large grains, is a major problem of this deposition method, OSPD is considered as the simplest and most important deposition method. The defects of this method can be overcome by solvent engineering (Nam et al., 2014), thermal annealing (Hsu et al., 2014) and additive treatment (Liang et al., 2014), and a PCE of up to 19.3% was achieved (Shaowei et al., 2015).

A sequential deposition process (SDP) was introduced by Liang et al. (1998) and later used by Burschka et al. (2013) in PSCs. In this process, the metal halide in DMF or DMSO was first spin coated on the porous or planar substrate and then allowed to react with a solution of MAI in 2-propanol by dipping the metal halide-coated substrates in the MAI solution. Though SDP improves the coverage, pore filling and morphology, which in turn leads to an enhancement in the PCE, the method is not as efficient in planar heterojunction (PHJ) solar cells as in nanostructured solar cells. In PHJ cells, the MAI cannot penetrate deep into the lead halide layer for the complete conversion of lead halide to perovskite crystals. Quick crystallization and rough perovskite film hamper the efficiency (Xiao et al., 2014).

Xiao et al. (2014) reported a TSSD method, by spin coating a lead halide solution in DMF and MAI solution in 2-propanol step by step. Thermal annealing at 100°C drives the interdiffusion of precursors and thus the formation of perovskites. The MAI concentration and the exposure time of lead halide to MAI solution before spin coating have a major role in deciding the perovskite structure and thus influence the efficiency of the system (Im et al., 2014).

In the SCD method, the substrates were coated with a precursor solution of MAI and $PbCl_2$ (3:1 molar ratio) in DMF or DMSO by using a spray head. Subsequent thermal annealing transformed it into $MAPbI_{3-x}Cl_x$. A planar device reported by Barrows et al. (2014) yielded a PCE of 11.1%. The obtained PCE is comparable to the devices fabricated *via* spin coating and shows the commercial potential of SCD, which can be used in the large-area, low-cost, efficient manufacture of PSCs.

5.6.2 Vapour-Based Deposition Methods

Dual-source vacuum deposition (DSVD) was first reported by Liu et al. (2014) by fabricating a $MAPbI_{3-x}Cl_x$-based PHJ perovskite solar cell and reporting a high PCE of 15.4%. In this method, the precursor sources, MAI and $PBCl_2$, were evaporated from separate sources simultaneously. This method ensured a uniform film of perovskite over a range of length scales.

SVD process is very similar to the precursor interdiffusion deposition method (TSSD) except that SVD uses vapour deposition. In the reported procedure, PbX_2

TABLE 5.2
Examples of Devices Via Different Perovskite Deposition Methods and their Corresponding Device Configurations and PCE

Deposition Process (DP)	Device Structure	J_{sc} (mA cm^{-2})	V_{oc} (V)	FF (%)	PCE (%)	References
OSPD	ITO/PEIE/bl-TiO_2(Y)/ $MAPbI_{3-x}Cl_x$/**spiro**/Au	22.75	1.13	75.01	19.3	Zhou et al. (2014)
SDP	ITO/bl-ZnO/$MAPbI_3$/**spiro**/Ag	20.4	1.03	74.9	15.7	Liu et al. (2014)
TSSD	ITO/bl-TiO_2/mp-TiO_2/ $MAPbI_3$/**spiro**/Au	21.64	1.056	74.1	17.01	Im et al. (2014)
DSVD	FTO/bl-TiO_2/$MAPbI_{3-x}Cl_x$/ **spiro**/Ag	21.5	1.07	67	15.4	Liu et al. (2014)
SVD	ITO/PEDOT:PSS/$MAPbI_{3-x}Cl_x$/ **C60**/Bphen/Ca/Ag	20.9	1.02	72.2	15.4	C.W. Chen et al. (2014)
VASP	FTO/bl-TiO_2/$MAPbI_{3-x}Cl_x$/ **spiro**/Ag	19.8	0.924	66.3	12.1	Q. Chen et al. (2014)
SCD	ITO/PEDOT:PSS/$MAPbI_{3-x}Cl_x$/ **PC60BM**/Ca/Al	16.8	0.92	72	11.1	Barrows et al. (2014)

(X = I and Cl) and MAI were vapour deposited on the substrates layer by layer (C.W. Chen et al., 2014). A thermal annealing following the deposition of layers completes the perovskite crystal transformation. Chen et al. achieved a high PCE of 15.4%.

The vapour-assisted solution (VASP) process, as reported by Chen et al. (2014), is a blended deposition method. At first, a solution of PbI_2 in DMF was spin coated on the substrates and dried at 110°C for 15 min. The PbI_2-coated substrates were then annealed in MAI vapour at 150°C in a N_2 atmosphere for the desired time to yield the perovskite films. It was then cooled down, washed with isopropanol and dried. A further annealing ensured a complete deposition. The perovskite film exhibited full-surface coverage, uniform grain structure with grain size up to micrometres and ~100% precursor transformation completeness. The different methods described above for the deposition of perovskite films are schematically represented in Figure 5.9.

5.7 SYNTHESIS OF PEROVSKITE NANOMATERIALS (PNMS)

The above-mentioned methods limit the perovskite into bulk/thin films. Compared to the bulk materials, the perovskite nanomaterials (PNMs) can be utilized for the fabrication of thin films and flexible devices (Shamsi et al., 2019). In addition to this, PNMs display high processability, and feature-rich and controllable facets and active sites. It also exhibits outstanding photo-electro-magnetic properties due to the small-size effect and quantum effect (Wang et al., 2017).

Nanomaterial synthesis can be broadly classified into bottom-up approach and top-down approach. Perovskite nanomaterial synthesis methods can be divided into two major categories: wet chemical synthesis methods and other synthetic approaches.

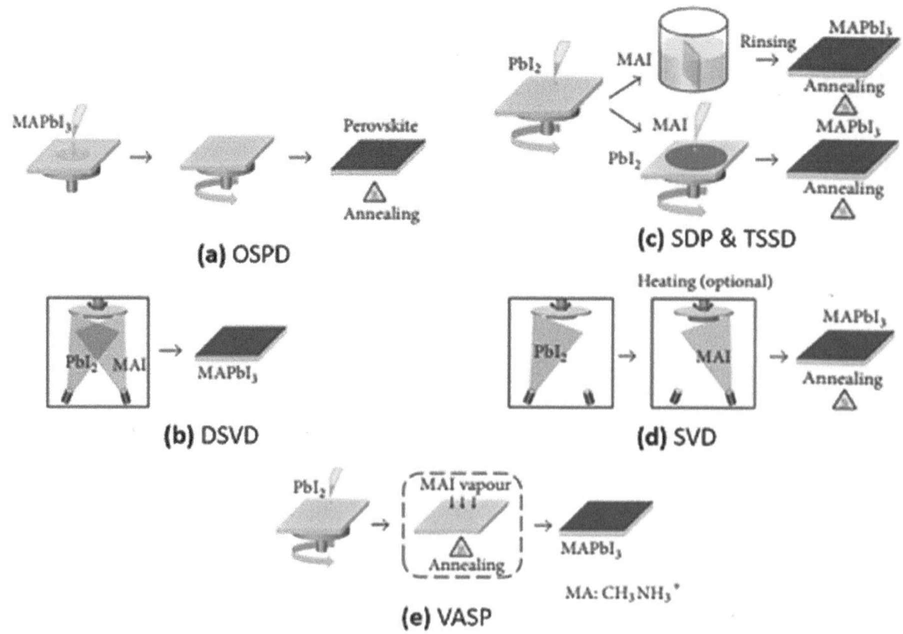

FIGURE 5.9 Illustration of different perovskite deposition methods: (a) one-step precursor deposition, (b) dual-source vapour deposition, (c) solution-based two-step method (SDP and TSSD), (d) sequential vapour deposition and (e) vapour-assisted solution process. Reprinted with permission from Zhou et al. (2018).

5.7.1 Wet Chemical Synthesis Methods

This method comes under the broad category of bottom-up approach. Hot injection (HI) method, thermal decomposition (TD) method and reprecipitation methods, which include emulsion-based synthesis and ligand-assisted reprecipitation (LARP), were widely used for the preparation of PNMs under this category (Dey et al., 2021).

5.7.1.1 Hot Injection (HI) Method and Thermal Decomposition (TD) Method

HI method was first reported by Protesescu and coworkers (2015) for the synthesis of all-inorganic halide perovskite quantum dots (PQDs). A certain amount of preheated (120–150°C) complex of A-site cation with oleate (Cs-oleate) was rapidly injected into a heated (150–180°C) solution containing B-site cation salt with oleic acid (OA) and oleylamine ligands (Chen et al., 2019). The reaction can be arrested by keeping the reaction vessel in ice-cold water. The product can be obtained by washing with nonpolar solvents. Instead of adding Cs-oleate solution, metal (Cs or Pb) acetates can be used and trimethylchlorosilane, trimethylbromosilane, acyl chloride, etc., can be injected at higher temperature (Wang et al., 2017). At higher temperature, halogen precursors will decompose and free halide ions will react with metal ions in solution to form perovskite nanocrystals. Nanoplatelets (NPLs) and nanowires (NWs) can be synthesized by tuning the temperature of the reaction. It was found that decreasing the temperature leads to the formation of NPLs and increasing the temperature and time leads to the formation of NWs in colloidal form (Imran et al., 2018). This indicates that the reaction time and temperature can be tuned to produce different dimensional perovskite materials.

HI method generally performs at anhydrous and anaerobic conditions because nucleation and growth of the PNMs are very fast and are very sensitive to the reaction conditions like temperature, ratio of surfactants and polarity of solvents. Since the halide perovskites are ionic and have less stability in polar solvents, care must be taken to ensure the purity of the precursor reagents and the post-synthesis washing reagents. The post-synthesis washing reagents should be dried and purified by distillation before being used in the HI method.

In the TD method, metal trifluoroacetate precursors were added to the mixture of OA, OM and ODE and heated at elevated temperature under vacuum (removal of O_2, H_2O, etc.). Decomposition of metal trifluoroacetate yields the perovskite nanocrystals. TD is mainly used for the synthesis of fluoride perovskite (Du et al., 2007).

5.7.1.2 Reprecipitation Methods—Emulsion-based Method and LARP

The first synthesis of hybrid organic–inorganic PQDs was reported by Schimidt et al. (2014) through the synthesis of $MAPbBr_3$ by the solvent precipitation method. In this particular synthesis, methylammonium bromide and $PbBr_2$ were added to a solution containing long-chain alkylammonium bromide and OA in ODE kept at 80°C. The addition of acetone to this solution precipitates out the PQDs. As in the case of the HI method, the first synthesis of PQDs was followed by the synthesis of NWs and NPLs (Tyagi et al., 2015).

LARP is also a similar synthetic method. It was first used by Zhang et al. (2015) for the synthesis of $MAPbX_3$QDs. In this particular synthesis, the perovskite salt of the targeted halide is dissolved in a solvent (DMF/DMSO) that is "good" for the precursor salt. The solution also contains specific quantities of n-octylamine and OA. It was then dropped to a bad solvent for the precursor salts with vigorous stirring. It then yields the desired nanomaterial, PQDs, in solution. Later, perovskite NWs and NPLs were synthesized by the LARP method (Kostopoulou et al., 2017). LARP is a highly convenient method for the synthesis of various PNMs by varying the A, B and X site compositions.

5.7.2 Other Synthetic Methods

Chemical vapour deposition (CVD) and epitaxial methods were used to prepare $MAPbX_3$ NWs by Xing et al. (2015) for the first time. PbX_2 was initially grown on the substrate. It was then converted to $MAPbX_3$ through a gas–solid reaction with MAX. All-inorganic perovskite $CsPbX_3$ was also grown by CVD by Park et al. (2016). Later, perovskite NWs and NPLs were grown by the CVD method (Liu et al., 2018). The advantage of the CVD method is that the perovskite can be grown on any substrate and thus can be used for heterostructure formation, which can then be used for novel optoelectronic applications.

As listed above, most of the synthesis methods adopted for the PNMs are bottom-up methods. Top-down methods like ball milling and mechanical exfoliation can be used for the fabrication of PNMs (Prochowicz et al., 2019). Though these methods have their own demerits, they are greener synthetic methods as they do not require any solvents (Rosales et al., 2019).

5.8 CURRENT RESEARCH IN PEROVSKITE PHOTOVOLTAICS

Hybrid organic–inorganic—more precisely—PSCs were first introduced to the scientific community in 2009 (Kojima et al., 2009). Since then, research in perovskite PV technologies has been witnessing an explosion and reached a laboratory-confirmed efficiency of 25.5% (NREL, 2019). Though the perovskite has this boom in the photovoltaic technology, it still has some limitations that have to be addressed before commercialization in the form of solar cells. It is found that the efficiency of the PSCs decreases with increasing area, which is a major problem in commercializing the PSCs. A lot of industrial research studies are currently being conducted to address this problem. Another area of intensive research in perovskite is the stability of the material. Scientists are trying to enhance the stability and efficiency by various methods; varying the composition of the perovskite material, either by using various organic cations or by intermixing these organic cations at A site, substituting/intermixing halide ions at X site (Cl, Br) (Jeon et al., 2015) and changing the focus from hybrid to all-inorganic perovskites (Tian et al., 2020) and 2D bulk/thin film structure to QDs, by surface engineering like passivation (Ma et al., 2021), 2D/3D, 2D@3D/2D concept (Liu et al., 2021), defect engineering (Alharbi et al., 2021) and 1D-3D mixed-dimensional perovskite material (Zhan et al., 2021), are some of the examples. Another area of concern is the environmental aspect.

Replacing the lead (Pb) with another environment-friendly metal like tin (Sn) is also getting large interest among the photovoltaic community (Xu et al., 2021). But the easy oxidation of Sn^{2+} to Sn^{4+} again creates the stability problem. The rapid crystallization process in tin halide perovskite may be reduced or controlled by controlling the crystallization process (Ghimire et al., 2022). The use of a cold precursor solution and additives like tin fluorides, g-C_3N_4 and graphene-based composite perovskite will lead to controlled crystal growth and suppress Sn^{2+} to Sn^{4+} oxidation (Rao et al., 2021). This results in an enhancement in the power conversion efficiency and stability of the cell. Mixing of the cations, Pb^{2+}/Sn^{2+}, is also found to enhance the stability of the Sn-based PSCs (Sahamir et al., 2022). Lim et al. (2021) explain the various strategies adapted for optimizing the PCE of Sn-based perovskite solar cell. It has been found that the Eu^{2+} acts as an electron donor, provides a reducing environment and helps to suppress the oxidation of Sn^{2+} to Sn^{4+} (Wang et al., 2021).

Perovskite solar cell efficiency has been improved by modifying the fabrication process by introducing optional layers into the solar cell device. They are electron-transporting medium (ETM) and hole-transporting medium (HTM). TiO_2 nanoparticle- and Al_2O_3 nanoparticle-based ETMs and several HTMs are already tested and found to have an appreciable increase in the efficiency of PSCs (Heo et al., 2013). The use of ITO/ZnO nanoparticles instead of FTO/TiO_2 nanoparticles as ETMs and inorganic HTMs instead of organic HTMs also shows reasonable results (Liu et al., 2014). Scientists around the world are trying out all the possible sorts of device engineering by different electron-transporting mediums/layers, introducing a very thin barrier layer between ETM and absorber, n-i-p and p-i-n structures, mesoporous electrode structure, doped and undoped hole-transporting mediums, organic and inorganic hole-transporting mediums, etc. (Hai et al., 2021; Ling et al., 2021). Researchers around the world are trying to reduce the processing temperature so that these PSCs can be used to make flexible solar cells (and thereby roll-to-roll production) by using polymers as substrates. Currently, the perovskite solar cell is getting widespread attention for using this material with conventional silicon or CIGS solar cell as a tandem cell. The result is so promising and has a confirmed laboratory-tested efficiency of 29.5% and 18.2%, respectively (NREL, 2019). Jonathan et al. from Stanford University attained an open-circuit voltage of 1.65 V for perovskite/Si tandem cell (Jonathan et al., 2015). Compared to the advancement in DSSCs over the last decade, the advances in PSCs are so promising and open up much space for research.

5.9 SUMMARY

In this chapter, we tried to give an overview of the PSCs by discussing the basic structure of 3D and 2D layered perovskites, the basic architecture of the solar cells made up of perovskite as the light-absorbing material and various strategies for the synthesis of perovskite material in bulk/film form and as quantum dots. The current research scenario of perovskite photovoltaics was discussed in brief, and it points to the various areas on which we have to focus and the fields where much advancement and further research are essential for the successful development of perovskite photovoltaics as that of silicon photovoltaics, which we are witnessing now.

ACKNOWLEDGEMENTS

The author S.K.S. Patel acknowledges the financial support from the IoE Seed Grant, Banaras Hindu University, Varanasi, and UGC-BSR Research Start-Up Grant, New Delhi.

The author Sajith Kurian acknowledges the financial support (YSS/2015/000975) from the Science and Engineering Research Board (SERB), DST, New Delhi.

REFERENCES

Alharbi, E. A. et al. (2021). Methylammonium triiodide for defect engineering of high-efficiency perovskite solar cells, *ACS Energy Lett.* 6, 3650–3660.

Azzopardi, B. et al. (2011). Economic assessment of solar electricity production from organic-based photovoltaic modules in a domestic environment, *Energy Environ. Sci.* 4, 3741.

Barrows, A. T. et al. (2014). Efficient planar heterojunction mixed-halide perovskite solar cells deposited via spray-deposition, *Energy Environ. Sci.* 7, 2944–2950.

Burschka, J. et al. (2013). Sequential deposition as a route to high-performance perovskite-sensitized solar cells, *Nature* 499, 316–319.

Chapin, D. M., Fuller, C. S., & Pearson, G. L. (1954). A new silicon p-n junction photocell for converting solar radiation into electrical power, *J. Appl. Phys.* 25, 676–677.

Chen, C. W. et al. (2014). Efficient and uniform planar-type perovskite solar cells by simple sequential vacuum deposition, *Adv. Mater.* 26, 6647–6652.

Chen, D., & Chen, X. (2019). Luminescent perovskite quantum dots: Synthesis, microstructures, optical properties and applications, *J. Mater. Chem. C* 7, 1413–1446.

Chen, J., Jia, D., Johansson, E. M. J., Hagfeldt, A., & Zhang, X. (2021). Emerging perovskite quantum dot solar cells: Feasible approaches to boost performance. *Energy Environ. Sci.*, 14, 224–261.

Chen, J. B.-X. et al. (2016). 3,4-phenylenedioxythiophene (PheDOT) based hole-transporting materials for perovskite solar cells, *Chem.: Asian J.* 11, 1043–1049.

Chen, Q. et al. (2014). Planar heterojunction perovskite solar cells via vapor-assisted solution process, *J. Am. Chem. Soc.*, 136, 622–625.

Chen, Z., Li, H., Tang, Y., Huang, X., Ho, D., & Lee, C.-S. (2014). Shape-controlled synthesis of organolead halide perovskite nanocrystals and their tunable optical absorption, *Mater. Res. Express* 1, 015034.

Dey, A. et al. (2021). State of the art and prospects for halide perovskite nanocrystals, *ACS Nano* 15, 10775–10981.

Du, Y.-P. et al. (2007). Single-crystalline and near-monodispersed $NaMF_3$ (M = Mn, Co, Ni, Mg) and $LiMAlF_6$ (M=Ca, Sr) nanocrystals from cothermolysis of multiple trifluoroacetates in solution, *Chem.: Asian J.* 2, 965–974.

Ghimire, S. et al. (2022). Structural reconstruction in lead-free two-dimensional tin iodide perovskites leading to high quantum yield emission, *ACS Energy Lett.* 7, 3, 975–983.

Gu, L. L. et al. (2016). 3D arrays of 1024-pixel image sensors based on lead halide perovskite nanowires, *Adv. Mater.* 28, 9713–9721.

Hai, J. et al. (2021). Dopant-free hole transport materials based on a large conjugated electron-deficient core for efficient perovskite solar cells, *Adv. Funct. Mater.* 2105458.

Heo, J. H. et al. (2013). Efficient inorganic-organic hybrid heterojunction solar cells containing perovskite compound and polymeric hole conductors, *Nat. Photonics* 7, 486–491.

Hsu, H.-L., Chen, C.-P., Chang, J.-Y., Yu, Y.-Y., & Shen, Y.-K. (2014). Two-step thermal annealing improves the morphology of spin-coated films for highly efficient perovskite hybrid photovoltaics, *Nanoscale* 6, 10281–10288.

Hussain, I. et al. (2018). Functional materials, device architecture, and flexibility of perovskite solar cell, *Emergent Mater.* 1, 133–154.

Im, J-H., Jang, I. H., Pellet, N. P., Gratzel, M., & Park, N. G. (2014). Growth of $CH_3NH_3PbI_3$ cuboids with controlled size for high-efficiency perovskite solar cells, *Nat. Nanotechnol.* 9, 927–932.

Im, J.-H., Kim, H.-S., & Park, N.-G. Morphology-photovoltaic property correlation in perovskite solar cells: One-step versus two-step deposition of CH3NH3PbI3. *APL Mater.* 2, 081510 (2014).

Imran, M. et al. (2018). Benzoyl halides as alternative precursors for the colloidal synthesis of lead-based halide perovskite nanocrystals, *J. Am. Chem. Soc.* 140, 2656–2664.

Jeon, N. J. et al. (2015). Compositional engineering of perovskite materials for high-performance solar cells, *Nature* 517, 476–480.

Jiang, X. et al. (2017). High-performance regular perovskite solar cells employing low-cost poly(ethylenedioxythiophene) as a hole-transporting material, *Sci. Rep.* 7, 42564.

Kakiage, K., Aoyama, Y., Yano, T., Oya, K., Fujisawab, J., & Hanaya, M. (2015). Highly-efficient dye-sensitized solar cells with collaborative sensitization by silyl-anchor and carboxy-anchor dyes, *Chem. Commun.* 51, 15894–15897.

Kamat, P. V. (2008). Quantum dot solar cells: Semiconductor nanocrystals as light harvesters, *J. Phys. Chem. C* 112(48), 18737–18753.

Knop, O., Wasylishen, R. E., White, M. A., Cameron, T. S., & Van Oort, M. J. M. (1990). Alkylammonium lead halides: Part 2. $CH_3NH_3PbI_3$ (X = Cl, Br, I) perovskites: Cuboctahedral halide cages with isotropic cation reorientation, *Can. J. Chem.* 68, 412–422.

Kojima, A., Teshima, K., Shirai, Y., & Miyasaka, T. (2009). Organometal halide perovskites as visible-light sensitizers for photovoltaic cells, *J. Am. Chem. Soc.* 131, 17, 6050–6051.

Kostopoulou, A., Sygletou, M., Brintakis, K., Lappas, A., & Stratakis, E. (2017). Low-temperature benchtop-synthesis of all-inorganic perovskite nanowires, *Nanoscale* 9, 18202–18207.

Lee, J.-W., Seol, D.-J., Cho, A.-N., & Park, N.-G. (2014). High efficiency perovskite solar cells based on the black polymorph of $HC(NH_2)_2PbI_3$, *Adv. Mater.* 26, 4991–4998.

Li, H., Tong, G., Chen, T., Zhu, H., Li, G., Chang, Y., Wang, L., & Jiang, Y. (2018). Interface engineering using a perovskite derivative phase for efficient and stable $CsPbBr_3$ solar cells, *J. Mater. Chem. A* 6, 14255–14261.

Liang, K., Mitzi, D. B., & Prikas, M. T. (1998). Synthesis and characterization of organic–inorganic perovskite thin films prepared using a versatile two-step dipping technique, *Chem. Mater.* 10, 403–411.

Liang, P-W. et al. (2014). Additive enhanced crystallization of solution-processed perovskite for highly efficient planar-heterojunction solar cells, *Adv. Mater.* 26, 3748–3754.

Lim, E. L., Hagfeldt, A., & Bi, D. (2021). Toward highly efficient and stable Sn^{2+} and mixed Pb^{2+}/Sn^{2+} based halide perovskite solar cells through device engineering, *Energy Environ. Sci.* 14, 3256–3300.

Ling, W., Liu, F., Li, Q., & Li, Z. (2021). The crucial roles of the configurations and electronic properties of organic hole-transporting molecules to the photovoltaic performance of perovskite solar cells, *J. Mater. Chem. A* 9, 18148–18163.

Liu, D., & Kelly, T. L. (2014). Perovskite solar cells with a planar heterojunction structure prepared using room-temperature solution processing techniques, *Nat. Photonics* 8, 133–138.

Liu, K. et al. (2021). Architecturing 1D-2D-3D multidimensional coupled $CsPbI_2Br$ perovskites toward highly effective and stable solar cells, *Small* 17, 2100888.

Liu, M. M. B., & Snaith, J. (2013). Efficient planar heterojunction perovskite solar cells by vapour deposition, *Nature* 501, 395–398.

Liu, Z. et al. (2018). Morphology-tailored halide perovskite platelets and wires: From synthesis, properties to optoelectronic devices, *Adv. Opt. Mater.* 6, 1800413.

Liu, Z. et al. (2020). Single-crystal hybrid perovskite platelets on graphene: A mixed-dimensional Van Der Waals heterostructure with strong interface coupling, *Adv. Funct. Mater.* 30, 1909672.

Ma, D. et al. (2021). An effective strategy of combining surface passivation and secondary grain growth for highly efficient and stable perovskite solar cells, *Small* 17, 2100678.

Mahmood, K., Sarwarb, S., & Mehran, M. T. (2017). Current status of electron transport layers in perovskite solar cells: Materials and properties, *RSC Adv.* 7, 17044.

Mailoa, J. P. et al. (2015). A 2-terminal perovskite/silicon multijunction solar cell enabled by a silicon tunnel junction, *Appl. Phys. Lett.* 106, 121105.

Marinova, N., Valero, S., & Delgado, J. L. (2017). Organic and perovskite solar cells: Working principles, materials and interfaces, *J. Colloid Interface Sci.* 488, 373–389.

Matsui, T., Seo, J.-Y., Saliba, M., Zakeeruddin, S. M., & Gratzel, M. (2017). Room-temperature formation of highly crystalline multi cation perovskites for efficient, low-cost solar cells, *Adv. Mater.* 29, 1606258.

Mitzi, D. B. (2001). Templating and structural engineering in organic-inorganic perovskites, *J. Chem. Soc. Dalton Trans.* 1, 1–12.

Mitzi, D. B. (2007). *Synthesis, Structure, and Properties of Organic-Inorganic Perovskites and Related Materials*. John Wiley & Sons, Inc., New York, pp. 1–121.

Morgano, M., Perez-Wurfl, I., & Binetti, S. (2011). Nanostructured silicon-based films for photovoltaics: Recent progresses and perspectives, *Sci. Adv. Mater.* 3, 388–400.

Nam, J. J., Noh, J. H., Kim, Y., Yang, W. S., Ryn, S., & Seok, S. I. (2014). Solvent engineering for high-performance inorganic-organic hybrid perovskite solar cells, *Nat. Mater.* 13, 897–903.

Nam, J. K. et al. (2017). Potassium incorporation for enhanced performance and stability of fully inorganic cesium lead halide perovskite solar cells, *Nano Lett.* 17, 2028–2033.

National Renewable Energy Laboratory. (2021). https://www.nrel.gov/pv/cell-efficiency.html, Golden, CO

Niezgoda, J. S., Foley, B. J., Chen, A. Z., & Choi, J. J. (2017). Improved charge collection in highly efficient CsPbBrI, *ACS Energy Lett.* 2, 1043–1049.

NREL. (2019). Best research-cell efficiencies. https://www.nrel.gov/pv/assets/pdfs/best-reserch-cell-efficiencies.20190411.pdf

O'Regan, B., & Gratzel, M. (1991). A low-cost, high-efficiency solar cell based on dye-sensitized colloidal TiO_2 films, *Nature* 353, 737–740.

Parida, B., Iniyan, S., & Goic, R. (2011). A review of solar photovoltaic technologies, *Renew. Sustain. Energy Rev.* 15, 1625–1636.

Park, K. et al. (2016). Light-matter interactions in cesium lead halide perovskite nanowire lasers, *J. Phys. Chem. Lett.* 7, 3703–3710.

Ping, L.-Q. et al. (2016). Copper-doped chromium oxide hole-transporting layer for perovskite solar cells: Interface engineering and performance improvement, *Adv. Mater. Interf.* 3, 1500799.

Prochowicz, D., Saski, M., Yadav, P., Grätzel, M., & Lewiński, J. (2019). Mechanoperovskites for photovoltaic applications: Preparation, characterization, and device fabrication, *Acc. Chem. Res.* 52, 3233–3243.

Protesescu, L. et al. (2015). Nanocrystals of cesium lead halide perovskites ($CsPbX_3$, X = Cl, Br, and I): Novel optoelectronic materials showing bright emission with wide color gamut, *Nano Lett.* 15, 3692–3696.

Rafique, S. et al. (2017). Bulk heterojunction organic solar cells with graphene oxide hole transport layer: Effect of varied concentration on photovoltaic performance, *J. Phys. Chem. C* 121, 140–146.

Rao, L. et al. (2021). Wearable tin-based perovskite solar cells achieved by a crystallographic size effect, *Angew. Chemie Int. Ed.* 60, 14693–14700.

Rosales, B. A., Wei, L., & Vela, J. (2019). Synthesis and mixing of complex halide perovskites by solvent-free solid-state methods, *J. Solid State Chem.*, 271, 206–215.

Sahamir, S. R. et al. (2022). Enhancing the electronic properties and stability of high-efficiency tin-lead mixed halide perovskite solar cells via doping engineering, *J. Phys. Chem. Lett.* doi:10.1021/acs.jpclett.2c00699

Saliba, M. et al. (2016). Cesium-containing triple cation perovskite solar cells: Improved stability, reproducibility and high efficiency, *Energy Environ. Sci.*, 9, 1989–1997.

Santosh, J. et al. (2021). Hole-transporting materials for perovskite solar cells employing an anthradithiophene core, *ACS Appl. Mater. Interf.* 13, 24, 28214–28221.

Schmidt, L. C. et al. (2014). Nontemplate synthesis of $CH_3NH_3PbBr_3$ perovskite nanoparticles, *J. Am. Chem. Soc.* 136, 850–853.

Shamsi, J., Urban, A. S., Imran, M., De Trizio, L., & Manna, L. (2019). Metal halide perovskite nanocrystals: Synthesis, post-synthesis modifications, and their optical properties, *Chem. Rev.* 119, 3296–3348.

Shaowei, S., Yongfang, L., Xiaoyu, L., & Wang, H. (2015). Advancements in all-solid-state hybrid solar cells based on organometal halide perovskites, *Mater. Horizons* 2, 378–405.

Shi, J. et al. (2014). Hole-conductor-free perovskite organic lead iodide heterojunction thin-film solar cells: High efficiency and junction property, *Appl. Phys. Lett.* 104, 063901.

Snaith, H. J. (2013). Perovskites: The emergence of a new era for low-cost, high-efficiency solar cells, *J. Phys. Chem. Lett.* 4, 3623–3630.

Stoumpos, C. C., Malliakas, C. D., & Kanatzidis, M. G. (2013). Semiconducting tin and lead iodide perovskites with organic cations: Phase transitions, high mobilities, and near-infrared photoluminescent properties, *Inorg. Chem.* 52, 9019–9038.

Subhani, W. S., Wang, K., Du, M., Wang, X., & Liu, S. (2019). Interface modification-induced gradient energy band for highly efficient $CsPbIBr_2$ perovskite solar cells, *Adv. Energy Mater.* 9, 1803785.

Swarnkar, A. et al. (2016). Quantum dot-induced phase stabilization of α-CsPbI3 perovskite for high-efficiency photovoltaics, *Science* 354, 92–95.

Tanaka, K., Sano, F., Takahashi, T., Kondo, T., Ito, R., & Ema, K. (2002). Two-dimensional wannier excitons in a layered-perovskite-type crystal $(C_6H1_3NH_3)_2PbI_4$, *Solid State Comm.* 122, 249–252.

Tian, J. et al. (2020). Inorganic halide perovskite solar cells: Progress and challenges, *Adv. Energy Mater.* 10, 2000183.

Tyagi, P., Arveson, S. M., & Tisdale, W. A. (2015). Colloidal organohalide perovskite nanoplatelets exhibiting quantum confinement, *J. Phys. Chem. Lett.* 6, 1911–1916.

Wang, X., Liu, L., Qian, Z., Gao, C., & Liang, H. (2021). Eu^{2+} ions as an antioxidant additive for Sn-based perovskite light-emitting diodes, *J. Mater. Chem. C* 9, 12079–12085.

Wang, Z. et al. (2017). Stability of perovskite solar cells: A prospective on the substitution of the A cation and X anion, *Angew. Chem. Int. Ed. Engl.* 56, 1190–1212.

Wu, Q. et al. (2015). Kesterite Cu_2ZnSnS_4 as a low-cost inorganic hole-transporting material for high-efficiency perovskite solar cells, *ACS Appl. Mater. Interf.* 7, 28466–28473.

Xiao, Z. et al. (2014). Efficient, high yield perovskite photovoltaic devices grown by interdiffusion of solution-processed precursor stacking layers, *Energy Environ. Sci.* 7, 2619–2623.

Xing, J. et al. (2015). Vapor phase synthesis of organometal halide perovskite nanowires for tunable room-temperature nanolasers, *Nano Lett.* 15, 4571–4577.

Xu, L. et al. (2021). Recent advances and challenges of inverted lead-free tin-based perovskite solar cells, *Energy Environ. Sci.* 14, 4292–4317.

Xu, L., Deng, L.-L., Cao, J., Wang, X., Chen, W.-Y., & Jiang, Z. (2017). Solution-processed $Cu(In,Ga)(S,Se)_2$ nanocrystal as inorganic hole-transporting material for efficient and stable perovskite solar cells, *Nanos. Res. Lett.* 12, 159.

Yan, L. et al. (2018). Interface engineering for all inorganic $CsPbI_2Br$ perovskite solar cells with efficiency over 14%, *Adv. Mater.* 30, 1802509.

Yi, Z., Ladi, N. H., Shai, X., Li, H., Shen, Y., & Wang, M. (2019). Will organic-inorganic hybrid halide lead perovskites be eliminated from optoelectronic applications? *Nanoscale Adv.* 1, 1276.

Yin, W.-J., Yang, J.-H., Kang, J., Yan, Y., & Wei, S.-H. (2015). Halide perovskite materials for solar cells: A theoretical review, *J. Mater. Chem. A* 3, 8926–8942.

Zeng, Z. et al. (2020). Rare-earth-containing perovskite nanomaterials: Design, synthesis, properties and applications, *Chem. Soc. Rev.* 49, 1109–1143.

Zhan, Y. et al. (2021). Elastic lattice and excess charge carrier manipulation in 1D-3D perovskite solar cells for exceptionally long-term operational stability, *Adv. Mater.* 2105170. doi:10.1002/adma.202105170

Zhang, F. et al. (2015). Brightly luminescent and color-tunable colloidal $CH_3NH_3PbX_3$ (X=Br, I, Cl) quantum dots: Potential alternatives for display technology, *ACS Nano* 9, 4533–4542.

Zhang, H., Wang, H., Chen, W., & Jen, A. K.-Y. (2017). $CuGaO_2$: A promising inorganic hole-transporting material for highly efficient and stable perovskite solar cells, *Adv. Mater.* 29, 1604984.

Zhou, D. et al. (2018). Perovskite-based solar cells: Materials, methods, and future perspectives, *J. Nanomater.* 1, 8148072.

Zhou, H., Chen, Q., & Li, G. (2014). Interface engineering of highly efficient perovskite solar cells, *Science* 345, 542–546.

6 Third-Generation Photovoltaic Solar Cells

Synthesis, Fabrication, and Applications

Varsha Yadav, Rahul Bhatnagar, and Upendra Kumar

6.1 INTRODUCTION

Energy is the primary input for almost all economic activities today, and it has become critical for improving the quality of life (Corkish et al., 2016; Yadav et al., 2020). Consequently, low-cost, highly scalable, and zero/low-emission energy production is one of humanity's most critical and pressing issues. Global energy demand is rapidly increasing in this period, with projections of a 30% rise in current energy consumption from 2015 to 2040 (Yadav et al., 2021). The world's economy and population are expanding, and the energy demand is also increasing. When examining the world's energy demand, we can split it into few categories such as natural gas, oil nuclear, renewable energy, and coal. As shown in Figure 6.1a, world energy use has been heavily reliant on fossil fuels like natural gas, coal, and oil, which is expected to continue (IEA, 2019).

More than half of the world's total energy demand is currently satisfied by fossil fuels like natural gas and other non-renewable energy sources, including oil, coal, and nuclear power. Figure 1b shows the historical and projected global energy consumption by various energy sources up to 2040, as well as the global energy consumption

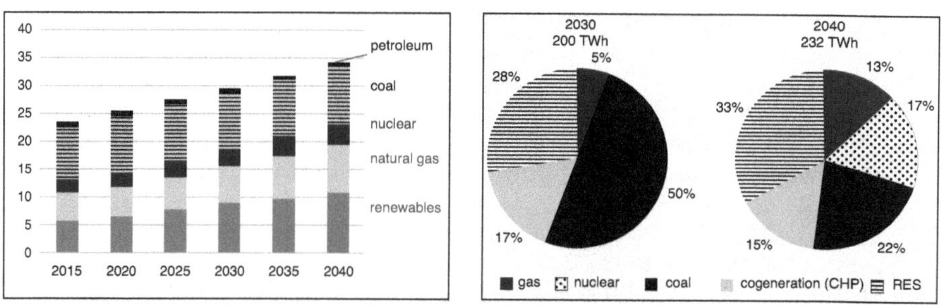

FIGURE 6.1 (a) Forecasts for global energy use by different energy sources from 2015 to 2040. (b) The amount of energy consumed by different sectors.

DOI: 10.1201/9781003481157-6

of different industries (Coyle and Simmons, 2014). Due to their size and other disadvantages, such as their restricted ability to reverse, emissions, and price rises, traditional sources have a number of problems that limit their utilization. Consumption of fossil fuels will decrease while technology and the development of renewable energy sources, such as wind turbines, solar panels, and fuel cells, should be enhanced to meet the world's energy needs. As a result, switching to alternative, unconventional, and renewable energy sources is crucial for society's long-term survival. By 2025, renewable energy sources will account for 29% of all energy, up from 22% in 2012. In 2040, 30% of the world's power would come from renewable energy if the Clean Power Plan (CPP) for the United States were considered (U.S. Energy Information Administration, 2019). Numerous renewable energy sources were used as substitute energy resources. Renewable energy sources like geothermal, wind, biomass, hydroelectric, and solar power will all be used to produce future renewable energy, as shown in Figure 6.2 because they are always available and have no adverse environmental effects.

FIGURE 6.2 Various sources of alternative renewable energy.

Green and sustainable solar energy has a lower location dependency, needs less maintenance, has greater control over power generation, and has a wide variety of applications than other renewable energy sources, making it a promising potential energy source. The cumulative amount of solar irradiation obtained on the earth's surface each year is approximately 10,000 times the energy required by the planet each year. The annual striking solar radiation at the earth's apex atmosphere is about 5.5 1024 J, with around 60% of this energy reaching the surface. This amount of energy is several times greater than our global society's average energy appetite (GEA, 2012).

At 6,000 K, solar radiation is released from the sun's photosphere, giving it a spectral distribution that resembles that of a black body at a similar temperature, as shown in Figure 6.3 (Widén and Munkhammar, 2019). Solar radiation is scattered by air molecules, aerosols, and dust particles and absorbed by air molecules, especially oxygen, ozone, water vapor, and carbon dioxide, as it passes through the earth's environment. The solar radiation spectrum's imprint on the earth's surface is unique. The latitude, altitude, and climatic form of a region, the season, time of day, and weather conditions at a particular time all affect the quantity of solar irradiation accessible there. Solar energy needs to take in a certain amount of energy from the solar spectrum to function.

The visible region of the solar spectrum ranges from 100 to 1 mm. Still, the majority of irradiance occurs between 250 and 2,500 nm, with the maximum for air mass (AM) 0 in the wavelength range of visible light (400–700 nm), meaning that solar cells will try to absorb as much as possible in this region (Scofield et al.). Photovoltaic (PV) technology, which generates energy from direct sunlight, has recently gained worldwide attention as a critical component of energy production. It has many benefits, including being environmentally friendly, turning light into high-level energy right away, requiring less maintenance, and being modular, thus serving both small and large power demands. As an outcome, solar energy can be used as a possible

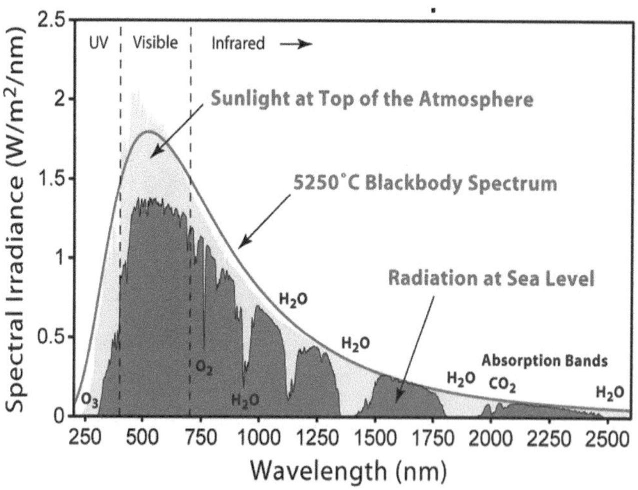

FIGURE 6.3 Solar radiation spectrum.

clean and sustainable energy source in the future, providing essential benefits to both the climate and the economies of both developed and developing countries.

6.1.1 Types of Solar Energy Conversions

There are currently a few different solar energy conversion methods available (Corkish et al., 2016). This is listed further down.

6.1.1.1 Thermal Solar Energy Conversion

This form of energy is directly used for water heating in the home. Its output is in the form of black panels on rooftops, but it is ineffective due to low radiation density in a few kinds of weather (Musunuri et al., 2007).

6.1.1.2 Chemical Solar Energy Conversion

This track depicts the transformation of solar energy into chemical energy. This is important due to its ability to solve long-term storage and energy transfer issues.

6.1.1.3 Thermoelectric Solar Energy Conversion

Turbines, rotated by steam provided by heated water with traditional fuels, are well known for producing electricity. Solar energy can be converted to heat, which is then used to heat water, which drives the steam turbine.

6.1.1.4 Photoelectric Solar Energy Conversion

Solar cells and other photovoltaic devices are examples of this form of energy conversion. This process directly transforms sunlight into electricity (Ranabhat et al., 2016). The increasing cost of solar cells is the primary factor restricting the widespread use of solar energy.

6.1.2 Solar Cell: An Overview

Photovoltaic is the process of converting sunlight into electricity using a semiconductor material. The terms "photo" and "voltaic" are derived from the Greek words "light" and "Volta," the name of an Italian physicist (Ranabhat et al., 2016). In illumination, the photovoltaic effect generates a voltage or electric current in a substance or system. In 1839, a French physicist discovered the photovoltaic effect when Becquerel discovered that light was shone on an electrode immersed in a conductive solution to produce an electric current (Tress, n.d.). He was ecstatic. Later that year, in 1873, W. Smith discovered selenium's photoconductivity, which led to the development of photovoltaic technology. Albert Einstein was awarded the Nobel Prize in Physics in 1921 for explaining the photoelectric effect. Photovoltaic devices or solar cells are devices that exhibit the photovoltaic effect. A solar cell, also known as a photogalvanic cell or a photoelectric cell, is an electronic device that transforms sunlight directly into electricity (Jao et al., 2016). A solar cell generates both a voltage and a current when sunlight falls on it, which is used to produce electricity. This process requires two things: first, a material that raises an electron's energy state due to light absorption, and second, the transfer of that electron with high energy from the

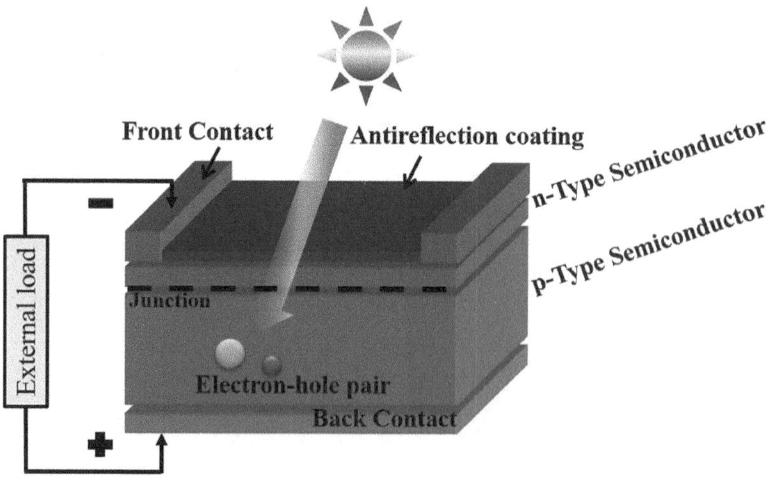

FIGURE 6.4 Solar cell system.

solar cell to an external circuit. The electron returns to the solar cell after its energy has been dissipated in the external circuit. The specifications for photovoltaic energy conversion can be met with several materials and methods, but in practice, p-n junction semiconductor materials are used in almost all photovoltaic energy conversion. A solar cell system is exhibited in Figure 6.4.

The following processes can be used to understand how a solar cell works.

- The absorption of photons in the materials that make up a junction occurs in the charge carrier generation.
- The charge carriers produced by photons are then segregated in the junction.
- The charge carriers produced by photons are collected at the junction's interfaces.

6.1.2.1 The Absorption of Photons in the Materials that Make Up a Junction Occurs in the Charge Carrier Generation

The initial stage of a solar cell's activity is the absorption of incident photons. The energy levels of the active layer (or absorber layer) and the energy of each incident photon are the main determinants. In solar cells, the active layer is created from a semiconductor material. The band gap can describe semiconductor materials by separating the energy gaps between the valence band (VB) and the conduction band (CB). The most prominent energy band occupied by electrons and the lowest energy band where no states are occupied, respectively, are depicted by the CB and VB. An electron is excited from the VB to the CB when a photon with energy more than or equal to the band gap interacts with a substance. The extra energy is then thermalized away. The photon will also be transmitted rather than absorbed by the material if its energy is lower than the band gap energy.

6.1.2.2 The Charge Carriers Produced by Photons Are then Segregated in the Junction

The creation of an excited electron (negatively charged, e^-) in the CB and the vacancy of an electron in the VB, also known as a hole (positively charged, h^+), occurs in the second stage of photon absorption. In inorganic semiconductors, these electron-hole pairs can be considered free charges at room temperature. Coulomb interaction binds the electron-hole pairs tightly together. The electron-hole pair is designated as an exciton due to the interaction, which contributes to the stabilizing energy balance. For the photovoltaic effect to occur, the exciton requires additional energy to produce free electrons and holes that enable it to pass through the active material freely. The separation and selection of free-charge carriers is the next step in solar cell service. Diffusion-based and drift-based transport are two transport mechanisms that lead to this procedure. The concentration gradient induces charge carrier transport in semiconductor materials in the diffusion-based transport mechanism, while the electric field induces charge carrier transport in the drift-based transport mechanism. As a result, these two mechanisms will be critical in transporting free-charge carriers by active material (Silicon and Cells, 2021).

6.1.2.3 The Charge Carriers Produced by Photons Are Collected at the Junction's Interfaces

The final stage of solar cell operation is the selection of photo-generated charge carriers at the electrode into the external circuit. An ohmic touch collects charge carriers between the electrode and the semiconductor. To avoid the Schottky barrier, the work function of the semiconductor should be greater than the work function of the electrode for electron collection. Furthermore, the electrode's work function should be higher than the semiconductor's work function for whole collection. The cycle is complete when an electron enters the electrode and dissipates the energy while producing electricity. The void on the opposite electrode then recombines with it. Proper electrode material and transport layer choices are essential for achieving overall high power conversion efficiency (PCE) in solar cells.

6.1.3 Photovoltaic Characteristics and Parameters

Solar cell parameters include fill factor, open-circuit voltage, short-circuit current, maximum power output, and overall PCE (Parasuraman, 2015). The current-voltage peculiarities of a solar cell in darkness and under light are shown in Figure 6.5. The *J–V* characteristics behavior of a solar cell can be analyzed by calculating the current difference as the applied voltage varies around the solar cell. It is the photo-generated current superimposed on the *I–V* curve of a solar cell diode in the dark. It can be done in both dark and light environments. In the dark, an *I–V* curve of a solar cell is similar to a diode's. The *I–V* curve moves into the fourth quadrant as light falls on a solar cell (illumination condition), and the cell begins to produce electricity. The greater the amount of change detected, the higher the light intensity on the cell. An ideal solar cell's *I–V* curve is located in the fourth quadrant (Devasia and Kurinec, 2011). The fourth quadrant equation of the *I–V* curve is represented by

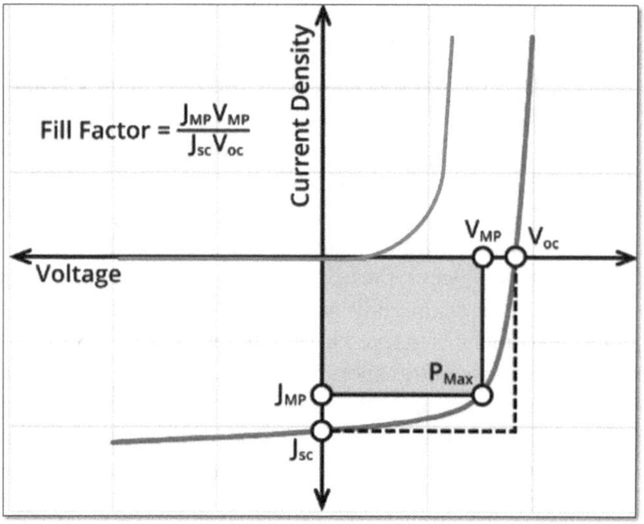

FIGURE 6.5 Schematic diagram of *I–V* peculiarities of a solar cell under light illumination.

$$I = I_0 - I_L\left[\exp\frac{qV}{\eta kT} - 1\right]$$

where I_L is photo-generated current, I_0 is the reverse saturation current of diode, q is an elementary charge, V is the applied voltage, n is the ideality factor of a diode, k is Boltzmann constant, and T is the temperature.

6.1.3.1 Open-circuit Voltage

The number of forwarding biases on the solar cell caused by the bias of the solar cell junction with the light-generated current is known as the open-circuit voltage. The maximum voltage present in a solar cell when no current is flowing through it is identified as the open-circuit voltage or V_{OC}. The open-circuit voltage can be seen on the *I–V* curve above. The V_{OC} equation is obtained by setting the current value to zero in the above solar cell equation.

$$V_{OC} = \frac{nkT}{q}\ln\left(\frac{I_L}{I_O} + 1\right)$$

According to this equation, the open-circuit voltage of a solar cell is determined by the saturation current and the light-produced light-generated current. The main effect is the saturation current, which is dependent on recombination in the solar cell and is caused by minor variations in short-circuit current. Furthermore, V_{OC} is influenced by several structural factors. While the charge carrier loss increases, the value of V_{OC} decreases (Lecture, 2015). It is also necessary to improve the compatibility between the semiconductor (deposited active layer) and the metal electrode since the morphology of the active layer material affects open-circuit voltage.

6.1.3.2 Short-Circuit Current

When the terminals are wired together, the current flow between the two terminals of a solar cell and light impinges on the cell is identified as a short-circuit current. Short-circuit current is proportional to the number of light-generated carriers, and I_{SC} stands for it. In an ideal case, the light-induced and short-circuit currents are similar. Consequently, the maximum amount of current that can be calculated with the solar cell is the short-circuit current. As an outcome, I_{SC} relies on the main factors mentioned below:

- Solar cell active region
- Light intensity's effect
- The spectrum of incident light
- Optical characteristics
- Possibility of collection

The diffusion length and surface passivation are essential parameters for distinguishing solar cells made of same materials (Wenham et al., 2007). The following formula is used to determine the short-circuit current of an ideal cell with uniform generation and a passivated surface:

$$I_{SC} = q_{AG}[L_n + L_P]$$

where A is the solar cell's region, q_{AG} is the generation rate, and L_n and L_p are the electron and hole diffusion lengths, respectively. "The high proportionality between diffusion length and generation rate can be seen in the short-circuit current."

6.1.3.3 Fill Factor

As shown in Figure 6.5, the point on the I–V curve where the solar cell generates its maximum power under light conditions is known as the maximum power point. The voltage and current at full power, denoted by V_m and I_m, are shown in the diagram. The product of this power P_{max}, the open-circuit voltage and the short-circuit current, is the fill factor.

$$\text{FF} = \frac{V_m I_m}{V_{OC} I_{SC}} = \frac{P_{max}}{V_{OC} I_{SC}}$$

The fill factor in an ideal solar cell battery must be unity, where all incident power is converted to energy. The fill factor (FF) is graphically classified as the "squareness" of the cell in Figure. Recombination processes charge carrier transport and charge dissociation, and all play a role in the FF of a solar cell (Devasia and Kurinec, 2011). These processes affect the overall transport capacity. A space charge region forms when hole and electron transport are unbalanced in the active layer, resulting in low fill factors.

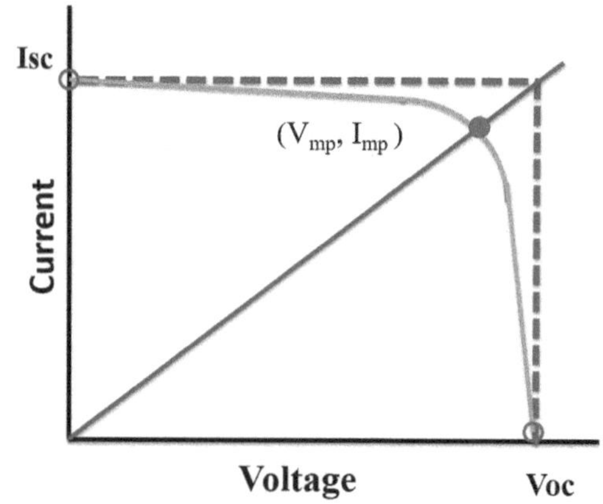

FIGURE 6.6 Characteristic resistance of a solar cell.

6.1.3.4 Maximum Power Output

Maximum power output (Pmax) of a solar cell is provided, as shown in Figure 6.6. It is calculated as $P_{max} = I_m\ V_m$, where I_m and V_m represent the heights of voltage and current, respectively, of a solar cell in the fourth quadrant of the I–V response. The point at which point the delivered power reaches its total value is called operating point of a solar cell.

6.1.3.5 Power Conversion Efficiency

The efficiency of a solar cell is a critical factor in comparing its performance and characteristics to those of other solar cells. The ratio of the power delivered at the operational point to the incident power is used to determine a solar cell's efficiency.

$$\eta = \frac{V_m I_m}{P_{in}} \times 100\%$$

Efficiency is connected to I_{SC} and V_{OC} using FF as given by,

$$\eta = \frac{I_{SC} V_{OC}}{P_{in}} \times \mathrm{FF}$$

Furthermore, performance is highly dependent on the strength of incident light, the solar spectrum, and the device's operating temperature [21]. As a result, the operating conditions should be meticulously monitored when comparing one solar cell to another. At a temperature of 25°C, the performance of terrestrial solar cells is calculated using the AM 1.5 G (air mass) standard. The solar elevation angle of $\psi = 42°$ is defined by the term AM 1.5, where AM = 1/Cos ψ.

6.1.4 Resistance Characteristic

The performance resistance of a solar cell at its highest power point, known as the solar cell's characteristic resistance, is exhibited in Figure 6.6. If the characteristic and load resistance are equal, the solar cell is ready to work, and the most power is transferred to the load at its optimum power point. It is an essential aspect of solar cell research, especially when considering parasitic loss mechanisms' effects (Smets et al., 2016).

The inverse of the line's slope is expressed by characteristic resistance, as seen in the diagram. Equation-based display is given as:

$$R_{CH} = \frac{V_{MP}}{I_{MP}} \approx \frac{V_{OC}}{I_{SC}}$$

6.1.4.1 Effect of Parasitic Resistances

By dissolving power in impact resistances, resistant factors in solar cells minimize the performance of solar cells. Series and shunt resistances are the most common parasitic resistances (Loulou et al., 2018). Figure 6.7 shows the anastomosis of the series and shunt resistances on the solar cell.

The main impact of parasitic resistance is that it reduces the FF in most situations and the average sum of both series and shunt resistance. The geometry of the solar cell determines this. The solar cell field determines the value of resistance at the operating point of the solar cell. After comparing the sequence, the resistance can have a different area; however, a typical resistance unit is Ω cm^2.

6.1.4.2 Series Resistance

The first is a current movement in the emitter and base, which induces series resistance in a solar cell. The second is the material-to-metal touch resistance, and the third is the rear metal contact and maximum resistance. The main aim of series resistance is to lower the FF, but a value that is too high can also lower the short-circuit current. The series resistance is depicted in Figure 6.8 and is part of the equation for a solar cell with a shunt resistance.

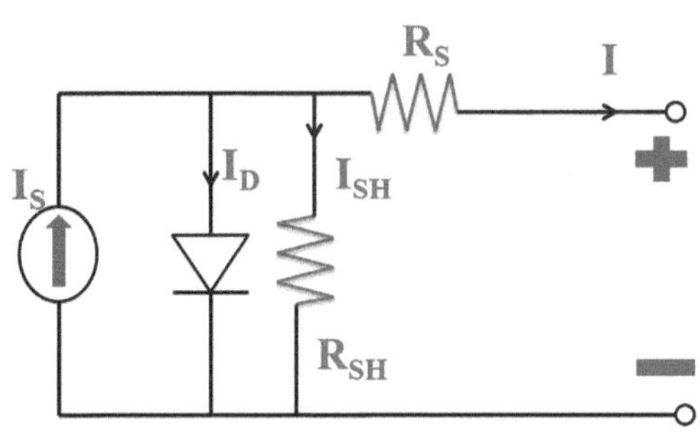

FIGURE 6.7 The anastomosis of the shunt and series resistances on the solar cell.

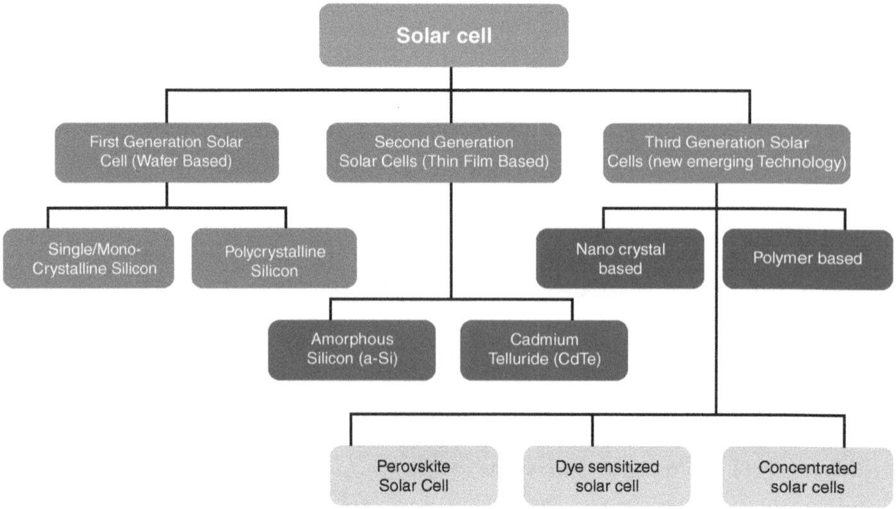

FIGURE 6.8 Generation of solar cells.

$$I = I_L - I_O \exp\left[\frac{q(V + IR_S)}{nkT}\right]$$

where I represents the cell out current, I_L represents the light-induced current, V represents the voltage across the cell terminals, T represents temperature, q and k represent constants, n represents the ideality factor, and R_s represents the cell series resistance.

Series resistance has no impact on the solar cell, since at open-circuit voltage, the total current flow through the solar cell, and consequently through the series resistance, is zero (Wolf and Rauschenbach, 1963). In contrast, near the open-circuit voltage, the series resistance significantly affects the I–V curve. An easy way to estimate the series resistance of a solar cell is to find the slope of the I–V curve at the open-circuit voltage stage.

6.1.5 Generation of Solar Cells

The most popular materials in photovoltaic solar cells are silicon (amorphous silicon, single crystal, and multi-crystalline), copper-indium-gallium-selenide (CIGS), cadmium-telluride, and copper-indium-gallium-sulfide. Photovoltaic solar cells are classified into different groups based on these materials, as shown in Figure 6.8.

The first-generation solar cells are classified as (a) single-crystal solar cells and (b) multi-crystal solar cells. Because of its high efficiency, this is the oldest and most widely used technology category. The first generation of solar cells is produced on wafers. The first generation of solar cells, produced on wafers, typically delivers a power output of 2 to 3 watts per wafer (Ranabhat et al., 2016). Solar modules made up of several cells increase strength. As shown in the flow chart, there are two types of first-generation solar cells. Their crystallization levels differentiate them. A single-crystal solar cell has identical crystals on its wafer throughout. Crystal grains on a wafer make up multi-crystal solar cells (Katja, 2014; Kibria et al., 2014).

The solar cell's grain borders are visible to everyone. While monocrystalline solar cells have a higher efficiency than multi-crystalline solar cells, multi-crystalline wafer processing is more straightforward and less expensive. As a result, they compete with monocrystals. A-Si thin-film solar cells and mc-Si solar cells were the main focus of the second generation of solar cells like CIS/CIGS (Copper indium gallium selenide solar cell) solar cell. CdTe solar cells' costs are cheaper, and their efficiencies are lower than those of the first generation. They also provide a visual aesthetic benefit (Simya et al., 2018). Thin-film solar cells are much more applicable on walls, vehicles, building integrations, and other surfaces, since there are no fingers in front of them for metallization. Flexible substrates may also be used to expand these thin films. Thin-film solar cells have the advantage of being able to expand on vast areas of up to 6 cm^2. On the other hand, wafer-based solar cells can only be manufactured in wafer dimensions (Kibria et al., 2014).

CIGS and cadmium-telluride/cadmium-sulfide (CdTe/CdS) solar cells are examples of third-generation solar cells that take into account both polymer- and nanocrystal-based solar cells. The solar cells were focused. Although they haven't yet been commercialized, some recent advances in dye-sensitized solar cells offer promise (Mohammad Bagher, 2015). Dye-sensitized and concentrated solar cells are the most evolved third-generation solar cell types. Electricity is produced by dye molecules sandwiched between electrodes in dye-sensitized solar cells (DSSCs; Carella et al., 2018). In dye molecules, electron-hole pairs are generated and transported by TiO_2 nanoparticles. Their cost is also meager, despite their low performance. In comparison with other technologies, their development is easy. There are numerous color options for DSSCs (sensitized and cells, n.d.). The concentrated PV solar cell is another intriguing invention. Concentrated cells work by concentrating a lot of solar radiation into a condensed space around the PV cell. This lowers the semiconductor content, which could be highly expensive. This system ought to include a flawless optical setup. The concentration ranges range from tens of thousands to tens of thousands of suns. As a result, the overall cost could be lower than the traditional systems. In the near future, CPVs are a promising technology (Khan, 2013).

6.1.6 State-of-the-Art: Dye-Sensitized Solar Cells

Dye-sensitized solar cells are a cost-effective and environmentally friendly substitute for traditional Si solid-state devices (DSSCs). DSSCs are photoelectrochemical cells that convert direct solar energy into electric energy. It is simple to manufacture, and the floats and versatile thin-film structures are well-suited to automated manufacturing, implying that it could be scaled up easily (Yang et al., 2012). Fast scale-up, low production costs, excellent dim/disperse light performance, and conformability with flexible substrates and building window glass are just some of its impressive characteristics (Selvaraj et al., 2018; Tomar et al., 2020). Consequently, DSSCs have been identified as a viable substitute for bulk silicon-based solar cells (Kannan Balasingam et al., 2013). Vogel et al. (1870) first introduced DSSC with the help of silver halide in gelatin medium, which is photoenergetic in the UV range. Then, Vogel (1873) used AgX sensitized with dye molecules to produce photoabsorption up to visible range. Hishiki (1965) and Gerisher (1968) used zinc oxide to sensitize rose bengal and cyanine dyes. Daltrozzo and Tributch studied the rhodamine B dye on ZnO. Spitler and Calvin used TiO_2 to replace ZnO in different studies in 1977 (Rokesh et al., 2014). Ruthenium dye sensitized with TiO_2 semiconductor has been commonly used for DSSC since Grätzel first introduced it in 1990. Dye sensitization has a lengthy history, and the photosensitization phenomena brought on by organic dyes were first observed in 1887 (Wei, 2010).

Photoelectrochemistry and photography depend on photo-induced charge division at the liquid-solid interface, and they've merged interestingly (Sharma et al., 2018). Michael Grätzel, who conducted groundbreaking research and invented the new structure of DSSC, was responsible for its conceptual revival until 1991. Introducing a TiO_2 nanoparticle-based photoanode with an extensive surface area was the breakthrough in Grätzel's cell. A small thin film of photoactive dye is placed on the morphology of the TiO_2 film, and this layer preferentially absorbs the dye's photo-generated electrons (O'Regan and Grätzel, 1991; Srinivasu et al., 2011). Under normal testing conditions, the PCE of the first Grätzel cell was up to 7.1% (O'Regan and Grätzel, 1991). DSSCs have been widely studied for the past two decades, spanning the fields of fundamental condensed matter physics, system operating dynamics, material innovation, and novel structure design. DSSCs' latest state-of-the-art high power conversion efficiencies are about 14% (Arifin et al., 2018). DSSCs differ from other semiconductor-based solar cells because sunlight is not absorbed primarily by the semiconductor (Yadav et al., 2020). The pair of electron and hole is not isolated at the *p–n* junction's built-in potential (Rokesh et al., 2014). It's been 20 years since O' Regan and Grätzel's widely cited 1991 *Nature* paper "A Low-Cost, High-Efficiency Solar Cell Based on Dye-Sensitized Colloidal TiO_2 Films" (O'Regan and Grätzel, 1991) has been published. Due to their high efficiency, low production cost, and environmental benefits, traditional solar cells are losing ground to DSSCs based on nanocrystalline semiconductors like TiO_2 (Ahliha et al., 2018; Ammar et al., 2019; O'Regan and Grätzel, 1991). They've seen an exponential enhancement in the number of papers published per year on different aspects of DSSCs, with about 1,000 papers published in the last few years (Jinchu et al., 2014). Over the last 20 years, it has been the focus of extensive academic and commercial study, owing to its potential as a low-cost solar energy power conversion technology (Liu et al., 2019). Many researchers are working harder to increase the performance and stability of solar cells (Atli and Yildiz, 2022). The molecular composition of the dye, counter electrode, electrolyte, structural morphology of semiconductor metal oxides (Tsui and Cheung, 2015), and conducting flexible transparent material can all be altered to enhance the performance and stability of DSSCs. In recent years, printing on flexible conductive substrate materials followed by a roll-to-roll manufacturing has made it possible to manufacture flexible devices on a large scale, and the purification processes have also become more cost-effective (Gong et al., 2012, Yadav et al., 2021).

6.1.6.1 DSSC Structure and Working Principle

A platinum-coated conductive glass substrate, dye molecules coated on a large band gap semiconductor surface, and an electrolyte make up the DSSC (redox couple). When subjected to solar irradiation, an incident photon produces a bound electron-hole pair in DSSCs. In the dye sensitizer, a large band portion of solar light generates electron-hole pairs, which anchor on the photoanode of large band gap semiconductor materials (nanoparticles; Nakade et al., 2003). Owing to the difference in energy levels, electron-hole pairs form in dye solution of molecules easily divided at the minor level (picoseconds scale). Electrons are pumped from the dye into the photoanode CB and transferred to the glass-coated transparent conduction oxide film (TCO). The maximum output voltage in this structure is determined by the difference between the Fermi energy of the semiconductor film and the redox potential of the electrolyte. As a working electrode, the nanocrystalline semiconductor is placed on a transparent conductive substrate (Jena et al., 2013). The two functions are separated in this system, unlike in conventional systems where the semiconductor performs light absorption and charge carrier transport. A sensitizer, applied to the surface of a wide band gap semiconductor, absorbs light. The most common compounds are TiO_2, SnO_2, and ZnO (Aliah et al., 2018; Ouyang, 2019; Park et al., 2013). Charge separation occurs at the interface between the two materials due to photo-induced electron injection from the dye into the CB of the solid. Carriers are transferred from the semiconductor's CB to the charge collector, normally the TCO. A mediator, such as an electrolyte, fills the mesoscopic pores in the nanocrystalline film (redox couple; Khan, 2013). To absorb light, the photosensitive dye is chemisorbed into the semiconductor surface. The counter electrode (CE) is separated from the photoanode (working electrode) by a thin spacer (Mehmood et al., 2014). The CE is made of Pt-coated TCO substrate.

The four basic steps in the operation of a DSSC are electron injection, light absorption, current selection, and carrier transportation (Mehmood et al., 2014). Figure 6.9b shows the conversion of photons into current via the following steps.

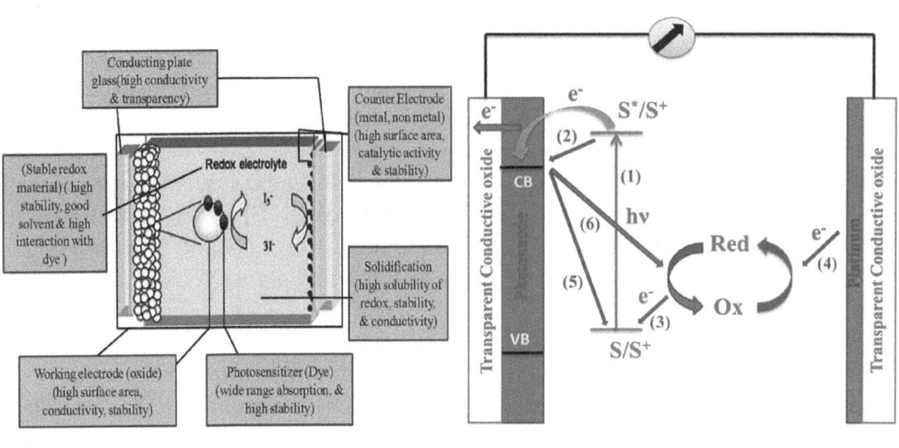

FIGURE 6.9 The working theory of DSSCs is depicted in (a) the schematic diagram and (b) the working principle of DSSCs.

- First, an electron–photon is generated by dye molecules, as in photosynthesis. A dye molecule is excited by photon (*hυ*) absorption, whereas an electron is excited from HOMO-D into LUMO-D [56].

$$D + h = D$$

- The free electron is inserted into the TiO_2 CB (with duration in the nanosecond range) and leaves the oxidized dye molecule D^+ (Jasim, 2011).

$$D^* = D^+ + e^-$$

- Then, the electron reaches the Pt catalyst layer (CE), where redox reactions occur by the recombination with holes within the electrolyte by reducing tri-iodide (I^{3-}) to iodide ion (I^-; Karthick et al., 2019).

$$I^{3-} + 2e^- = 3I^-$$

- However, in the final step, the negative charge of I diffuses back to the dye molecules and will react with the oxidized molecule D^+. Thus, it completes the electrical cycle and repeats.

$$3I^- + 2D^+ = I3^- + 2D$$

6.1.6.2 Components of DSSCs

In the fabrication of DSSCs, four main components are used: sensitizer (dye), photoanode (working electrode), CE, and electrolyte (redox couple). The following are the elements, as well as a description of them.

6.1.6.2.1 Conductive and Transparent Substrate

The two transparent conductive materials sheets are used to assemble DSSCs for catalyst and semiconductor deposition and serve as current collectors. Aside from these two main characteristics, the DSSC method has a few more (sensitized and cells, n.d.). To allow optimal sunlight to reach the cell's weighty region, it should first have nearly 80% transparency. Finally, high electrical conductivity is needed for able charge transport and minimal energy deficit in the cell. The conductive substrates in DSSCs are primarily fluorine-doped tin oxide (FTO, SnO_2: F) and indium-doped tin oxide (ITO, In_2O_3: Sn; Mahalingam and Abdullah, 2016). ITO and FTO layers are applied to these conductive substrates made of soda-lime glass (Yeoh et al., 2017).

6.1.6.2.2 Photoanode (Working Electrode)

The photoanode (working electrodes) is created by coating a transparent conducting substrate made of FTO or ITO with a thin layer of oxide semiconducting materials with a large band gap, such as TiO_2, NiO, ZnO, Nb_2O_5, and SnO_2. Due to a wide range of applications in sensors, photocatalytic systems, water splitting, medical implants, solar cells, and fuel cells, nano-sized TiO_2 has generated much interest

(Yeoh and Chan, 2017). Due to its commercial availability at low cost, nontoxic design, chemical stability, general reactivity, biocompatibility, and higher PCE achieved in the cell, it is one of the most extensively studied semiconductor materials used in many applications.

TiO_2 is a transition metal oxide that belongs to the transition metal oxide family. Titanium is the ninth extremely abundant element in the earth's crust and can be found in various mineral sands and rocks. It's an inorganic solid material with a white color that's insoluble in water, non-flammable, amphoteric, and readily accessible. Figure 6.10 depicts the unit cell of TiO_2 polymorphs such as rutile, anatase, and brookite (Mo and Ching, 1995). The tetragonal crystal structure of TiO_2's anatase and rutile phases, whereas brookite's crystal structure is orthorhombic. The band gap of anatase is 3.2 eV, while the band gaps of brookite and rutile are 3.3 and 3.0 eV, respectively.

I41/amd, P42/mnm, and Pbca are the space groups corresponding to anatase, rutile, and brookite stages, respectively. The rutile phase of TiO_2 has a higher density (4.25 g/cm^3) than the anatase phase (3.894 g/cm^3), and its melting point (MP) is 1,817°C. Anatase is more stable than the other polymorphs in the bulk, rutile, and nanorange. From the perspective of the oxygen octahedral picture, the crystal structure of TiO_2 can be seen (Reyes-Coronado et al., 2008). The Ti cation occupies the middle, while the six oxygen anions occupy the corner. The three octahedral corners share each oxygen corner in an ideal case, resulting in a 1:2 (Ti:O) stoichiometric ratio. The anatase phase differs from rutile in that it has a more refined crystal structure that allows for quick electron mobility and a high band gap that improves photo-catalysis and solar-electric conversion efficiency.

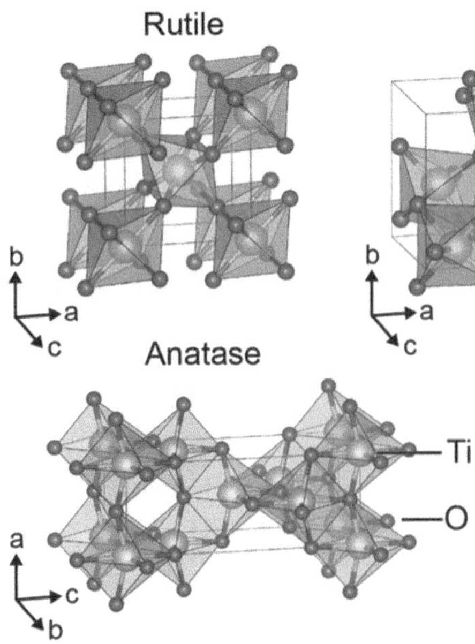

FIGURE 6.10 Unit cells of TiO_2 polymorphs: (a) rutile, (b) anatase, and (c) brookite.

6.1.6.2.3 Counter Electrode

In DSSCs like photoanode, the CE is also essential. Its research focuses on reducing tri-iodide to iodide at the FTO/electrolyte interface using a conductive substrate and a suitable catalyst (Kumar et al., 2019). Since the charge transfer resistance between the ITO surface and the electrolyte is high, using a suitable catalyst lowers the charge transfer resistance at the CE, allowing the reaction to proceed. As a result, one of the most critical characteristics of the catalyst is low charge transfer resistance. In addition, the CE catalyst must be chemically stable in the presence of the electrolyte. Platinum (Pt) is the most popular CE catalyst in DSSCs (Huang et al., 2007). Additional materials, such as biomass, conducting polymers, and other materials, have been used in DSSCs and work reasonably well. Pt has the best output of all the catalysts, but it comes at an exorbitant price. As a result, making the best use of platinum is a technical challenge. As a result, many alternatives to Pt in DSSCs have emerged, including biomass, carbonyl sulfide Au/GNP (CoS), and alloy CEs like CoNi0.25 and FeSe, despite the various kind of CEs.

6.1.6.2.4 Dye Sensitizer or Photosensitizer

The dye, responsible for the maximum absorption of the incident light, is the most significant component of DSSC. Some essential characteristics of the dye should include luminescence and absorption in the visible field (Batniji et al., 2016). The lowest unoccupied molecular orbital (LUMO) and the highest occupied molecular orbital (HOMO) should be close to and far from the surface of TiO_2's CB, respectively. For dye regeneration to be successful, the oxidized condition of the dye must have a higher potential than the redox electrolyte. The photosensitizer species must not accumulate, be chemically and thermally stable, and be photostable. DSSCs have been used and characterized with a variety of dye sensitizers. Metal-free organic dye molecules with a high extinction coefficient, porphyrin and phthalocyanine systems with good thermal and chemical stability and strong absorption in the near-infrared region, and metal-free metal complexes with large absorption spectra are the most common (Mozaffari et al., 2017). Ru-based metal-complex dyes (e.g., Z907, N3, N719) have more than 10% efficiencies, strong absorption spectra, suitable excited and ground state energy levels, and good chemical stability but low molar extinction coefficient. Organic dyes, like indoline, coumarin, tetrahydroquinoline, triarylamine, heteroanthracene, and natural dyes, are simpler to make and have a high molar absorption coefficient (Xu et al., 2008). These are less costly and environmentally friendly than "metal-complex" dyes, but they have broad absorption spectra and have demonstrated remarkable DSSC sensitizer performance.

6.1.6.2.5 Electrolyte (Redox Couple)

Aside from being an electrically conducting medium, the critical function of DSSCs is to donate electrons to oxidized dye molecules to prevent the dye from resuming the excited electrons. It should be transparent enough to allow light to pass through and have a quick redox reaction and good conductivity (Su'ait et al., 2015). Redox couples, chemicals, cations, solvents, and ionic liquids are all essential components of an electrolyte. In general, electrolyte materials should possess specific characteristics to function as an electrolyte in DSSCs.

- Electrolyte redox potential should be negative when compared to the dye oxidation potential.
- The electrolyte should be high-conducting, electrochemically stable, thermally stable, and chemically capable for a long time.
- Chemically inert against all other DSSC components, coupled with much reversibility to simple kinetics of rapid electron transfer.
- High solubility in the solvent ensures high charge carrier concentrations in the electrolyte and high diffusion coefficients in the used solvent for effective mass transport.

6.2 MATERIALS SYNTHESIS METHODS

6.2.1 Solution Processing Techniques

One of the most common solution processing techniques is a magnetic stirrer. In this present thesis work, we have prepared the solutions of dye and electrolyte materials with the help of a magnetic stirrer with a hot plate. Two other techniques are used for dye extraction: Soxhlet and ultrasonic *bath.* A detailed description of both techniques is given below.

6.2.2 Magnetic Stirrer with a Hot Plate

A magnetic stirrer is an electromechanical system that is used to combine liquid and bulk materials, as well as to prepare a uniform liquid mixture. It is commonly used in chemistry and biology laboratories and industries for sample preparation and investigation. Figure 6.11 shows a photograph of the magnetic stirring apparatus.

It's a flexible and long-lasting laboratory device that generates a spinning magnetic field using a rotating magnet and stationary electromagnets. This agitates/mixes the liquid by rapidly spinning a stir bar immersed in it. This system also includes a heater to keep the liquid warm when mixed. Stirrers are commonly used in almost any lab, including chemistry, physics, bioscience, and others (Phuse and Khan, 2017). Since they are more effective at mixing and quieter than gear-driven motorized stirrers, they are preferred. The magnetic bars are most suitable for glass vessels used for chemical reactions because glass has little effect on a magnetic field. When interacting with viscous liquids or dense suspensions, stir bars are obscure. Mechanical stirring is commonly needed for larger volumes or more viscous liquids.

In contrast to other stirring devices, the stirring bars need less cleaning and sterilization due to their small size. They don't need lubricants that could contaminate the reaction vessel or the final product. A hot plate can also be used in magnetic stirrers to heat the liquid or for other reasons. Many features, such as speed and temperature control, are included. Powerful motors and softer magnets produce exceptional magnetic power, and microprocessors regulate speed precisely over the entire range in many models.

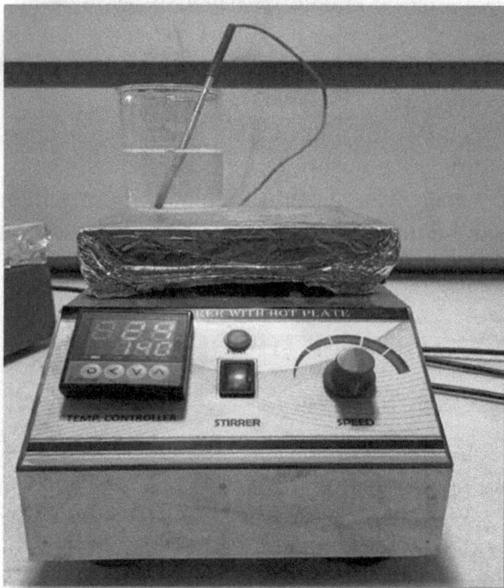

FIGURE 6.11 Photograph of magnetic stirrer.

6.2.2 Ultrasonic Bath

A water bath is a large piece of laboratory equipment that uses an ultrasonic electric generator to generate a signal that powers a transducer. The photograph of the ultrasonic bath used in this thesis work is shown in Figure 6.12. Ultrasonic bathing is a technique that involves agitating a fluid with sound waves (usually between 20 and 40 kHz). Even if these frequencies are outside your hearing range, ear protection is still recommended during sonication because the process produces a loud screeching noise. The heavier the agitation of particles, the higher the frequency is advantageous. While ultrasound can be used only with water, using a solvent suitable for the object to be cleaned and the form of soiling improves the effect (Ellery and Schleyer, 1984). Cleaning takes between 5 and 10 minutes on average but can take up to 30 minutes depending on the item to be cleaned.

Thousands of microscopic vacuum bubbles in the solution produce pressure cycles during sonication. Cavitations are the mechanism by which the bubbles fall into the solution (Majid et al., 2015). This generates powerful vibrational waves in the cavitation field, disrupting molecular interactions such as those between water molecules, separating clumps of particles and facilitating mixing. For example, the gas bubbles come together and leave the solution more rapidly in dissolved gas vibrations (Cao et al., 2017).

FIGURE 6.12 Picture of ultra-sonication bath cleaner.

6.3 DEVICE FABRICATION TECHNIQUES

We have fabricated various novel dyes and TiO_2-based devices in this thesis work. Figure 6.13 shows a flow chart of the steps undertaken from solution preparation to the final step of film deposition for the fabrication of the device. In this section, we explained every process step-by-step in detail.

6.4 CONCLUSIONS

We have fabricated various novel dyes and TiO_2-based devices in this thesis work. Figure 6.13 shows a flow chart of the steps undertaken from solution preparation to the final step of film deposition for the fabrication of the device.

Latest innovations in the field of DSSCs have raised hopes for developing more powerful and long-lasting devices at a low cost. One of the main goals of the current research is to synthesize and characterize low-cost, environmentally friendly materials and cost-effective devices based on these materials for dye-sensitized solar cell applications. This chapter offers an overview of DSSCs, their elements, and how they function. It also explains the concept behind using materials as components of DSSCs. This chapter also discusses the different experimental methods and equipment descriptions used to characterize thin films and devices to assess their various properties and the instruments and techniques applied to synthesizing materials and fabricating dye-sensitized solar-based devices.

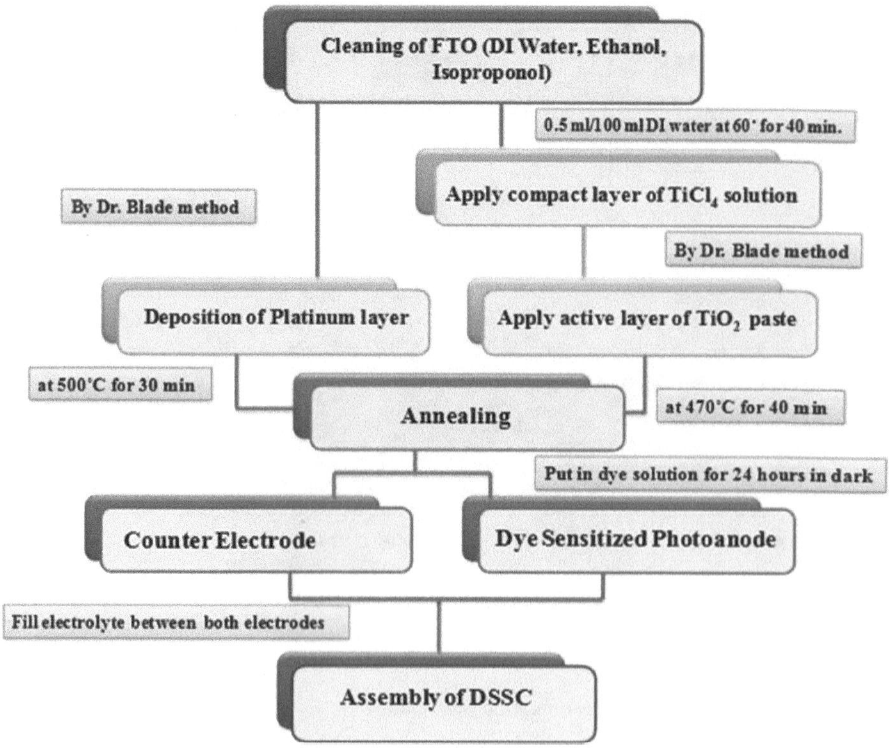

FIGURE 6.13 Flow chart of DSSC fabrication process.

REFERENCES

Ahliha, A. H., F. Nurosyid, A. Supriyanto, and T. Kusumaningsih. 2018. "Optical Properties of Anthocyanin Dyes on TiO_2 as Photosensitizers for Application of Dye-Sensitized Solar Cell (DSSC)." *IOP Conference Series: Materials Science and Engineering* 333 (March): 012018. https://doi.org/10.1088/1757-899X/333/1/012018.

Aliah, H., B. Bernando, F. Puspitasari, A. Setiawan, P. Pitriana, B. W. Nuryadin, and M. A. Ramdhani. 2018. "Dye Sensitized Solar Cells (DSSC) Performance Reviewed from the Composition of Titanium Dioxide (TiO_2)/Zinc Oxide (ZnO)." *IOP Conference Series: Materials Science and Engineering* 288 (1). https://doi.org/10.1088/1757-899X/288/1/012070.

Ammar, A. M., H. S. H. Mohamed, M. M. K. Yousef, G. M. Abdel-Hafez, A. S. Hassanien, and A. S.G. Khalil. 2019. "Dye-Sensitized Solar Cells (DSSCs) Based on Extracted Natural Dyes." *Journal of Nanomaterials* 2019. https://doi.org/10.1155/2019/1867271.

Andualem, A., and S. Demiss. 2018. "Review on Dye-Sensitized Solar Cells (DSSCs)." *Edelweiss Applied Science and Technology* 2: 145–50.

Arifin, Z., S. Suyitno, S. Hadi, and B. Sutanto. 2018. "Improved Performance of Dye-Sensitized Solar Cells with TiO_2 Nanoparticles/Zn-Doped TiO_2 Hollow Fiber Photoanodes." *Energies* 11 (11). https://doi.org/10.3390/en11112922.

Batniji, A., M. S. Abdel-Latif, T. M. El-Agez, S. A. Taya, and H. Ghamri. 2016. "Dyes Extracted from Trigonella Seeds as Photosensitizers for Dye-Sensitized Solar Cells." *Journal of Theoretical and Applied Physics* 10 (4): 265–70. https://doi.org/10.1007/s40094-016-0225-9.

Cao, Q., J. Cheng, Q. Feng, S. Wen, and B. Luo. 2017. "Surface Cleaning and Oxidative Effects of Ultrasonication on the Flotation of Oxidized Pyrite." *Powder Technology* 311: 390–97.

Carella, A., F. Borbone, and R. Centore. 2018. "Research Progress on Photosensitizers for DSSC." *Frontiers in Chemistry* 6 (SEP): 1–24. https://doi.org/10.3389/fchem.2018.00481.

Corkish, R., W. Lipiński, and R. J. Patterson. 2016. "Introduction to Solar Energy." *Solar Energy* 2: 1–29. https://doi.org/10.1142/9637.

Coyle, E. D., and R. A. Simmons. 2014. *Understanding the Global Energy Crisis*. https://doi.org/10.26530/oapen_469619.

Devasia, A., and S. K. Kurinec. 2011. "Teaching Solar Cell I-V Characteristics Using SPICE." *American Journal of Physics* 79 (12): 1232–39. https://doi.org/10.1119/1.3636525.

Dou, J., Y. Li, F. Xie, X. Ding, and M. Wei. 2016. "Metal–Organic Framework Derived Hierarchical Porous Anatase TiO_2 as a Photoanode for Dye-Sensitized Solar Cell." *Crystal Growth & Design* 16 (1): 121–25.

Duarte, C. A., L. R. Goulart, L. S. C. Filice, I. L. De Lima, E. Campos-Fernándes, N. O. Dantas, A. C. A. Silva, et al. 2020. "Characterization of Crystalline Phase of TiO_2." *Materials* 13: 4071.

EIA. 2009. "World Energy Demand and Economic Outlook- Chapter1" 2016 (May 2015): 7–17. https://www.eia.doe.gov/oiaf/ieo/pdf/world.pdf.

Ellery, W. N., and M. H. Schleyer. 1984. "Comparison of Homogenization and Ultrasonication as Techniques in Extracting Attached Sedimentary Bacteria." *Marine Ecology Progress Series. Oldendorf* 15 (3): 247–50.

GEA. 2012. "Global Energy Assessment-Renewable Energy," 761–900. https://www.iiasa.ac.at/web/home/research/Flagship-Projects/Global-Energy-Assessment/Home-GEA.en.html.

Gong, J., J. Liang, and K. Sumathy. 2012. "Review on Dye-Sensitized Solar Cells (DSSCs): Fundamental Concepts and Novel Materials." *Renewable and Sustainable Energy Reviews* 16 (8): 5848–60. https://doi.org/10.1016/j.rser.2012.04.044.

Huang, Z., X. Liu, K. Li, D. Li, Y. Luo, H. Li, W. Song, L. Q. Chen, and Q. Meng. 2007. "Application of Carbon Materials as Counter Electrodes of Dye-Sensitized Solar Cells." *Electrochemistry Communications* 9 (4): 596–98.

IEA. 2019. "World Energy Outlook 2019 エグゼクティブサマリー." *World Energy Outlook* 2019, 1. https://www.iea.org/reports/world-energy-outlook-2019; https://webstore.iea.org/download/summary/2467?fileName=Japanese-Summary-WEO2019.pdf.

Jao, M. H., H. C. Liao, and W. F. Su. 2016. "Achieving a High Fill Factor for Organic Solar Cells." *Journal of Materials Chemistry A* 4 (16): 5784–5801. https://doi.org/10.1039/c6ta00126b.

Jasim, K. E.. 2011. "Dye Sensitized Solar Cells-Working Principles, Challenges and Opportunities." *Solar Cells-Dye-Sensitized Devices* 8: 172–210.

Jena, A., S. P. Mohanty, P. Kumar, J. Naduvath, P. Lekha, J. Das, H. K. Narula, S. Mallick, P. Bhargava, and V. Gondane. 2013. "Transactions of the Indian Ceramic Society," no. April: 1–16.

Jinchu, I., C. O. Sreekala, and K. S. Sreelatha. 2014. "Dye Sensitized Solar Cell Using Natural Dyes as Chromophores-Review." *Materials Science Forum* 771: 39–51. https://doi.org/10.4028/www.scientific.net/MSF.771.39.

Kannan Balasingam, S., M. Lee, M. G. Kang, and Y. Jun. 2013. "Improvement of Dye-Sensitized Solar Cells toward the Broader Light Harvesting of the Solar Spectrum." *Chemical Communications* 49 (15): 1471–87. https://doi.org/10.1039/c2cc37616d.

Karthick, S. N., K. V. Hemalatha, S. K. Balasingam, F. M. Clinton, S. Akshaya, and H. -J. Kim. 2019. "Dye-Sensitized Solar Cells: History, Components, Configuration, and Working Principle." *Interfacial Engineering in Functional Materials for Dye-Sensitized Solar Cells*, 1–16.

Katja, V. 2014. "Seminar -1st Year, 2nd Cycle Solar Cells." *Nature Energy*, no. May: 2–3. https://doi.org/10.1038/s41560-019-0463-6.

Khan, M. I. 2013. "A Study on the Optimization of Dye-Sensitized Solar Cells," no. January.

Kibria, M. T., A. Ahammed, S. M. Sony, and F. Hossain. 2014. "A Review : Comparative Studies on Different Generation Solar Cells Technology." *International Conference on Environmental Aspects of Bangladesh*, no. March: 51–53.

Kumar, D. K., S. K. Swami, V. Dutta, B. Chen, N. Bennett, and H. M. Upadhyaya. 2019. "Scalable Screen-Printing Manufacturing Process for Graphene Oxide Platinum Free Alternative Counter Electrodes in Efficient Dye Sensitized Solar Cells." *FlatChem* 15 (April): 100105. https://doi.org/10.1016/j.flatc.2019.100105.

Labat, F., I. Ciofini, H. P Hratchian, M. J. Frisch, K. Raghavachari, and C. Adamo. 2011. "Insights into Working Principles of Ruthenium Polypyridyl Dye-Sensitized Solar Cells from First Principles Modeling." *The Journal of Physical Chemistry C* 115 (10): 4297–4306.

Lecture. 2015. "Fundamental Properties on Today's Menu Fundamental Solar Cell Properties Dark Current Efficiency Internal and External QE" 8 (Phys 4400): 25–32.

Liu, J., Y. Li, S. Yong, S. Arumugam, and S. Beeby. 2019. "Flexible Printed Monolithic-Structured Solid-State Dye Sensitized Solar Cells on Woven Glass Fibre Textile for Wearable Energy Harvesting Applications." *Scientific Reports* 9 (1): 1–11. https://doi.org/10.1038/s41598-018-37590-8.

Loulou, M., M. K. Al Turkestan, N. Brahmi, and M. Abdelkrim. 2018. "Current Dependence of Series and Shunt Resistances of Solar Cells." *2018 9th International Renewable Energy Congress, IREC 2018*, no. March 2018: 1–5. https://doi.org/10.1109/IREC.2018.8362494.

Mahalingam, S., and H. Abdullah. 2016. "Electron Transport Study of Indium Oxide as Photoanode in DSSCs: A Review." *Renewable and Sustainable Energy Reviews* 63: 245–55. https://doi.org/10.1016/j.rser.2016.05.067.

Majid, I., G. A. Nayik, and V. Nanda. 2015. "Ultrasonication and Food Technology: A Review." *Cogent Food & Agriculture* 1 (1): 1071022.

Mehmood, U., S. U. Rahman, K. Harrabi, I. A. Hussein, and B. V. S. Reddy. 2014. "Recent Advances in Dye Sensitized Solar Cells." *Advances in Materials Science and Engineering* 2014: 974782.

Mo, S. D., and W. Y. Ching. 1995. "Electronic and Optical Properties of Three Phases of Titanium Dioxide: Rutile, Anatase, and Brookite." *Physical Review B* 51 (19): 13023–32. https://doi.org/10.1103/PhysRevB.51.13023.

Mohammad Bagher, A. 2015. "Types of Solar Cells and Application." *American Journal of Optics and Photonics* 3 (5): 94. https://doi.org/10.11648/j.ajop.20150305.17.

Mozaffari, S., M. R. Nateghi, and M. B. Zarandi. 2017. "An Overview of the Challenges in the Commercialization of Dye Sensitized Solar Cells." *Renewable and Sustainable Energy Reviews* 71 (December 2016): 675–86. https://doi.org/10.1016/j.rser.2016.12.096.

Musunuri, R. K., D. Sánchez, and R. Rodriguez. 2007. "Solar Thermal Energy Engineering." *Renewable Energy*, no. October.

Nakade, S., Y. Saito, W. Kubo, T. Kitamura, Y. Wada, and S. Yanagida. 2003. "Influence of TiO_2 Nanoparticle Size on Electron Diffusion and Recombination in Dye-Sensitized TiO_2 Solar Cells." *Journal of Physical Chemistry B* 107 (33): 8607–11. https://doi.org/10.1021/jp034773w.

Ouyang, J. 2019. "Applications of Carbon Nanotubes and Graphene for Third-Generation Solar Cells and Fuel Cells." *Nano Materials Science* 1 (2): 77–90. https://doi.org/10.1016/j.nanoms.2019.03.004.

Parasuraman, S. 2015. "Lecture 19: Solar Cells," 17.

Park, K., Q. Zhang, D. Myers, and G. Cao. 2013. "Charge Transport Properties in TiO_2 Network with Different Particle Sizes for Dye Sensitized Solar Cells." *ACS Applied Materials and Interfaces* 5 (3): 1044–52. https://doi.org/10.1021/am302781b.

Phuse, S. S, and D. Khan. 2017. "Influence of Extraction Methods Using Different Solvents on Caesalpinia Pulcherrima Leaves." *International Journal of Pharma and Bio Sciences* 8 (2): 829–37.

Pujahari, R. M. (2021). Crystalline silicon solar cells. In "*Energy Materials*." Elsevier. pp. 107–138.

Raj, C. C., and R. Prasanth. 2016. "A Critical Review of Recent Developments in Nanomaterials for Photoelectrodes in Dye Sensitized Solar Cells." *Journal of Power Sources* 317: 120–32. https://doi.org/10.1016/j.jpowsour.2016.03.016.

Ranabhat, K., L. Patrikeev, A. Antal evna Revina, K. Andrianov, V. Lapshinsky, and E. Sofronova. 2016. "An Introduction to Solar Cell Technology." *Journal of Applied Engineering Science* 14 (4): 481–91. https://doi.org/10.5937/jaes14-10879.

Regan, B. O, and M. Gratzelt. 1991. "A Low-Cost, High-Efficiency Solar Cell Based on Dye-Sensitized Colloidal TiO_2 Films." *Nature* 353 (October): 737–40.

Reyes-Coronado, D., G. Rodríguez-Gattorno, M. E. Espinosa-Pesqueira, C. Cab, R. De Coss, and G. Oskam. 2008. "Phase-Pure TiO_2 Nanoparticles: Anatase, Brookite and Rutile." *Nanotechnology* 19 (14). https://doi.org/10.1088/0957-4484/19/14/145605.

Rokesh, K., A. Pandikumar, and K. Jothivenkatachalam. 2014. "Dye Sensitized Solar Cell: A Summary." *Materials Science Forum* 771: 1–24. https://doi.org/10.4028/www.scientific.net/MSF.771.1.

Scofield, J. H. 2009. "The Solar Spectrum, the Stefan-Boltzmann Law, the Planck Distribution." *Optics and Lasers in Engineering* 1: 1–9

Selvaraj, P., H. Baig, T. K. Mallick, J. Siviter, A. Montecucco, W. Li, M. Paul, et al. 2018. "Enhancing the Efficiency of Transparent Dye-Sensitized Solar Cells Using Concentrated Light." *Solar Energy Materials and Solar Cells* 175 (October 2017): 29–34. https://doi.org/10.1016/j.solmat.2017.10.006.

Sharma, K., V. Sharma, and S. S. Sharma. 2018. "Dye-Sensitized Solar Cells: Fundamentals and Current Status." *Nanoscale Research Letters* 13. https://doi.org/10.1186/s11671-018-2760-6.

Simya, O. K., P. Radhakrishnan, A. Ashok, K. Kavitha, and R. Althaf. 2018. "Engineered Nanomaterials for Energy Applications." *Handbook of Nanomaterials for Industrial Applications*, 751–767. https://doi.org/10.1016/B978-0-12-813351-4.00043-2.

Smets, A. H., Jäger, K., Isabella, O., van Swaaij, R. A., & Zeman, M. (2016). "Solar Cell Parameters and Equivalent Circuit." Solar Energy*: Physics, Engineering, and Photovoltaic Conversion Technology and Systems*, 113–121.

Srinivasu, P., S. P. Singh, A. Islam, and L. Han. 2011. "Novel Approach for the Synthesis of Nanocrystalline Anatase Titania and Their Photovoltaic Application." *Advances in Optoelectronics* 2011: 539382. https://doi.org/10.1155/2011/539382.

Su'ait, M. S., M. Y. A. Rahman, and A. Ahmad. 2016. "Review on Polymer Electrolyte in Dye-Sensitized Solar Cells (DSSCs)." *Solar Energy* 115: 452–471.

Tomar, N., A. Agrawal, V. S. Dhaka, and P. K. Surolia. 2020. "Ruthenium Complexes Based Dye Sensitized Solar Cells: Fundamentals and Research Trends." *Solar Energy* 207: 59–76. https://doi.org/10.1016/j.solener.2020.06.060.

Tress, W., and W. Tress. 2014. *Organic Solar Cells*. Springer International Publishing. pp. 67–214.

U.S. Energy Information Administration. 2019. "*International Energy Outlook 2019*." U.S. Energy Information Administration, vol. September, no. 09: 25–150.

Wenham, S. R., M. A. Green, M. E. Watt, and R. Corkish. 2007. "The Behaviour of Solar Cells." *Applied Photovoltaics*, 43–56.

Widén, J., and J. Munkhammar. 2019. "*Solar Radiation Theory*." Uppsala University.

Wolf, M., and H. Rauschenbach. 1963. "Series Resistance Effects on Solar Cell Measurements." *Advances in Energy Conversion* 3 (2): 455–479. https://doi.org/10.1016/0365-1789(63)90063-8.

Xu, W., B. Peng, J. Chen, M. Liang, and F. Cai. 2008. "New Triphenylamine-Based Dyes for Dye-Sensitized Solar Cells." *Journal of Physical Chemistry C* 112 (3): 874–880. https://doi.org/10.1021/jp076992d.

Yadav, V., S. Chaudhary, C. M. S. Negi, and S. K. Gupta. 2020. "Textile Dyes as Photo-Sensitizer in the Dye Sensitized Solar Cells." *Optical Materials* 109 (2020): 110306. https://doi.org/10.1016/j.optmat.2020.110306.

Yadav, V., C. M. S. Negi, D. K. Kumar, and S. K. Gupta. 2021. "Fabrication of Eco-Friendly, Low-Cost Dye Sensitized Solar Cells Using Harda Fruit-Based Natural Dye." *Optical Materials* 122 (2021): 111800. https://doi.org/10.1016/j.optmat.2021.111800.

7 Carbon-Based Nanomaterials

Carbon Nanotubes and Graphene for Energy Storage Applications

Nandita Singh, Uday Pratap Azad, Ashish Kumar Singh, Sunil Kumar Singh, Bharat Lal Sahu, Shobhana Ramteke, and Divya Pratap Singh

7.1 INTRODUCTION

Presently, most of the research studies are focused on green and clean energy production, healthy environmental development, and affordable healthcare development, as these are the main global concerns. Electrochemistry plays a vital role and can be considered as the heart of biological and chemical sensor development, energy storage, and power generation (Simon and Gogotsi, 2008; Tarascon and Armand, 2001). It's very difficult to secure green and effective energy sources, which is one of the major challenges that we face in the 21st century. Presently, most of the research studies focus on sustainable and green energy production, green environmental development, and healthcare development, as these are the main global concerns (Broto and Kirshner, 2020; Choi et al., 2018). As we know, natural energy resources are very limited, and in the near future, they will be run out very soon. At the same time, due to their regular use, day by day, the environment is getting more and more polluted, which is responsible for the birth of various kinds of fatal diseases. So, it's the responsibility of everyone to work for better and cleaner energy resources to get rid of these energy and environmental and health-related issues. The advancement in highly stable electrochemical energy storage materials is a better alternative to tackle the global energy crisis problem. Carbon is one of the basic and fourth most plentiful elements in the universe by mass. The atoms of carbon can bind together in numerous ways and produce various allotropes such as diamond, graphite, fullerenes, and amorphous carbon. Under normal conditions, all the allotropes of carbon are solid at room temperature. As per structure, the different forms of carbon are responsible for different physical properties. Since the last 36 years, our standing has become enriched about sp^2 carbons. Fullerene (C60) was the first identified nanostructure of carbon,

DOI: 10.1201/9781003481157-7

whereas carbon nanotube was the second identified nanostructure of carbon, which come under the categories of zero-dimensional and one-dimensional materials, respectively.

With the increasing call for sustainable energy resources, countless research studies have been done for the development of advanced conversion and energy storage materials. Carbon nanotube- and graphene-derived materials are considered to be the best alternatives to tackle the energy crisis problem. Due to their inimitable properties of large surface area, high chemical stability, and superior electric and thermal conductivities (Tasis et al., 2006; Yuan and He, 2015), they have been extensively used for electrode material fabrication in electrochemical energy storage systems.

Carbon nanotubes (CNTs), which are the allotropes of carbon, are one of the best cylindrical tube-shaped novel classes of nanostructured materials that have been frequently used in various research fields for different purposes. They come under the category of one-dimensional material. Mainly, two types of carbon nanotubes have been discovered, namely single-walled carbon nanotubes (SWCNTs) and multiwalled carbon nanotubes (MWCNTs). SWCNTs are mainly formed by a single layer of graphite, which is arranged in a cylindrical shape, and they are intermediate between fullerene and flat graphene. They have been prepared by rolling up sheets of single-layer carbon atoms (graphene) and are believed to be intermediate between fullerene cases and flat graphene. MWCNTs consist of multiple carbon nanotubes nested within one another, and depending on the number of tubes, sometimes they are called as double-walled carbon nanotubes, triple-walled carbon nanotubes, etc. The number of nanotubes in MWCNTs can vary from 3 to 20 (Rathinavel et al., 2021). The pictorial illustration for the preparation of carbon nanotubes from graphene and various types of graphene is given in Figure 7.1.

Graphene is a well-known two-dimensional (2D) nanomaterial having sp^2-bonded single-layer carbon atoms and a honeycomb-type lattice structure. Apart from layered 2D graphene, porous 2D and 3D graphenes (such as graphene aerogel, graphene hydrogen, and graphene flowers) have been widely used for energy storage applications, which have been prepared by adopting numerous synthetic strategies (Ambrosi et al.,

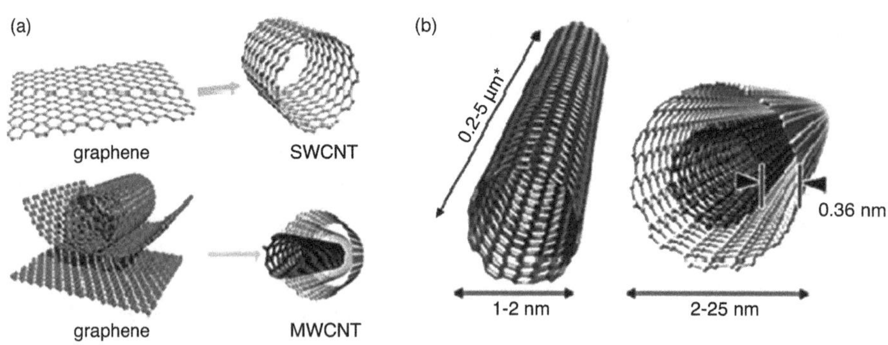

FIGURE 7.1 (a) Schematic illustration for the formation of carbon nanotubes from graphene sheets. Reproduced with permission from Vidu et al. (2014). (b) Single- and double-walled carbon nanotubes and their dimensions. Reproduced with permission from Roldo et al. (2013).

2014). Numerous heteroatoms, metal nanoparticles, metal oxide-doped metal sulfides, conducting polymers, and metal-organic framework-decorated graphenes have been used for energy storage applications. Due to its advantageous properties like very high surface area, better chemical and mechanical stability, and very good electrical and thermal conductivity, it has been widely used in various subjects such as chemistry, biotechnology, physics, and material science (Guo et al., 2011; Shao et al., 2010).

This chapter covers the recent advancement in carbon nanotube- and graphene-based electrode materials for various energy storage applications, i.e., supercapacitors, lithium-sulfur batteries, and lithium-ion batteries, and we are mainly focused on the carbon nanotube- and graphene-derived materials that have been utilized for the development of efficient, stable, durable, and high-energy storage materials.

7.2 GRAPHENE AND ITS NANOCOMPOSITES FOR ANODE FABRICATION IN LITHIUM-ION BATTERIES

Increasing the capacity of the anode is the key factor in developing a lithium-ion battery (LIB) with higher power and energy densities. Yoo et al. achieved a large improvement in the storage capacity of lithium by using graphene nanosheets (GNS). They have used the exfoliation and resemble process to control the layered structure of the GNS and achieved the specific capacitance of 540 mA h g^{-1} for GNS. Additionally, they have used CNT and C60 for anode fabrication; the specific capacity was increased to 730 and 784 mA h g^{-1}, for CNT and C60, respectively (Yoo et al., 2008); and the improved performance could be due to dissimilar electronic structure of graphite and graphene sheets and extra sites for the lodging of lithium ions due to increased d-spacing value of graphene layers (Pan et al., 2009; Yin et al., 2011; Bhardwaj et al., 2010). It has been reported that the reversible capacity of chemically derived graphene can be achieved up to 1,264 mA h g^{-1}, which is almost two times higher as compared to conventional graphite, which was used as anode material in LIB at a slower charge-discharge rate. Wu et al. (2011) reported that a high charge-discharge rate and large-capacity graphene electrode can be constricted by doping the graphene with other heteroatoms such as boron and nitrogen. Different heteroatom-doped graphenes (Figure 7.2a) and their corresponding energy densities (Figure 7.2b) are depicted in Figure 7.2.

At a low charge/discharge rate, the capacity for nitrogen- and boron-doped graphene was found to be 15,401,043 mA h g^{-1}. Various metal oxide-doped graphene oxides such as SiO_2 and Co_3O_4 have also been used as anode materials to improve the charge-discharge rate by using some organic molecules as a binder. Here, the role of graphene is to control the aggregation of metal oxide nanoparticles, and in the case of metal oxide-doped graphene oxide material, higher capacity and cycling stability were obtained as compared to undoped graphene oxide. Dong et al. (2012) have reported Co_3O_4 nanorods on 3D graphene foam by adopting a very simple hydrothermal synthetic route and using the same as an electrode material for supercapacitors (SCs), and they were able to achieve a specific capacity of ~1,100 F g^{-1} by using the current density of 10 A g^{-1} with outstanding cycling stability. Metal hydroxides such as $Ni(OH)_2$ (Yan et al., 2012; Wang et al., 2010), $Co(OH)_2$ (Chen et al., 2010), and

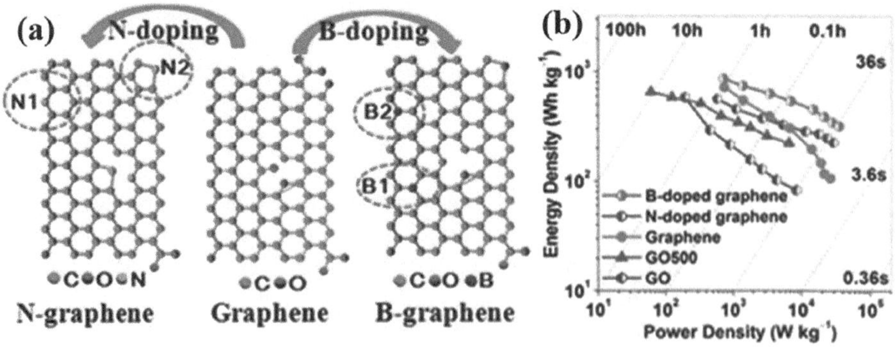

FIGURE 7.2 (a) Graphene and nitrogen- and boron-doped graphene. (b) Ragone plots for nitrogen- and boron-doped graphene, pristine graphene, graphene oxide, and thermally reduced graphene. Reproduced with permission from Wu et al. (2011). Copyright © 2011, American Chemical Society.

double-layered hydroxides (Jiang et al., 2011) have also been used to prepare composite materials with graphene, and they have been used as an electrode material for SCs. Dong et al. (2012) prepared $Ni(OH)_2$ nanoplates on graphene, and they were able to achieve a specific capacity of 1,335 F g^{-1} by using a current density of 2.8 A g^{-1}.

7.3 GRAPHENE-DERIVED MATERIALS FOR CATHODE FABRICATION IN LITHIUM-ION BATTERIES

Over the past decades, extensive research has been carried out to increase the capacity and the energy density of previously developed cathode materials for their possible use as an alternative to fulfill the high-energy call in the impending electronics market. Graphene and graphene-derived materials have been widely used as cathode materials to increase the conductivity, specific capacitance, lithium-ion transportation, and electron transfer kinetics. Rangappa et al. (2010) have synthesized $LiMPO_4$ (M = Fe and Mn) colloidal nanocrystals by a supercritical ethanol process and utilized them in rechargeable lithium-ion batteries as promising cathode materials due to their high capacity and good cycle life, but the main problem with this cathode material was its low ionic conductivity and electrical conductivity, which restricted its use for practical application. Furthermore, to overcome this shortcoming, the composite of $LiMPO_4$ with reduced graphene oxide and porous 3D graphene has also been utilized (Yang et al., 2012; Shi et al., 2012). Yang et al. (2013) have prepared VO_2-doped graphene ribbon and used it as cathode material for the ultrafast lithium storage. They have found that highly ordered and crystalline VO_2-graphene nanoribbons were capable of boosting discharging and charging abilities with high capacity and longer cycle concert even at high temperatures. Furthermore, in another report, vanadium pentoxide (V_2O_5) has been reported as a promising cathode material for LIBs, but the sluggish electron transfer kinetics and slow transportation of lithium were the major problems with this cathode material (Wang et al., 2011). To improve the electron transfer kinetics and lithium-ion transportation, V_2O_5 was further used with some carbonaceous materials (Wang et al., 2011; Rui et al., 2011), but the long cycle stability still remains a challenging issue, and to rectify this issue, V_2O_5 and graphene sheet composites were used, and they were able to achieve a current density of 10,000 mA g^{-1} for 100,000 continuous cycles (Lee et al., 2012).

7.4 GRAPHENE-DERIVED MATERIALS FOR SUPERCAPACITORS

As we know, electrochemical supercapacitors (ECs) are another type of energy capture and storage device that are superior to batteries in terms of power density and cyclability and are considered as a better alternative for the renewable energy resources. The electric double-layer capacitors (EDLCs) are also ECs that accumulate the charge electrostatically through reversible adsorption of ions of the electrolyte onto active materials (Simon and Gogotsi, 2008). Large specific surface area and higher electrical conductivity are the key requirements for high capacitance, and graphene-based materials are the best alternatives to fulfill these requirements as they possess very high surface area and very high electrical conductivity. Zhu et al. (2011) have reported a facile chemical activation strategy for the production of higher-grade microwave-exfoliated graphene oxide with very high surface area (a-MEGO), and the specific capacitances achieved by using this material were found to be 165, 166, and 166 F g^{-1} by using current densities of 1.4, 2.8, and 5.7 A g^{-1}, respectively. Reduced graphene oxide, which had been doped with nitrogen, has also been extensively used for the cathode fabrication in lithium-ion batteries. Jeong et al. (2011) have used such types of material and were able to achieve a specific capacitance of 282 F g^{-1} at a higher current density of 1 A g^{-1}, which is almost four times higher as compared to the specific capacitance of pristine graphene (69 F g^{-1}). Dreyer and Bielawski (2012) used graphene oxide as a heterogeneous catalyst to polymerize a large number of olefin monomers in the presence of heat, and they found that the materials synthesized in this way were quite suitable for supercapacitive applications due to their high capacitance (25–120 F g^{-1}). Table 7.1 summarizes the various types of graphene-based materials used for various energy storage applications.

7.5 APPLICATION OF GRAPHENE AND GRAPHENE-DERIVED MATERIALS IN LITHIUM-SULFUR BATTERIES

As we know, in the current scenario, the lithium-sulfur battery is the most demanding field of research, and the study in this direction had already started in 1940, and since then, numerous advancements have been made to increase its capacity. Generally, in lithium-sulfur batteries, lithium works as an anode, whereas sulfur works as a cathode and some kind of separator (solid or liquid) (Figure 7.3a). The basic electrode reaction of the Li-S cells in a simplified way can be given by the equation:

$$S + 2Li \leftrightarrow Li_2S \quad (7.1)$$

At the cathode, Li is oxidized to Li^+ ion, whereas at the anode, sulfur is reduced to S^{2-} ions. At the start of the discharge process, the elemental sulfur, which has a ring structure (S_8 phase), is gradually reduced and gets converted into long-chain

TABLE 7.1
Wide Range of Graphene-Based Materials for Various Energy Storage Applications

Materials	Application	Current density	Capacitance	Cycling stability	References
3D porous graphene	EDLCs	0.05 mA cm^{-2}	8.66 mF cm^{-2}	92.23% for 3,500 cycles	Wang et al. (2019)
3D porous rGO	EDLCs	1 A g^{-1}	245 F g^{-1}	92% over 10,000 cycles	Zhang et al. (2018)
Reduced holey rGO	EDLCs	0.2 A g^{-1}	260 F g^{-1}	91% after 5,000 cycles	Liu et al. (2018)
3D porous graphene	EDLCs	560 A L^{-1}	27.5 mF L^{-1}	94.6% after 20,000 cycles	Strauss et al. (2018)
N-doped rGO foams	EDLCs	0.1–20 A g^{-1}	260–173 F g^{-1}	92.8% after 5,000 cycles	Liu et al. (2019)
Porous CNT-rGO	EDLCs	0.05 A cm^{-3}	60.75 F cm^{-3}	94% after 10,000 cycles	Park et al. (2019)
3D porous rGO hydrogel/carbon dots	EDLCs	1 A g^{-1}	264 F g^{-1}	91% after 5,000 cycles	Feng et al. (2018)
RuO_2/holey rGO	PCs	–	199 F cm^{-3}	88.9% after 5,000 cycles	Zhai et al. (2018)
3D iron oxide/nanoporous N-doped graphene	PCs	–	High specific capacitance of 409 mA h g^{-1}	91% after 10,000 cycles	Liu et al. (2018a)
$Ni_xCo_{1-x}O$/porous rGO	PCs	20 A g^{-1}	697.8 F g^{-1}	86% after 10,000 cycles	Zhou et al. (2018)
$NiCo_2S_4$/holey defect rGO hydrogel	PCs	0.5–6 A g^{-1}	1,000–800 F g^{-1}	87% after 5,000 cycles	Tiruneh et al. (2018)
NiCoS/porous rGO framework	PCs	–	405 F g^{-1}	–	Tiruneh et al. (2018a)
Polyaniline nanotube/rGO aerogels	PCs	1 A g^{-1}	117.4 F g^{-1}	94.9% after 1,000 cycles	Bora et al. (2018)
Polyaniline/rGO aerogels	PCs	10 A g^{-1}	396 F g^{-1}	72.3% after 1,000 cycles	Li et al. (2019)
Porous rGO	LIBs	–	770 mA h g^{-1}	98.0% after 110 cycles	Xing et al. (2019)
MgO/N-doped graphene	LIBs	–	1,138–440 mA h g^{-1}	99.9% after 500 cycles	Mo et al. (2019)
N/Cl dual-doped porous graphene	LIBs	0.1–1.0 A g^{-1}	1,200–1,010 mA h g^{-1}	95% after 1,800 cycles	Liu et al. (2018b)
3D CoO/rGO aerogels	LIBs	8 A g^{-1}	624.7 mA h g^{-1}	1142.8 mA h g^{-1} after 100 cycles at 0.5 A g^{-1}	Yao et al. (2019)
NiO/GF	LIBs	1 A g^{-1}	330 mA h g^{-1}	640 mA h g^{-1} after 50 cycles at 100 mA g^{-1}	Shao et al. (2019)

(Continued)

TABLE 7.1 (*Continued*)
Wide Range of Graphene-Based Materials for Various Energy Storage Applications

Materials	Application	Current density	Capacitance	Cycling stability	References
3D Nb_2O_5/holey rGO	LIBs	–	139 mA h g^{-1}	125 mA h g^{-1} after 10,000 cycles at 10C	Sun et al. (2017)
3D CoP/rGO aerogel	LIBs	10 A g^{-1}	351.8 mA h g^{-1} after 4,000 cycles	805.3 mA h g^{-1} after 200 cycles at 200 mA g^{-1}	Gao et al. (2019)
Porous rGO	LSBs	–	127.1–510.3 mA h g^{-1}	70.9% over 300 cycles	Yeon et al. (2019)
Porous rGO/graphene crumples	LSBs	–	713 mA h g^{-1}	524 mA h g^{-1} after 100 cycles at 0.2 C	Chong et al. (2018)
3D porous N-doped rGO	LSBs	–	1,311 mA h g^{-1}	714 mA h g^{-1} at 1.5 mA cm^2 after 400 cycles	Cheng et al. (2019)
3D polypyrrole/porous rGO	LSBs	–	1,020 mA h g^{-1}	802 mA h g^{-1} over 200 cycles at 0.1 C	Zhang et al. (2018a)
Fe_3O_4/porous graphene	LSBs	–	1,423–589 mA h g^{-1}	732 mA h g^{-1} after 500 cycles at 1 C	Liu et al. (2018c)
3D porous graphene grids	SIBs	50 mA h g^{-1}	160 mA h g^{-1}	112 mA h g^{-1} capacity during 1,000 cycles at 1 A h g^{-1}	Zhang et al. (2019)
3D hierarchical porous rGO	SIBs	100 mA g^{-1}	369.4 mA h g^{-1}	253 mA h g^{-1} at of 1 A g^{-1} after 500 cycles	Wang et al. (2019a)
Holey rGO	SIBs	0.1–10 A g^{-1}	365–131 mA h g^{-1}	163 mA h g^{-1} after 3,000 cycles at 2 A g^{-1}	Zhao et al. (2018)
MoS_2-rGO sponges	SIBs	1 A g^{-1}	208.6 mA h g^{-1}	372.0 mA h g^{-1} at 100 mA g^{-1} after 50 cycles	Li et al. (2018)

(*Continued*)

TABLE 7.1 (*Continued*)
Wide Range of Graphene-Based Materials for Various Energy Storage Applications

Materials	Application	Current density	Capacitance	Cycling stability	References
3D porous Ni foam/rGO/SnS_2	SIBs	50 mA g^{-1}	1503.7 mA h g^{-1}	561.9 mA h g^{-1} after 160 cycles at 1 A g^{-1}	Ye et al. (2018)
N/P dual-doped rGO aerogel	SIBs	50 mA g^{-1}	330 mA h g^{-1}	195 mA h g^{-1} at 1 A g^{-1} after 1,000 cycles	Li et al. (2018a)
3D N-doped holey rGO/transition metal oxide	SIBs	1 A g^{-1}	403 mA h g^{-1}	510 mA h g^{-1} at the 100th cycle at 0.1 A g^{-1}	Yang et al. (2019)
Et al. (2018) Fe^{3+}/V_2O_5@graphene quantum dots	PCs	2 A g^{-1}	761 F g^{-1}	95% after 10 000 cycles	Shi et al. (2019)
N-doped graphene/phosphorus composite	LIBs	130 mA g^{-1}	2522.6 mA h g^{-1}	470.1 mA h g^{-1} at 1,300 mA g^{-1} for 300 cycles	Jiao et al. (2019)
Sulfiphilic few-layered $MoSe_2$/rGO	LSBs	–	1,608 mA h g^{-1}	870 mA h g^{-1} after 120 cycles at 0.2 C	Tian et al. (2019)

Reproduced with permission from Zhang et al. (2019a).

Notes: EDLCs, electrochemical double-layer capacitors; PCs, pseudocapacitors; LIBs, lithium-ion batteries; SIBs, sulfur-ion batteries.

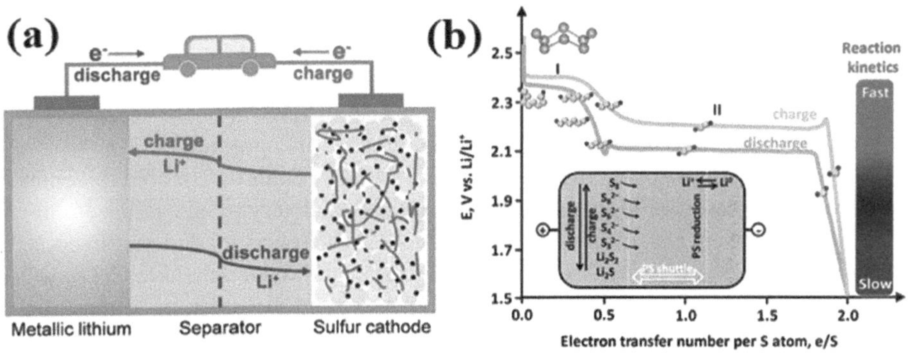

FIGURE 7.3 (a) Pictorial illustration of a lithium-sulfur battery. Reproduced with permission from Zheng et al. (2019). (b) Sulfur charge-discharge profile and insert show polysulfide (PS) shuttle. Reproduced with permission from Wang et al. (2013).

polysulfides, and later on, the chain-length polysulfide is reduced and gets converted into lithium sulfides. This is the reason behind two or three voltage plateau regions in the discharging graph of Li-S battery, which are related to various reduction stages of sulfur based on the amount of carbon present on the sulfur electrode and on the discharge rate (Figure 7.3b). Low toxicity and lower cost of sulfur are the positive aspects of Li-S battery, whereas low conductivity of sulfur, dissolution of polysulfide anion during the charge-discharge process, and stability of the electrode are the negative aspects. So, to reduce these negative aspects of sulfur, graphene, and carbon nanotubes, other nanomaterials have been used to improve the performance of Li-S batteries. Zhang et al. (2012) have prepared nano-sulfur onto the surface of graphene oxide in a microemulsion by adopting a chemical method, and as a result, they were able to achieve a reversible capacitance of 954 mA h g^{-1} and a high coulombic efficiency of approximately 96.7% above 50 cycles with insignificant capacity loss. Wang et al. (2011a) have also reported the synthesis of sulfur nanoparticles. In this synthesis, firstly, they have coated the sulfur nanoparticles with poly(ethylene glycol), and secondly, these nanoparticles were enfolded with carbon black, which was decorated with graphene oxide sheets, and by using this nanocomposite, they have achieved capacity up to 600 mA h g^{-1} over 100 cycles.

7.6 CARBON NANOTUBES AND ITS NANOCOMPOSITES FOR ANODE FABRICATION IN LITHIUM-ION BATTERIES

Carbon nanotubes (CNTs) are considered as one-dimensional nanomaterials. They have better thermal and electrical conductivity as well as superior mechanical properties. CNTs have strong intercalation properties for lithium, which makes them auspicious anode material for application in lithium-ion batteries. Both SWCNTs and MWCNTs have been used as anode materials in lithium-ion batteries. In the case of SWCNTs, the lithium-ion accumulation capacity of metallic CNTs is almost five times higher than that of semiconducting CNTs (Kawasaki et al., 2008).

MWCNTs have multiple rolled layers of graphene sheets, which are suitable for the insertion of Li ions in a way similar to graphite, making them an auspicious candidate as an anode material for LIBs. In MWCNTs, not only the spacing of graphitic sheets between two layers but also the hollow pores of MWCNTs are accessible for lithium-ion intercalation/deintercalation. Kang et al. (2012) prepared a 3D anode of MWCNTS for LIBs by using higher loading of MWCNTs and were able to achieve better concert in terms of higher specific capacitance and very good cycling stability as compared to their two-dimensional (2D) anode, which was loaded on 2D copper foil. As compared to pristine CNTs, the anodes prepared from defective CNTs were found to be more operative for the easier adsorption and diffusion of lithium ions, and the holes on the sidewalls of CNTs are considered as defects, which are quite helpful for the intercalation and diffusion of lithium ions. Ren et al. (1998) have used the short-length and long-length CNTs in LIBs for anode fabrication and performed the comparative study and found that short and long CNTs show the reversible capacity of 266 and 170 mA h g^{-1} at a current density of 0.2 and 0.8 mA cm^{-2}, respectively. The capacity of short CNTs was found to be almost two times higher as compared to long CNTs, and the most probable reason behind this could be easier intercalation and deintercalation of lithium ions in short CNTs. As we know, the lithium ions inserted into CNTs undergo a one-dimensional random movement inside the CNT, and if the tube is too long, in that case, the effective diffusion of lithium ions will be decreased, which will be responsible for the lower capacity. Thus, we can say that while making anodes of CNTs for LIBs, the types of CNTs, defects, and lengths of the CNTs play a very important role.

7.7 CARBON NANOTUBE-DERIVED MATERIALS FOR CATHODE FABRICATION IN LITHIUM-ION BATTERIES

Due to their advantageous properties such as high electrical conductivity, better mechanical properties, and superior intercalation properties, carbon nanotube-derived materials have also been used in cathode fabrication for lithium-ion batteries. Le et al. (2015) have prepared $LiMn_2O_4$/CNT and $LiNi_{0.5}Mn_{1.5}O_4$/CNT nanocomposites and used the same for the construction of cathode material for LIBs, and the specific capacity was found to be 145 and 120 A h kg^{-1} for $LiMn_2O_4$/CNT (10 wt%) and $LiNi_{0.5}Mn_{1.5}O_4$/CNT (10 wt%), respectively, which was measured at a charging/discharging rate of 0.1 C. Jiang et al. (2006) have also reported the preparation of $LiMn_2O_4$/CNT nanocomposite by adopting hydrothermal synthesis route and using the same as flexible and binder-free cathode in LIBs. They have found that these stretchy cathode materials exhibit a high capacity and a very stable cycling concert, which suggests the potential application of this flexible cathode in lithium-ion batteries. Wu et al. (2013) have prepared nano-$LiFePO_4$ and furthered its nanocomposites with carbon nanotube and amorphous carbon denoted as LFP@C/CNT, where LFP, C, and CNT stand for $LiFePO_4$, amorphous carbon, and CNTs, respectively. It was found that as compared to original LFP, single-carbon ornamented LFP (LFP@C), and carbon nanotube-decorated LFP (LFP/CNT), the carbon and carbon nanotube-decorated LFP (LFP@C/CNT) shows the best performance in terms of high-rate capability (~59% capacity retention), the best capacity performance during cycling (98.5% of original capacity after 500 cycles), and outstanding negative-temperature performance (about 71.4% of original capacity, when discharged at –25°C). Zhu et al. (2013) have prepared the $CuCrO_2$-CNT nanocomposites by adopting the hydrothermal synthesis route. They have used these nanocomposites as cathodes in LIBs and were able to achieve outstanding reversible specific capacity and cyclic concert; at the same time, very good capacity retention (530 mA h g^{-1} after 40 cycles) was achieved at a higher charging/discharging rate (1 C).

7.8 CARBON NANOTUBE-DERIVED MATERIALS FOR SUPERCAPACITORS

Supercapacitors are mainly known for their very high-power density, good reversibility, and longer cycle life. They have been designed to deliver power pulses for numerous applications. As per the current report given by the United States Department of Energy, supercapacitors and batteries have been given equal attention. The international market for supercapacitors has been rising progressively and quickly. To advance the concert of presently used supercapacitors, research work is still going on for new electrode materials with superior supercapacitive properties. To fulfill the requirements of superior capacitive performance, high-surface-area carbon nanotubes are still the major electrode materials for commercial supercapacitors. Sakamoto et al. (2002) have prepared V_2O_5 aerogel and SWCNT nanocomposite and used the same composite for the supercapacitive application, and this electrode material shows a specific capacity of 452 mA h g^{-1} and still retains up to 65% of initial capacity when the discharging rate is fastened from 112 mA g^{-1} (0.2 C) to 2,800 mA g^{-1} (5 C). Kim et al. (2006) have also prepared V_2O_5/CNT nanocomposite by electrochemically depositing V_2O_5 thin film onto the surface of CNT. They have used V_2O_5/CNT, a binder-free electrode, for the supercapacitive application. This electrode shows very high electrical conductivity and high capacity (680 mA h g^{-1} when applying a current density of 5 A g^{-1}). In this case, the improved performance could be due to the increased loading of V_2O_5 in V_2O_5/CNT film and the double-layer capacitance of V_2O_5/CNT film.

Fan et al. (2006) have prepared MnO_2/CNT nanocomposite by dispersing MnO_2 in CNT matrix and utilizing the same material for electrode fabrication, and this nanocomposite material shows a very large specific capacitance of 568 F g^{-1} when applying a charge/discharge current density of 1 mA cm^{-2}. Zhu et al. (2014) have prepared nitrogen-doped carbon nanotube and polyaniline composite. The electrode was fabricated by using this composite material, which shows a very high specific capacitance of 365.9 F g^{-1} at a current density of 0.1 A g^{-1} and excellent stability over 1,000 cycles. De Oliveira et al. (2016) have prepared CNT@MnO_2@polypyrrole composites and utilized this material for the electrode fabrication, and the fabricated electrode showed the capacitance of 272.7 F g^{-1} with reasonable cycling performance. Apart from the above discussion, research is still going on to develop other nanocomposites of carbon nanotubes that can significantly improve the capacity with increased cycling stability.

Jia et al. (2018) reported a novel mesostructured carbon nanotube and MnO_2 (CNT/MnO_2) nanocomposite, and they found that the electrode fabricated by this nanocomposite showed a very high specific capacitance of 1,229 F g^{-1} and very long-term cycling stability and retained 94.4% capacitance after 100,000 cycles.

7.9 CARBON NANOTUBE-DERIVED MATERIALS FOR LITHIUM-SULFUR BATTERIES

As we know, due to their very large theoretical capacity (1,672 mA h g^{-1}) and very high theoretical energy (2,600 W h kg^{-1}), lithium-sulfur batteries have potential applications in various fields such as advanced rechargeable batteries for electric vehicles and electronic devices. As we have previously discussed, low electrical conductivity and discharge product of sulfur, large sulfur volume expansion during discharge, and polysulfide shuttling effect restrict the commercialization of lithium-sulfur batteries, and to solve these problems, researchers have used graphene and graphene-derived composites. Here, in this section, we are mainly focused on carbon nanotubes and their composites with other nanomaterials. We will discuss how can we improve the performance of lithium-sulfur batteries. A pictorial presentation of carbon nanotube utilization in lithium-sulfur batteries is shown in Figure 7.4. Xiao et al. synthesized N-doped carbon nanotubes via a very simple and catalyst-less method, in which they have used CH_5N_3S as the source of nitrogen, and the concert of the cathode prepared from nitrogen-doped CNTs was found to be superior to that of the cathode prepared from ordinary CNTs. Figure 7.4 shows the application of CNTs in lithium-sulfur batteries (Xiao et al., 2015).

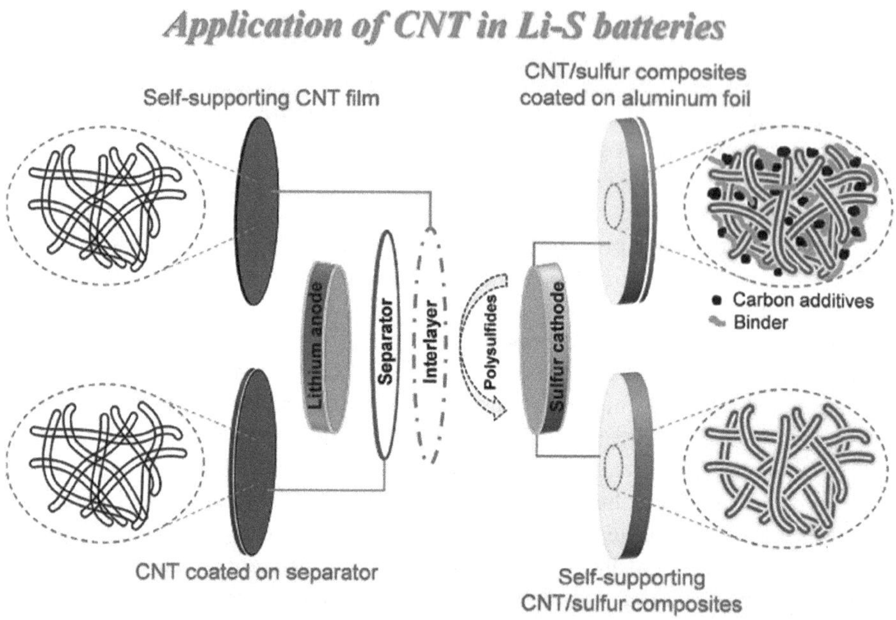

FIGURE 7.4 The various application modes of carbon nanotube-derived materials in lithium-sulfur batteries. Reproduced with permission from Zheng et al. (2019).

Zhang et al. (2017) reported a three-dimensional GN-CNT matrix that was prepared by a simple one-step pyrolysis process and used the same matrix as the host for sulfur. They have prepared the cathode for lithium-sulfur batteries by using this sulfur-doped GN-CNT material (S@GN-CNT) and found that this cathode material shows the reversible capacity of 758.4 mA h g^{-1} by applying 0.5 C. Zhao et al. (2015) prepared the sulfur/nitrogen-doped CNT and used the same material for cathode fabrication in lithium-sulfur batteries and achieved the specific capacity of 1,267 mA h g^{-1}. Guo et al. (2011) have prepared disordered CNTs by impregnating sulfur into CNTs; furthermore, they have utilized this material as cathode in lithium-sulfur batteries for better capacity and superior cycling performance. Various types of CNT-based materials utilized in lithium-sulfur batteries have been tabulated in Table 7.2.

TABLE 7.2
Various Types of CNT-Based Material Used in Lithium-Sulfur Batteries and Their Electrochemical Performance

CNT-based materials	Current density	Cycle number	Initial capacity (mA h g^{-1})	Capacity after cycling (mA h g^{-1})	References
MWCNT	C/5	150	1,324	881	Chung et al. (2014)
DWCNT	0.1 C	100	598	508	Sun et al. (2020)
o-MWCNT	0.1 A g^{-1}	50	1,105	925	Cheng et al. (2016)
PAN/SiO_2-MWCNT	0.2 C	100	1,182	741	Zhu et al. (2016)
MWCNT/GO/MWCNT	0.2 C	200	1,290	620	Chang et al. (2017)
CNT	1 C	500	684	361	Liu et al. (2017)
aCNT	0.5 C	500	1,306	621	Huang et al. (2018)
CNTOH	0.5 C	400	1,056	591	Ponraj et al. (2017)
OCNT	0.5 C	400	1034.9	732	Kim et al. (2018)
CNT	1 C	200	968	670	Manoj et al. (2018)
AB/MWCNT	1.6 A g^{-1}	200	1,500	662	Tian et al. (2018)
CNT/PVP	0.1 C	100	1,226	887	Li et al. (2019c)
Ni@NG-CNTs	1 C	800	1144.2	655.5	Zuo et al. (2019)
Co-NCNTs	0.2 C	200	1251.4	857.7	Wei et al. (2020)
MoS_2@CNT	1 C	500	690	670	Jeong et al. (2017)
Sb_2S_3/CNT	1 C	1,000	720	373	Yao et al. (2018)
SnS_2/CNT	2 C	800	782	555	Jiang et al. (2019)
NSCNTs/MoS_2	1 C	1,000	1,024	814	Xiang et al. (2019)
SWCNT	0.2 C	100	953	713	Chang et al. (2016)
PANiNF/MWCNT	0.2 C	100	1,020	709	Chang et al. (2015)
MCNT@PEG	0.5 C	200	1,283	727	Wang et al. (2015)
MWCNT/PEG	0.2 C	300	1,206	630	Luo et al. (2016)
PEI/MWCNT)	2 C	300	819	590	Lee et al. (2018)
PEDOT: PSS-CNT	0.2 C		950		Manoj et al. (2018a)
PAMAM-CNTs	2 C	1,200	1,050	573	Li et al. (2020c)

(Continued)

TABLE 7.2 (*Continued*)
Various Types of CNT-Based Material Used in Lithium-Sulfur Batteries and Their Electrochemical Performance

CNT-based materials	Current density	Cycle number	Initial capacity ($mA\ h\ g^{-1}$)	Capacity after cycling ($mA\ h\ g^{-1}$)	References
MWCNT/SPANI	0.1 A g^{-1}	100	1,127	913	Shi et al. (2018)
CNT/AC	0.2C	200	1495.6	742	Guo et al. (2017)
G@CNT	0.2C	200	1231.3	935.1	Wu et al. (2017)
CNT/CH	0.5C	200	894	826	Jiang et al. (2018)
MWCNT/NCQD	0.5C	1000	1330.8	507.9	Pang et al. (2018)
PC/MWCNT	0.5C	200	911	659	Tan et al. (2018)
G/CNT	0.5C	100	913.5	813	Gao et al. (2019a)
GO/CNT	0.2C	50	1591.56	1003	Lee et al. (2019)
Co/NCNS/CNT	2C	1,000	972.4	486	Song et al. (2020)
CNT/Al_2O_3	0.2C	100	1,282	807.8	Xu et al. (2015)
Al_2O_3/CNT	0.2C	100	1,096	760.4	Chen et al. (2020)
Fe_3O_4@C/CNTO	2C	1,000	833	335	Du et al. (2020)
CNTs@FeOOH	3.2 A g^{-1}	350	1121.9	556	Li et al. (2020e)
$MgBO_2(OH)$/CNT	0.5C	200	924	785	Kong et al. (2017)
CNT@TiO_2	0.1C	200	1,351	803	Yang et al. (2017)
TiO_2/CNTs	1C	250	1036.9	763	Guan et al. (2019)
MWCNTs @TiO_2	838 mA g^{-1}	600	1,083	610	Ding et al. (2018)
Ni-MOF/MWCNTs	0.5C	300	1,358	1183	Lee et al. (2018)
CNT/ZrO_2	0.1C	120	1,207	685	Liu et al. (2018)
MoO_3@CNT	0.3C	200	1,251	755	Luo et al. (2018)
CNT/MoP_2	0.2C	100	1,223	905	Luo et al. (2018)
CNT@ZIF	0.2C	100	1588.4	870.3	Wu et al. (2018)
COF-CNT	1 A g^{-1}	500	1,068	621	Wang et al. (2020)
MoS_2/CNT	0.5C	500	1,237	648	Yan et al. (2018)
MWCNTs/MnO_2	1C	500	880	610	Yao et al. (2018)

Reproduced with permission from Wei et al. (2020).

Notes: o-MWCNT, oxidized multiwalled carbon nanotube; PAN/SiO_2-MWCNT, poly-acrylonitrile/silica-MWCNT; aCNT, activated carbon nanotube; AB/MWCNT, acetylene black/MWCNT; SCL, sandwich-structure composite carbon layer (a N-doped mesoporous carbon sandwiched between two carbon nanotubes); NSCNTs/MoS_2, nitrogen and sulfur co-doped carbon nanotubes/MoS_2; SWCNT, single-walled carbon nanotube; MWCNT/PEG, MWCNT/polyethylene glycol; MWCNT/SPANI, MWCNT/sulfonated polyaniline; SMAP, nano-SnO_2/MWCNTs/aramid paper; PDAAQ/K-FGF/MWCNT/CTAB, poly 1,5-diaminoanthraquinone/potassium-functionalized graphene/MWCNT/cetyltrimethylammonium bromide; CNT/AC, CNT/active carbon; PAMAM-CNTs, poly(amidoamine)-modified multiwalled carbon nanotubes; COF-CNT, covalent organic framework-CNT.

7.10 CONCLUSION

In this chapter, we have covered the various types of graphene- and carbon nanotube-based materials and their recent progress for various energy storage systems, i.e., supercapacitors, lithium-sulfur batteries, and lithium-ion batteries, and we are mainly focused on the carbon nanotube- and graphene-based materials that have been used for the development of efficient, stable, durable, and high-energy storage materials. The potential use of CNT- and graphene-based materials has been covered for the anode and cathode fabrication of lithium-ion batteries. We have discussed the improvements that have been made to improve the performance of lithium-ion and lithium-sulfur batteries by using these two carbon-based materials in terms of reversible capacity and cycling stability. We have also covered the potential use of graphene- and CNT-based materials for supercapacitive applications.

ACKNOWLEDGMENTS

UPA acknowledges SERB-DST and UGC New Delhi for providing financial assistance in the form of National Post-Doctoral Fellowship (File Number: PDF/2017/002942) and start-up grant (No. F.30-551/2021 (BSR)), respectively.

REFERENCES

Ambrosi, A., Chua, C. K., Bonanni, A., Pumera, M. (2014) Electrochemistry of graphene and related materials. *Chem. Rev.* 114, 14, 7150–7188.

Bhardwaj, T., Antic, A., Pavan, B., Barone, V., Fahlman, B. D. (2010) Enhanced electrochemical lithium storage by graphene nanoribbons. *J. Am. Chem. Soc.* 132, 12556–12558.

Bora, A., Mohan, K., Doley, S., Dolui, S. K. (2018) Flexible asymmetric supercapacitor based on functionalized reduced graphene oxide aerogels with wide working potential window. *ACS Appl. Mater. Interfaces* 10, 7996–8009.

Broto, V. C., Kirshner, J. (2020) Energy access is needed to maintain health during pandemics. *Nano Energy* 5, 419–421.

Chang, C.-H., Chung, S.-H., and Manthiram, A. (2015) Ultra-lightweight PANiNF/MWCNT-functionalized separators with synergistic suppression of polysulfide migration for Li-S batteries with pure sulfur cathodes. *J. Mater. Chem. A* 3, 18829–18834.

Chang, C.-H., Chung, S.-H., Manthiram, A. (2016) Effective Stabilization of a High-Loading Sulfur Cathode and a Lithium-Metal Anode in Li-S Batteries Utilizing SWCNT-Modulated Separators. *Small* 12, 174–179.

Chang, C.-H., Chung, S.-H., Nanda, S., Manthiram, A. (2017) A rationally designed polysulfide-trapping interface on the polymeric separator for high-energy Li-S batteries. *Mater. Today Energy* 6, 72–78.

Chen, S., Zhu, J., Wang, X. (2010) One-step synthesis of graphene-cobalt hydroxide nanocomposites and their electrochemical properties. *J. Phys. Chem. C* 114, 11829–11834.

Chen, X., Huang, Y., Li, J., Wang, X., Zhang, Y., Guo, Y. et al. (2020) Bifunctional separator with sandwich structure for high-performance lithium-sulfur batteries. *J. Colloid Interface Sci.* 559, 13–20.

Cheng, D., Wu, P., Wang, J., Tang, X., An, T., Zhou, H., Zhang, D., Fan, T. (2019) Synergetic pore structure optimization and nitrogen doping of 3D porous graphene for high performance lithium sulfur battery. *Carbon* 143, 869–877.

Cheng, X., Wang, W., Wang, A., Yuan, K., Jin, Z., Yang, Y., et al. (2016) Oxidized multiwall carbon nanotube modified separator for high performance lithium-sulfur batteries with high sulfur loading. *RSC Adv.* 6, 89972–89978.

Choi, W., Azad, U. P., Choi, J.-P., Lee, D. (2018) Electrocatalytic oxygen reduction by dopant-free, porous graphene aerogel. *Electroanalysis* 30, 1472–1478.

Chong, W. G., Xiao, Y., Huang, J., Yao, S., Cui, J., Qin, L., Gao, C., Kim, J. K. (2018) Highly conductive porous graphene/sulfur composite ribbon electrodes for flexible lithium-sulfur batteries. *Nanoscale* 10, 21132.

Chung, S.-H., and Manthiram, A. (2014) High-performance Li-S batteries with an ultra-lightweight MWCNT-Coated separator. *J. Phys. Chemi. Lett.* 5, 1978–1983.

de Oliveira, A. H. P., Nascimento, M. L. F., de Oliveira, H. P. (2016) Carbon nanotube@ MnO_2@polypyrrole composites: Chemical synthesis, characterization and application in supercapacitors. *Mater. Res.* 19, 1080–1087.

Ding, H., Zhang, Q., Liu, Z., Wang, J., Ma, R., Fan, L., et al. (2018) TiO_2 quantum dots decorated multi-walled carbon nanotubes as the multifunctional separator for highly stable lithium sulfur batteries. *Electrochim. Acta* 284, 314–320.

Dong, X.-C., Xu, H., Wang, X.-W., Huang, Y.-X., Chan-Park, M. B., Zhang, H., Wang, L-H., Huang, W., Chen, P. (2012) 3D Graphene-cobalt oxide electrode for high-performance supercapacitor and enzymeless glucose detection. *ACS Nano* 6, 3206–3213.

Dreyer, D. R., Bielawski, C. W. (2012) Graphite oxide as an olefin polymerization carbocatalyst: Applications in electrochemical double layer capacitors. *Adv. Funct. Mater.* 22, 3247–3253.

Du, J., Ahmed, W., Xu, J., Zhang, M., Zhang, Z., Zhang, X., et al. (2020) Chainmail Catalyst of Fe_3O_4@C/CNTO-modified celgard separator with low metal loading for high-performance lithium-sulfur batteries. *Chemistryselect* 5, 3757–3762.

Fan, Z., Chen, J., Wang, M., Cui, K., Zhou, H., Kuang, Y. (2006) Preparation and characterization of manganese oxide/CNT composites as supercapacitive materials. *Diamond Relat. Mater.* 15, 1478–1483.

Feng, H., Xie, P., Xue, S., Li, L., Hou, X., Liu, Z., Wu, D., Wang, L., Chu, P. K. (2018) Synthesis of three-dimensional porous reduced graphene oxide hydrogel/carbon dots for high-performance supercapacitor. *J. Electroanal. Chem. 808*, 321–328.

Gao, F., Yan, X., Wei, Z., Qu, M., and Fan, W. (2019) Graphene/carbon nanotubes composite as a polysulfide trap for lithium-sulfur batteries. *Int. J. Electrochem. Sci.* 14, 3301–3314.

Gao, H., Yang, F., Zheng, Y., Zhang, Q., Hao, J., Zhang, S., Zheng, H., Chen, J., Liu, H., Guo, Z. (2019) Three-dimensional porous cobalt phosphide nanocubes encapsulated in a graphene aerogel as an advanced anode with high coulombic efficiency for high-energy lithium-ion batteries. *ACS Appl. Mater. Interfaces* 11, 5373–5379.

Guan, Y., Li, W., Xie, X., Qu, W., Shen, J., Fu, K. et al. (2019) Preparation of TiO2/CNTs composite coated separator and its application in Li-S battery. *Chem. J. Chin. Univ. Chin.* 40, 536–541.

Guo, J., Xu, Y., Wang, C. (2011) Sulfur-impregnated disordered carbon nanotubes cathode for lithium-sulfur batteries. *Nano Lett.* 11, 4288–4294.

Guo, Y., Xiao, J., Hou, Y., Wang, Z., Jiang, A. (2017) Carbon nanotube doped active carbon coated separator for enhanced electrochemical performance of lithium-sulfur batteries. *J. Mater. Sci. Mater. Electron.* 28, 17453–17460.

Huang, J.-Q., Chong, W. G., Zheng, Q., Xu, Z.-L., Cui, J., Yao, S. et al. (2018) Understanding the roles of activated porous carbon nanotubes as sulfur support and separator coating for lithium-sulfur batteries. *Electrochim. Acta* 268, 1–9.

Jeong, H. M., Lee, J. W., Shin, W. H., Choi, Y. J., Shin, H. J., Kang, J. K., Choi, J. W. (2011) Nitrogen-doped graphene for high-performance ultracapacitors and the importance of nitrogen-doped sites at basal planes. *Nano Lett.* 11, 2472–2477.

Jeong, Y. C., Kim, J. H., Kwon, S. H., Oh, J. Y., Park, J., Jung, Y. et al. (2017) Rational design of exfoliated 1T MoS_2@CNT-based bifunctional separators for lithium sulfur batteries. *J. Mater. Chem. A* 5, 23909–23918.

Jia, H., Cai, Y., Zheng, X., Lin, J., Liang, H., Qi, J., Cao, J., Feng, J., Fei, W. (2018) Mesostructured carbon nanotube-on-MnO_2 nanosheet composite for high-performance supercapacitors. *ACS. Appl.Mater. Inter.* 10, 38963–38969.

Jiang, C., Hosono, E., Zhou, H. (2006) Nanomaterials for lithium-ion batteries. *Nano Today* 1, 28–33.https://pubs.rsc.org/en/content/articlelanding/2012/jm/c2jm16543k

Jiang, J., Liu, J., Ding, R., Zhu, J., Li, Y., Hu, A., Li, X., Huang, X. (2011) Large-scale uniform α-$Co(OH)_2$ long nanowire arrays grown on graphite as pseudocapacitor electrodes. *ACS Appl. Mater. Interfaces* 3, 99–103.

Jiang, S., Chen, M., Wang, X., Wu, Z., Zeng, P., Huang, C. et al. (2018) MoS_2-Coated N-doped Mesoporous Carbon Spherical Composite Cathode and CNT/Chitosan modified separator for advanced lithium sulfur batteries. *ACS Sustain. Chem. Eng*. 6:16828.

Jiang, S., Chen, M., Wang, X., Zeng, P., Li, Y., Liu, H. et al. (2019) A tin disulfide nanosheet wrapped with interconnected carbon nanotube networks for application of lithium sulfur batteries. *Electrochim. Acta* 313, 151–160.

Jiao, X., Liu, Y., Li, T., Zhang, C., Xu, X., Kapitanova, O. O., He, C., Li, B., Xiong, S., Song, J. (2019) Crumpled nitrogen-doped graphene-wrapped phosphorus composite as a promising anode for lithium-ion batteries. *ACS Appl. Mater. Interfaces* 11, 30858–30864.

Kang, C., Lahiri, I., Baskaran, R., Kim, W.-G., Sun, Y.-K., Choi, W. (2012) 3-dimensional carbon nanotube for Li-ion battery anode. *J. Power Sour.* 219, 364–370.

Kawasaki, S., Hara, T., Iwai, Y., Suzuki, Y. (2008) Metallic and semiconducting single-walled carbon nanotubes as the anode material of Li ion secondary battery. *Mater. Lett.* 62, 2917–2920.

Kim, I-H., Kim, J-H., Cho, B.-W., Kim, K.-B. (2006) Synthesis and electrochemical characterization of vanadium oxide on carbon nanotube film substrate for pseudocapacitor applications. *J. Electrochem. Soc.* 153, A989–A996.

Kim, P. J. H., Kim, K., Pol, V. G. (2018). Towards highly stable lithium sulfur batteries: Surface functionalization of carbon nanotube scaffolds. *Carbon* 131, 175–183.

Kong, L., Peng, H.-J., Huang, J.-Q., Zhu, W., Zhang, G., Zhang, Z.-W. et al. (2017) Beaver-dam-like membrane: A robust and sulphifilic MgBO2(OH)/CNT/PP nest separator in Li-S batteries. *Energy Storage Mater*. 8, 153–160.

Le, T. V., Le, M. L. P., Tran, M. V., Nguyen, N. M. T., Luu, A. T., Nguyen, H. T. (2015) Fabrication of cathode materials based on $LiMn_2O_4$/CNT and $LiNi_{0.5}Mn_{1.5}O_4$/CNT nanocomposites for lithium -Ion batteries application. *Materials Research*. 18, 1044–1052.

Lee, D. H., Ahn, J. H., Park, M.-S., Eftekhari, A., Kim, D.-W. (2018) Metal-organic framework/carbon nanotube-coated polyethylene separator for improving the cycling performance of lithium-sulfur cells. *Electrochim. Acta* 283, 12911299.

Lee, J. W., Lim, S. Y., Jeong, H. M., Hwang, T. H., Kang, J. K., Choi J. W. (2012) Extremely stable cycling of ultra-thin V_2O_5 nanowire-graphene electrodes for lithium rechargeable battery cathodes. *Energy Environ. Sci.* 5, 9889–9894.

Lee, Y.-H., Kim, J.-H., Kim, J.-H., Yoo, J.-T., Lee, S.-Y. (2018) Spiderweb-mimicking anion-exchanging separators for Li-S batteries. *Adv. Funct. Mater*. 28:1870293.

Li, C., Q. Fu, K. Zhao, Wang, Y., Tang, H., Li, H., Jiang, H., Chen, L. (2018a) Nitrogen and phosphorous dual-doped graphene aerogel with rapid capacitive response for sodium-ion batteries. *Carbon* 132, 1117–1125.

Li, J., Qin, W., Xie, J., Lin, R., Wang, Z., Pan, L., Mai, W. (2018) Rational design of MoS_2-reduced graphene oxide sponges as free-standing anodes for sodium-ion batteries. *Chem. Eng. J.* 332, 260–266.

Li, N., Ma, X., Ye, H., Wang, S., Han, K. (2019c) Carbon nanotube-modified separator for lithium-sulfur batteries: Effects of mass loading and adding polyvinylpyrrolidone on electrochemical performance. *J. Phys. Chem. Solids* 134, 69–76.

Li, R., Yang, Y., Wu, D., Li, K., Qin, Y., Tao, Y., Kong, Y. (2019) Covalent functionalization of reduced graphene oxide aerogels with polyaniline for high performance supercapacitors. *Chem. Commun.* 55, 1738-1741.

Li, S., Zhang, H., Chen, W., Zou, Y., Yang, H., Yang, J. et al. (2020c) Toward commercially viable Li-S batteries: Overall performance improvements enabled by a multipurpose interlayer of hyperbranched polymer-grafted carbon nanotubes. *ACS Appl. Mater. Interfaces* 12, 25767–25774.

Li, Y., Li, X., Hao, Y., Kakimov, A., Li, D., Sun, Q. et al. (2020e) b-FeOOH interlayer with abundant oxygen vacancy toward boosting catalytic effect for lithium sulfur batteries. *Front. Chem.* 8:309.

Liu, B., Wang, S., Wu, X., Liu, Z., Gao, Z., Li, C. et al. (2018) Carbon nanotube/zirconia composite-coated separator for a high-performance rechargeable lithium-sulfur battery. *AIP Adv.* 8:105315.

Liu, B., Wu, X., Wang, S., Tang, Z., Yang, Q., Hu, G.-H. et al. (2017) Flexible carbon nanotube modified separator for high-performance lithium-sulfur batteries. *Nanomaterials* 7:196.

Liu, B., Zhao, M., Han, L., Lang, X., Wen, Z., Jiang, Q. (2018a) Three-dimensional nanoporous N-doped graphene/iron oxides as anode materials for high-density energy storage in asymmetric supercapacitors. *Chem. Eng. J. 335*, 467-474.

Liu, D., Li, Q., Li, S., Hou, J., Zhao, H. (2019) A confinement strategy to prepare N-doped reduced graphene oxide foams with desired monolithic structures for supercapacitors. *Nanoscale* 11, 4362–4368.

Liu, D., Li, Q., Zhao, H. (2018) Electrolyte-assisted hydrothermal synthesis of holey graphene films for all-solid-state supercapacitors. *J. Mater. Chem. A* 6, 11471–11478.

Liu, H., Tang, Y., Zhao, W., Ding, W., Xu, J., Liang, C., Zhang, Z., Lin, T., Huang, F. (2018b) Facile Synthesis of Nitrogen and Halogen Dual-Doped Porous Graphene as an Advanced Performance Anode for Lithium-Ion Batteries. *Adv. Mater. Interfaces* 5, 1701261.

Liu, Y., Qin, X., Zhang, S., Liang, G., Kang, F., Chen, G., Li, B. (2018a) Fe_3O_4-Decorated porous graphene interlayer for high-performance lithium-sulfur batteries. *ACS Appl. Mater. Interfaces* 10, 26264–26273.

Luo, L., Chung, S.-H., Manthiram, A. (2016) A trifunctional multi-walled carbon nanotubes/polyethylene glycol (MWCNT/PEG)-coated separator through a layer-by-layer coating strategy for high-energy Li-S batteries. *J. Mater. Chem. A* 4, 16805–16811.

Luo, L., Qin, X., Wu, J., Liang, G., Li, Q., Liu, M. et al. (2018) An interwoven MoO_3@CNT scaffold interlayer for high-performance lithium-sulfur batteries. *J. Mater. Chem. A* 6, 8612–8619.

Luo, Y., Luo, N., Kong, W., Wu, H., Wang, K., Fan, S. et al. (2018) Multifunctional interlayer based on molybdenum diphosphide catalyst and carbon nanotube film for lithium-sulfur batteries. *Small* 14:1702853.

Manoj, M., Jasna, M., Jayalekshmi, S. (2018a) "Activated carbon-sulfur composite with PEDOT: PSS-CNT Interlayer as Cathode Material for Lithium-Sulfur Batteries," In: *Proceedings of the PIE 10725 Low-Dimensional Materials and Devices*, Vol. 1075, eds N. P. Kobayashi, A. A. Talin, M. S. Islam, A. V. Davydov (San Diego, CA: SPIE).

Manoj, M., Jasna, M., Anil Kumar, K. M., Abhilash, A., Jinisha, B., Pradeep, V. S. et al. (2018) Sulfur-polyaniline coated mesoporous carbon composite in combination with carbon nanotubes interlayer as a superior cathode assembly for high-capacity lithium-sulfur cells. *Appl. Surf. Sci.* 458, 751–761.

Mo, R., Li, F., Tan, X., Xu, P., Tao, R., Shen, G., Lu, X., Liu, F., Shen, L., Xu, B., Xiao, Q., Wang, X., Wang, C., Li, J., Wang, G., Lu, Y. (2019) High-quality mesoporous graphene particles as high-energy and fast-charging anodes for lithium-ion batteries. *Nat. Commun.* 10, 1474.

Pan, D. Y., Wang, S., Zhao, B., Wu, M. H., Zhang, H. J., Wang, Y., Jiao, Z. (2009) Li storage properties of disordered graphene nanosheets. *Chem. Mater.* 21, 3136–3142.

Pang, Y., Wei, J., Wang, Y., and Xia, Y. (2018) Synergetic Protective Effect of the Ultralight MWCNTs/NCQDs modified separator for highly stable lithium-sulfur batteries. *Adv. Energy Mater.* 8:e1702288.

Park, H., Ambade, R. B., Noh, S. H., Eom, W., Koh, K. H., Ambade, S. B., Lee, W. J., Kim, S. H., Han, T. H. (2019) Porous graphene-carbon nanotube scaffolds for fiber supercapacitors. *ACS Appl. Mater. Interfaces* 11, 9011–9022.

Ponraj, R., Kannan, A. G., Ahn, J. H., Lee, J. H., Kang, J., Han, B. et al. (2017) Effective Trapping of Lithium Polysulfides using a functionalized carbon nanotube-coated separator for lithium-sulfur cells with enhanced cycling stability. *ACS Appl. Mater. Interfaces* 9, 38445–38454.

Rangappa, D., Sone, K., Ichihara, M., Kudo, T., Honma, I. (2010) Rapid one-pot synthesis of LiMPO4 (M = Fe, Mn) colloidal nanocrystals by supercritical ethanol process. *Chem. Commun.* 2010, 46, 7548–7550.

Rathinavel, S., Priyadharshini, K., Panda, D. (2021) A review on carbon nanotube: An overview of synthesis, properties, functionalization, characterization, and the application. *Mater. Sci. Eng. B* 268, 115095.

Ren, Z.F., Huang, Z.P., Xu, J.W., Wang, J.H., Bush, P., Siegal, M.P., Provencio, P.N. (1998) Synthesis of large arrays of well-aligned carbon nanotubes on glass. *Science* 282, 1105–1107.

Roldo, M., Fatouros, D. G. (2013) Biomedical applications of carbon nanotubes. *Annu. Rep. Prog. Chem. C Phys. Chem.* 109, 10–35.

Rui, X., Zhu, J., Sim, D., Xu, C., Zeng, Y., Hng, H. H., Lim, T. M., Yan, Q. (2011) Reduced graphene oxide supported highly porous V_2O_5 spheres as a high-power cathode material for lithium-ion batteries. *Nanoscale* 3, 4752–4758.

Sakamoto, J.S., Dunn, B. (2002) Vanadium oxide-carbon nanotube composite electrodes for use in secondary lithium batteries. *J. Electrochem. Soc.* 149, A26–A30.

Shao, J., Zhou, H., Feng, J., Zhu, M., Yuan, A. (2019) Facile synthesis of MOF-derived hollow NiO microspheres integrated with graphene foam for improved lithium-storage properties. *J. Alloys Compd.* 784, 869–8.

Shao, Y., Zhang, S., Engelhard, M. H., Li, G., Shao, G., Wang, Y., Liu, J., Aksay, I. A., Lin, Y. (2010) Nitrogen-doped graphene and its electrochemical applications. *J. Mater. Chem.* 20, 7491–7496.

Shi, L., Zeng, F., Cheng, X., Lam, K. H., Wang, W., Wang, A. et al. (2018) Enhanced performance of lithium-sulfur batteries with high sulfur loading utilizing ion selective MWCNT/SPANI modified separator. *Chem. Eng. J.* 334, 305–312.

Shi, Y., Chou, S.-L., Wang, J.-Z., Wexler, D., Li, H.-J., Liu., H.-K., Wu, Y. (2012) Graphene wrapped $LiFePO_4$/C composites as cathode materials for Li-ion batteries with enhanced rate capability. *J. Mater. Chem.* 22, 16465–16470.

Shi, Z., Chu, W., Hou, Y., Gao, Y., Yang, N. (2019) Asymmetric supercapacitors with high energy densities. *Nanoscale* 11, 11946–11955.

Simon, P., Gogotsi, Y. (2008) Materials for electrochemical capacitors. *Nat. Mater.* 7, 845–854.

Song, C.-L., Li, G.-H., Yang, Y., Hong, X.-J., Huang, S., Zheng, Q.-F. et al. (2020) 3D catalytic MOF-based nanocomposite as separator coatings for high-performance Li-S battery. *Chem. Eng. J.* 381:122701.

Strauss, V., Marsh, K., Kowal, M. D., El-Kady, M., Kaner, R. B. (2018) A simple route to porous graphene from carbon nanodots for supercapacitor applications. *Adv. Mater.* 30, 1704449.

Sun, H., Mei, L., Liang, J., Zhao, Z., Lee, C., Fei, H., Ding, M., Lau, J., Li, M., Wang, C., Xu, X., Hao, G., Papandrea, B., Shakir, I., Dunn, B., Huang, Y., Duan, X. (2017) Three-dimensional holey-graphene/niobia composite architectures for ultrahigh-rate energy storage. *Science* 356, 599–604.

Sun, Z., Xie, C., Fan, Z., Shen, F., Yin, Y., Niu, C. et al. (2020) Ultrathin dense double-walled carbon nanotube membrane for enhanced lithium-sulfur batteries. *J. Nanopart. Res.* 22:160.

Tan, L., Li, X., Wang, Z., Guo, H., Wang, J., and An, L. (2018) Multifunctional separator with porous carbon/multi-walled carbon nanotube coating for advanced lithium-sulfur batteries. *Chemelectrochem* 5, 71–77.

Tarascon, J.-M., Armand M. (2001) Issues and challenges facing rechargeable lithium batteries. *Nature* **414**, 359–367.

Tasis, D., Tagmatarchis, N., Bianco, A., Prato, M. (2006) Chemistry of carbon nanotubes. *Chem. Rev.* 106, 3, 1105–1136.

Tian, W., Xi, B., Feng, Z., Li, H., Feng, J., Xiong, S. (2019) Sulfiphilic few-layered $MoSe_2$ nanoflakes decorated rGO as a highly efficient sulfur host for lithium-sulfur batteries *Adv. Energy Mater.* 2, 1901896.

Tian, W., Xi, B., Mao, H., Zhang, J., Feng, J., Xiong, S. (2018). Systematic Exploration of the Role of a Modified Layer on the Separator in the Electrochemistry of Lithium-Sulfur Batteries. *ACS Appl. Mater. Interfaces* 10, 30306–30313.

Tiruneh, S. N., Kang, B. K., Kwag, S. H., Humayoun, U. B., Yoon, D. H. (2018a) Nickel cobalt sulfide anchored in crumpled and porous graphene framework for electrochemical energy storage. *Curr. Appl. Phys.* 18, S37–S43.

Tiruneh, S. N., Kang, B. K., Kwag, S. H., Lee, Y., Kim, M., Yoon, D. H. (2018) Synergistically active $NiCo2S_4$ nanoparticles coupled with holey defect graphene hydrogel for high-performance solid-state supercapacitors. *Chem. Eur. J.* 24, 3263–3270.

Vidu, R., Rahman, M., Mahmoudi, M., Enachescu, M., Poteca, T. D., Opris, I. (2014) Nanostructures: A platform for brain repair and augmentation. *Front. Syst. Neurosci.* 8, 91.

Wang, D.-W., Zeng, Q., Zhou, G., Yin, L., Li, F., Cheng, H.-M., Gentle, I. R., Lu, G. Q. M. (2013) Carbon-sulfur composites for Li-S batteries: Status and prospects. *J. Mater. Chem. A* 1, 9382–9394.

Wang, F., Mei, X., Wang, K., Dong, X., Gao, M., Zhai, Z., Zhu, Duan, J. Lv, C. W., Wang, W. (2019) Rapid and low-cost laser synthesis of hierarchically porous graphene materials as high-performance electrodes for supercapacitors. *J. Mater. Sci.* 54, 5658–5670.

Wang, G., Lai, Y., Zhang, Z., Li, J., and Zhang, Z. (2015) Enhanced rate capability and cycle stability of lithium-sulfur batteries with a bifunctional MCNT@PEG-modified separator. *J. Mater. Chem. A* 3, 7139–7144.

Wang, H., Casalongue, H. S., Liang, Y., Dai, H. (2010) $Ni(OH)_2$ nanoplates grown on graphene as advanced electrochemical pseudocapacitor materials. *J. Am. Chem. Soc.* 132, 7472–7477.

Wang, H., Yang, Y., Liang, Y., Robinson, J. T., Li, Y., Jackson, A., Cui, Y., Dai, H. (2011a) Graphene-wrapped sulfur particles as a rechargeable Lithium-sulfur battery cathode material with high capacity and cycling stability. *Nano Lett.* 11, 2644–2647.

Wang, J., Qin, W., Zhu, X., and Teng, Y. (2020) Covalent organic frameworks (COF)/CNT nanocomposite for high performance and wide operating temperature lithium-sulfur batteries. *Energy* 199:117372.

Wang, S., Li, S., Sun, Y., Feng, X., Chen, C. (2011) Three-dimensional porous V_2O_5 cathode with ultra-high-rate capability. *Energy Environ. Sci.* 4, 2854–2857.

Wang, Y., Liu, X., Yang, C., Li, N., Yan, K., Ji, T., Chi, H., Sun, F., Zhao, J., Li, Y. (2019a) Low-cost fabrication of three-dimensional hierarchical porous graphene anode material for sodium ion batteries application. *Surf. Coat. Technol.* 360, 110–115.

Wei, H., Liu, Y., Zhai, X., Wang, F., Ren, X., Tao, F., Li, T., Wang, G., Ren, F. (2020) Application of carbon nanotube-based materials as interlayers in high-performance lithium-sulfur batteries: A review. *Front. Energy Res.* 8, 585795.

Wei, L., Li, W., Zhao, T., Zhang, N., Li, L., Wu, F. et al. (2020) Cobalt nanoparticles shielded in N-doped carbon nanotubes for high areal capacity Li-S batteries. *Chem. Commun.* 56, 3007–3010.

Wu, F., Zhao, S., Chen, L., Lu, Y., Su, Y., Jia, Y. et al. (2018) Metal-organic frameworks composites threaded on the CNT knitted separator for suppressing the shuttle effect of lithium sulfur batteries. *Energy Storage Mater.* 14, 383–391.

Wu, H., Huang, Y., Zhang, W., Sun, X., Yang, Y., Wang, L. et al. (2017) Lock of sulfur with carbon black and a three-dimensional graphene@carbon nanotubes coated separator for lithium-sulfur batteries. *J. Alloys Compd.* 708, 743–750.

Wu, X.-L., Guo, Y.-G., Su, J., Xiong, J.-W., Zhang, Y.-L., Wan, L.-J. (2013) Carbon-nanotube-decorated nano-$LiFePO_4$@C cathode material with superior high-rate and low-temperature performances for lithium-ion batteries. *Adv. Energy Mater.* 3, 1155–1160.

Wu, Z. S., Ren, W., Xu, L., Li, F., Cheng, H. M. (2011) Doped graphene sheets as anode materials with superhigh rate and large capacity for lithium-ion batteries. *ACS Nano* 5, 5463–5471.

Xiang, K., Wen, X., Hu, J., Wang, S., and Chen, H. (2019) Rational fabrication of nitrogen and sulfur Codoped Carbon Nanotubes/MoS_2 for high-performance lithium-sulfur batteries. *Chemsuschem* 12, 3602–3614.

Xiao, J., Wang, H., Li, X., Wang, Z., Ma, J., Zhao, H. (2015) N-doped carbon nanotubes as cathode material in Li-S batteries. *J. Mater. Sci.: Mater. Electron.* 26, 7895–7900.

Xing, B., Zeng, Huang, H. G., Zhang, C., Yuan, R., Cao, Y., Chen, Z., Yu, J. (2019) Porous graphene prepared from anthracite as high-performance anode materials for lithium-ion battery applications. *J. Alloys Compd. 779*, 202–211.

Xu, Q., Hu, G. C., Bi, H. L., and Xiang, H. F. (2015) A trilayer carbon nanotube/Al2O3/ polypropylene separator for lithium-sulfur batteries. *Ionics* 21, 981–986.

Yan, J., Fan, Z., Sun, W., Ning, G., Wei, T., Zhang, Q., Zhang, R., Zhi, L., Wei, F. (2012) Advanced asymmetric supercapacitors based on $Ni(OH)_2$/graphene and porous graphene electrodes with high energy density. *Adv. Funct. Mater.* 22, 2632–2641.

Yan, L., Luo, N., Kong, W., Luo, S., Wu, H., Jiang, K. et al. (2018) Enhanced performance of lithium-sulfur batteries with an ultrathin and lightweight MoS_2/carbon nanotube interlayer. *J. Power Sources* 389, 169–177.

Yang, D., Xu, B., Zhao, Q., Zhao, X. S. (2019) Three-dimensional nitrogen-doped holey graphene and transition metal oxide composites for sodium-ion batteries. *J. Mater. Chem. A* 7, 363–371.

Yang, J., Wang, J., Wang, D., Li, X., Geng, D., Liang, G., Gauthier, M., Li, R., Sun, X. (2012) 3D porous $LiFePO_4$/graphene hybrid cathodes with enhanced performance for Li-ion batteries. *J. Power Sources* 208, 340–344.

Yang, L., Li, G., Jiang, X., Zhang, T., Lin, H., and Lee, J. Y. (2017) Balancing the chemisorption and charge transport properties of the interlayer in lithium-sulfur batteries. *J. Mater. Chem. A* 5, 12506–12512.

Yang, S., Gong, Y., Liu, Z., Zhan, L., Hashim, D. P., Ma, L., Vajtai, R., Ajayan P. M. (2013) Bottom-up approach toward single-crystalline VO_2-Graphene ribbons as cathodes for ultrafast lithium storage. *Nano Lett.* 13, 1596–1601.

Yao, M., Wang, R., Zhao, Z., Liu, Y., Niu, Z., and Chen, J. (2018) A flexible all-in-one lithium-sulfur battery a flexible all-in-one lithium-sulfur battery. *ACS Nano* 12, 12503–12511.

Yao, S., Cui, J., Huang, J.-Q., Lu, Z., Deng, Y., Chong, W. G. et al. (2018) Novel 2D Sb2S3 Nanosheet/CNT coupling layer for exceptional polysulfide recycling performance. *Adv. Energy Mater.* 8:1800710.

Yao, W., Zhang, F., Qiu, W., Xu, Z., Xu, J., Wen, Y. (2019) General synthesis of uniform three-dimensional metal oxides/reduced graphene oxide aerogels by a nucleation-inducing growth strategy for high-performance lithium storage. *ACS Sustainable Chem. Eng.* 7, 847–857.

Ye, J., Chen, Z., Liu, Q., Xu, C. (2018) Tin sulphide nanoflowers anchored on three-dimensional porous graphene networks as high-performance anode for sodium-ion batteries. *J. Colloid Interface Sci.* 516, 1–8.

Yeon, J. S., Yun, S., Park, J. M., Park, H. S. (2019) Surface-modified sulfur nanorods immobilized on radially assembled open-porous graphene microspheres for lithium-sulfur batteries. *ACS Nano* 13, 5163–5171.

Yin, S., Zhang, Y., Kong, J., Zou, C., Li, C. M., Lu, X., Ma, J., Boey, F. Y., Chen, X. (2011) Assembly of graphene sheets into hierarchical structures for high-performance energy storage. *ACS Nano* 5, 3831–3838.

Yoo, E. J., Kim, J., Hosono, E., Zhou, H. S., Kudo, T., Honma, I. (2008) Large reversible Li storage of graphene nanosheet families for use in rechargeable lithium-ion batteries. *Nano Lett.* 8, 2277–2282.

Yuan, H., He, Z. (2015) Graphene-modified electrodes for enhancing the performance of microbial fuel cells. *Nanoscale* 7, 7022–7029.

Zhai, S., Wang, C., Karahan, H. E., Wang, Y., Chen, X., Sui, X., Huang, Q., Liao, X., Liao, X., Wang, X., Chen, Y. (2018) Nano-RuO_2-decorated holey graphene composite fibers for micro-supercapacitors with ultrahigh energy density. *Small* 14, 1800582.

Zhang, F.-F., Zhang, X.-B., Dong, Y.-H., Wang, L.-M. (2012) Facile and effective synthesis of reduced graphene oxide encapsulated sulfur via oil/water system for high performance lithium sulfur cells. *J. Mater. Chem.* 22, 11452–11454.

Zhang, H., Guo, H., Li, A., Chang, X., Liu, S., Liu, D., Wang, Y., Zhang, F., Yuan, H. (2019) High specific surface area porous graphene grids carbon as anode materials for sodium ion batteries. *J. Energy Chem.* 31, 159–166.

Zhang, Q., Wang, Y., Zhang, B., Zhao, K., He, P., Huang, B. (2018) 3D superelastic graphene aerogel-nanosheet hybrid hierarchical nanostructures as high-performance supercapacitor electrodes. *Carbon* 127, 449–458.

Zhang, Y., Bakenov, Z., Tan, T., Huang, J. (2018a) Three-dimensional hierarchical structure of PPy/porous-graphene to encapsulate polysulfides for lithium/sulfur batteries. *Nanomaterials* 8, 606.

Zhang, Y., Wan, Q., Yang, N. (2019a) Recent advances of porous graphene: Synthesis, functionalization, and electrochemical applications. *Small* 15, 1903780.

Zhang, Z., Kong, L.-L., Liu, S., Li, G.-R., Gao, X.-P. (2017) A High-efficiency sulfur/carbon composite based on 3D graphene nanosheet@carbon nanotube matrix as cathode for lithium-sulfur battery. *Adv. Energy Mater.* 7, 1602543.

Zhao, J., Zhang, Y., Zhang, F., Liang, H., Ming, F., Alshareef, H., Gao, Z. (2018) Partially reduced holey graphene oxide as high-performance anode for sodium-ion batteries. *Adv. Energy Mater.* 2, 1803215.

Zhao, Y., Yin, F., Zhang, Y., Zhang, C., Mentbayeva, A., Umirov, N., Xie, H., Bakenov, Z. (2015) A free-standing sulfur/nitrogen-doped carbon nanotube electrode for high-performance lithium/sulfur batteries. *Nanoscale Res. Lett.* 10, 450.

Zheng, M., Chi, Y., Hu, Q., Tang, H., Jiang, X., Zhang, L., Zhang, S., Pang, H., Xu, Q. (2019) Carbon nanotube-based materials for lithium-sulfur batteries. *J. Mater. Chem. A* 7, 17204–17241.

Zhou, Y., Wen, L., Zhan, K., Yan, Y., Zhao, B. (2018) Three-dimensional porous graphene/nickel cobalt mixed oxide composites for high-performance hybrid supercapacitor. *Ceram. Int. 44*, 21848–21854.

Zhu, J., Yildirim, E., Aly, K., Shen, J., Chen, C., Lu, Y. et al. (2016) Hierarchical multi-component nanofiber separators for lithium polysulfide capture in lithium-sulfur batteries: An experimental and molecular modeling study. *J. Mater. Chem. A* 4, 13572–13581.

Zhu, T., Zhou, J., Li, Z., Li, S., Si, W., Zhuo, S. (2014) Hierarchical porous and N-doped carbon nanotubes derived from polyaniline for electrode materials in supercapacitors. *J. Mater. Chem. A* 2, 12545–12551.

Zhu, X. D., Tian, J., Le, S. R., Chen, J. R., Sun, K. N. (2013) Enhanced electrochemical performances of $CuCrO_2$-CNTs nanocomposites anodes by in-situ hydrothermal synthesis for lithium-ion batteries. *Mater. Lett.* 107, 147–149.

Zhu, Y., Murali, S., Stoller, M. D., Ganesh, K. J., Cai, W., Ferreira, P. J., Pirkle, A., Wallace, R. M., Cychosz, K. A., Thommes, M., Su, D., Stach, E. A., Ruoff, R. S. (2011) Carbon-based supercapacitors produced by activation of graphene. *Science* 332, 1537–1541.

Zuo, X., Zhen, M., Wang, C. (2019) Ni@N-doped graphene nanosheets and CNTs hybrids modified separator as efficient polysulfide barrier for high-performance lithium sulfur batteries. *Nano Res.* 12, 829–836.

8 Mesoporous Carbon Materials

Mandakini Gupta and Vijay Lakshmi Mishra

8.1 INTRODUCTION

Mesoporous materials are flexible, disordered or ordered materials containing pores with diameters between 2 and 50 nm (Rouquerol et al., 1994). Typical examples of mesoporous materials are generally alumina, silica (Direnzo et al., 1997, Innocenzi, 2022) and mesoporous oxides of copper, iron (Wang et al., 2022), tin, niobium, tantalum, titanium, zirconium and cerium (Wang et al., 2012) that have similarly sized mesopores. Recently, carbon-based mesoporous materials called mesoporous carbons (MCs) have been deeply studied and achieved great attention across the globe as another family of carbon-based nanomaterials after their unique flexibility in forming nanostructures like fullerenes, carbon nano tubes and graphenes. On account of its uniform and synthetically controllable pore sizes, highly distinct surface area, chemical inertness, thermostability and biocompatibility and even the presence of monodispersed mesopores (Fang et al., 2010), this type of porous materials have gained the attention of scientists across the world to use these materials for various applications, and as a result, the research in this field has rapidly grown. On account of these unique properties, MCs are reported to have vast industrial applications such as catalysis (Xu et al., 2014), gas sensing (Luo et al., 2016), photovoltaics (Alon et al., 2021), ion exchange techniques (Wang et al., 2022), separation of large biomolecules (Vinu et al., 2006), supercapacitors (Yan et al., 2022), and batteries (Ali Eftekhari 2017a). Moreover, MCs control and manage the oral drug delivery system of the body and also reported to adsorb toxic and heavy metal from waste water and many other compounds from an aqueous media and help in waste water treatment (Wang et al., 2022). On the other hand, the significant development in the field of clean energy technology has increased the interest of researchers to explore these porous materials usable for energy storage devices (Eftekhari, 2017a).

8.2 SYNTHESIS

Earlier, porous carbon materials like activated carbon as well as carbon molecular sieves, on a large scale, were reported to be synthesized by treating fruit shells, polymers or wood by physical or chemical methods at high temperatures or simple pyrolysis of these substances. However, porous carbon materials made by these methods have comparatively bigger pore size, i.e. both in micropores as well as in mesoporous range and are less significant for application purposes. Further, the process of carbonization removes heteroatom from the framework which ultimately results

 DOI: 10.1201/9781003481157-8

in defects in the carbon materials synthesized by this route. Moreover, the porous carbons synthesized by these routes have poor conductivity, robustness, transport of matter via pores due to non-uniform distribution of similar sized pores followed by short range mass transport via these materials.

The further advancement of these porous carbon materials with respect to improved morphology, tunable porosity and better properties are achieved after the discovery of porous nanomaterials called mesoporous carbon with highly ordered mesostructures having pore size between 2 and 50 nm (Innocenzi, 2022).

The first report of highly ordered MCs (OMCs) using silica template was reported (Ryoo et al., 1999), and later, several methods have been used for the synthesis of OMCs (Xin and Song, 2013). MCs with appropriate physical properties are usually synthesized by template carbonization processes which involve both hard template and soft template methodology. The soft templates involve the use of surfactants such as cetyltrimethylammonium bromide (CTAB) (Liang et al., 2004, Zhang et al., 2021) as well as amphiphilic block copolymers, e.g. commercial Pluronic P123 (Meng et al., 2005) while the hard templates require rigid frameworks, e.g. ordered carbon, mesoporous silica and colloidal crystals and are called as nanocasting (Koyuncu and Okur, 2021). The preparation of OMCs by hard template is done in several steps which involve (a) preparation of controlled architecture of mesoporous silica; (b) deposition of an appropriate carbon precursor into the mesopores either by chemical vapor deposition (CVD) or wet impregnation or (c) organic-inorganic composite interaction by polymerization; (d) high temperature treatment of the material formed in the step (c) called carbonization; (e) removal of silica template by treatment of the product obtained by alkaline dissolution or etching in HF (Figure 8.1).

As we have discussed above, during the synthesis of OMCs, carbon has been deposited over a hard silica template, and after removal of the host silica template, the hollow shell of carbon material remains undisturbed and left over space by the template after removal, contains mesopores in the newly synthesized carbon material, i.e. ordered mesoporous carbon materials. However, the preparation of OMCs via the hard templates method involves a multistep procedure with the loss of costly template and makes it unsuitable for large-scale production (Koyuncu and Okur, 2021). On the other hand, the soft template approach is more acceptable and is suitable for mass production. The soft template method also involves the preparation of mesopore structure with an easily adjustable shape and pore size by simply modifying conditions involved in synthesis such as ionic strength, pH and nature of template like hydrophobic/hydrophilic

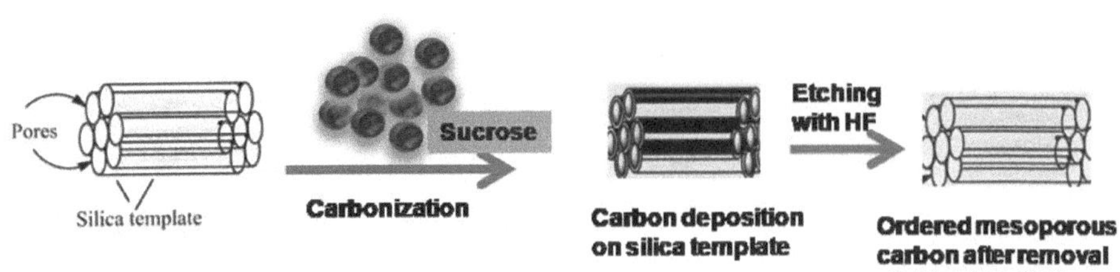

FIGURE 8.1 Schematic representation of hard template-based ordered mesoporous materials (OMCs) formation.

volume ratio, temperature and hydrophobic length. The basic mechanism of OMCs via soft templating involves the polymerizing process of the carbon frameworks over a soft template and further followed by subsequent pyrolysis resulting in the decomposition of the template molecule, leaving behind OMCs of uniform pore sizes. The resulting OMCs produced by the above technique crystallize in fcc structure with a bigger mesopore size of about 2–25 nm (Ishihara, 2019) (Figure 8.2).

The predominant soft template route involves evaporation-induced self-assembly (EISA) (Nanaji et al., 2019), but it is an unsuitable method for the commercial production of OMCs with respect to difficulties faced during its preparation such as sample collection and evaporation of a huge amount of solvents while the hydrothermal method may be more appropriate for mass production. The large-scale production of OMCs is also done by carbonization of organic sources, e.g. sucrose on hard silica template followed by acid treatment (Sudipta et al., 2015). Recently, OMCs are also reported to be synthesized using supramolecular templates using the following steps: (a) selection or preparation of supramolecular template; (b) polymerization at high temperature for getting better inorganic-organic cross linking; (c) decomposition of template; and (d) carbonization. As a result, synthesis of more precise OMCs involving carbon precursor and self-assembly of various copolymer started growing and became the promising area of research among scientific community. Thus, functionalization of OMCs by introducing inorganic substances either to the mesoporous walls or to the surface of the channels alters not only the morphology but also its property and can further be applied for various purposes (Trewyn et al., 2007, Stein, 2020, Deng et al., 2010, Zbair et al., 2020). Therefore, functionalized OMCs are valuable for both carbon-based electrode chemistry and catalyst support. The rise of age of nanotechnology has further enriched this field to another level of advancement and opened the door to the introduction of new synthesis techniques and to achieve improved physicochemical properties (Shuangzhu et al., 2020). Recently, the production of OMCs from cheap natural precursors through an eco-friendly approach has been reported and has great achievement in modern materials science research since it also opens the way to reduce horticulture waste for producing OMCs and can play a key role in waste management in the future (Jain et al., 2014). Some of the methods of this category involve chemical aviation (Jain et al., 2014) and pyrolysis. Owing to

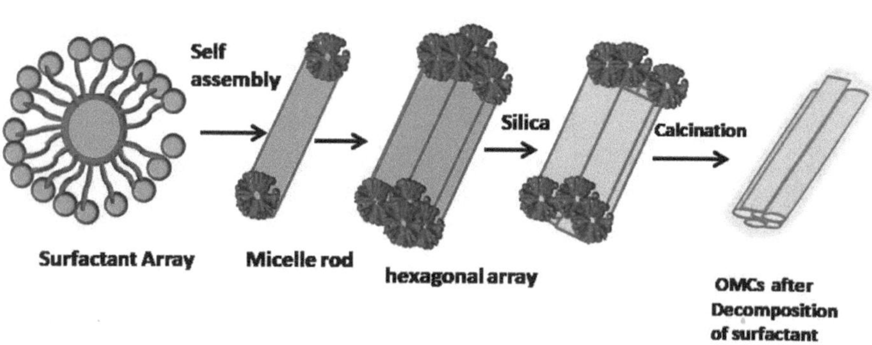

FIGURE 8.2 Schematic illustration of soft template-based ordered mesoporous materials (OMCs) formation.

further advancement in the production of OMCs, incorporation of inorganic compounds such as Fe, Co and Ni during polymerization is quite beneficial because this helps in the production of MCs bearing magnetic property. These materials containing magnetic behavior can be used as catalysts as well as adsorbents which can be removed from liquid solvents after use by applying magnetic field (Wang et al., 2021). OMCs can be functionalized and modified to further improve its physicochemical properties for various applications. The functionalization involves either direct synthesis or post synthesis reaction which may involve surface treatment or heteroatom doping (Liu et al., 2020).

8.3 APPLICATIONS

MCs have been reported to be prominent materials for the past two decades due to the increased surface areas, porous morphology and the presence of various functional groups on the walls of channels (Zhang et al., 2014). In this section, we will be discussing only some important applications, which include the adsorption property of OMCs for the removal of solid or gas from waste water; MCs as electrochemical sensors and biosensors, energy storage; production of catalyst or catalyst support.

8.3.1 OMCs as Adsorbent for Removal of Solid or Gas from Waste Water

Although, the removal of solid or gaseous pollutants from the environment can be processed in many ways, adsorption has been found to be a simple, timesaving and successful technology (Lian et al., 2020, Jeong et al., 2020). The presence of mesopores (2~50 nm) removes pollutants like toxic metals, complicated organic compounds, e.g. drugs and dyes and so on via phenomena adsorption. The adsorption capacity of OMCs is greatly affected by molecular mass, geometry, size, polarity and functional groups, e.g. organic dyes such as methylthionine chloride, fuchsin, rhodamine B, methyl orange have become the most serious environmental pollutants. In the new era of adsorbents, OMCs have showed remarkable improvement as an adsorbent with a large adsorption rate (>99.9%) even for very diluted dye solutions, extensively high performance in discolorations regardless of the nature of dye, i.e. acidic, basic or azo dyes.A

8.3.2 MCs as Electrochemical Capacitors, Batteries or Electrochemical Detection

A lot of research has been reported for OMCs as excellent materials bearing electrochemical applications (Eftekhari, 2017a, 2017b, Yan et al., 2022), and the main advantages are (a) high conductivity; (b) high surface area; (c) well-distributed pores with ordered channel, improving and enhancing the capacitance power and supporting the ions diffusion via these well-organized distinct porous architectures; and (d) large surface functionalities which can increase the abilities for associating with active sites, thereby speeding up the rate of redox reactions.

8.3.3 Sensor

MCs are recently utilized for the modification of electrode surface and can be applied in electrochemical detection as it can alter the surface property of the recognition layer of electrochemical sensing devices by improving conductivity, surface area, flow of matter and mass transport through the mesopores and chemical robustness of modified electrode, thereby improving sensitivity and selectivity. Modified electrodes with high surface area and porous morphology can have improved sensitivity which is one of the key requirements for electrochemical sensing devices. Tan and coworkers developed OMCs nanoparticles modified with MIP sensors (MCNs@MIP) for the detection of the antibiotic ofloxacin (Tan et al., 2014). There are extensive reports available in the literature for utilizing OMCs-modified electrodes for sensing applications (Luo et al., 2016).

8.3.4 Catalyst Supporter or Catalyst

OMCs have a highly organized and hierarchical nanoporous architecture with a high surface area and long-range order porosity, which is responsible for making them one of the more sustainable materials as the support for catalysts or active heterogeneous catalysts. The developed MCs-based catalyst is reported to be used in various organic reactions, including the base-catalyzed reactions, dehydrogenation, photocatalysis, electrocatalysis and selective oxidation. To enhance the use of catalyst, modified OMCs surface with functional groups or heteroatoms can be used. Ordered mesoporous carbon materials can be used as a good catalytic support also since the mesopores modified with functional groups have specific catalytic activity and behave as catalytic active sites and have a high surface area with extensive electrical conductivity. SO3H-bering amorphous carbons show extraordinary catalytic performance (Suganuma et al., 2011) for several acid catalyzed reactions with hydrophilic reactants, such as esterification (Liu at al., 2008), transesterification (Hara, 2009) and hydration reaction (Okamura et al., 2006). The catalytic activity and selectivity are affected by the pore structure of OMCs, e.g. nitrogen or transition metals oxides-doped OMCs can further increase the oxygen reduction reaction (ORR) performance. The pore size favoring large organic molecules diffusion enhances the catalytic property of OMCs.

8.4 CONCLUSION

This chapter aims to provide an overview of mesoporous carbon materials and its advanced version, i.e. modified mesoporous carbon material; synthesis of mesoporous carbon materials by different methodologies, including hard template and soft template and nanotechnology; and its application in various scientific research fields, including adsorption, sensors, catalysts as well as catalytic support and drug delivery (Huang et al., 2016). This chapter has also discussed the advantage of classical soft template-based materials over the hard template-based materials as well as the significance of modified mesoporous carbon materials over their classical version of these materials, e.g. use of nanotechnology for synthesis, incorporation of inorganic-organic nanoparticles to the walls of mesoporous carbon, and the use of various carbon precursors during the carbonization step which not only improve surface area and porosity but also show better properties. Hence, it opens the door for further research to achieve advanced mesoporous carbon in the future. Therefore, we hope that the above text would help researchers in the future.

REFERENCES

Alon, A., Sohmer, M., Pathak, C. S., Visoly-Fisher, I., Etgar, L. (2021). Photovoltaic recovery of all printable mesoporous-carbon-based perovskite solar cells, *RRL Solar*, 5, 2100028.

Deng, Y., Cai, Y., Sun, Z., Gu, D., Wei, J., Li, W., Guo, X., Yang, J., Zhao, D. (2010). Controlled synthesis and functionalization of ordered large-pore mesoporous carbons, *Adv. Funct. Mater.*, 20, 3658–3665.

Direnzo, F., Cambon, H., Dutartre, R. (1997). A 28-year-old synthesis of micelle-templated mesoporous silica, *Micropor. Mater.*, 10, 283–286.

Eftekhari, A. (2017a). Ordered mesoporous materials for lithium-ion batteries, *Micropor. Mesopor. Mater.*, 243, 355–369.

Eftekhari, A. (2017b). Ordered mesoporous carbon and its applications for electrochemical energy storage and conversion, *Mater. Chem. Front.*, 1, 1001–1027.

Fang, Y., Gu, D., Zou, Y., Wu, Z., Li, F., Che, R., Deng, Y., Tu, B., Zhao, D. (2010). A low-concentration hydrothermal synthesis of biocompatible ordered mesoporous carbon nanospheres with tunable and uniform size, *Angew. Chem. Int. Ed.*, 49, 7987–7991.

Hara, M. (2009). Environmentally benign production of biodiesel using heterogeneous catalysts, *ChemSusChem*, 2, 129–135.

Huang, X., Wu, S., Du, X. (2016). Gated mesoporous carbon nanoparticles as drug delivery system for stimuli-responsive controlled release, *Carbon*, 101, 135–142.

Innocenzi, P. (2022). *Mesoporous Ordered Silica Films: From Self-Assembly to Order*. Springer.

Ishihara, A. (2019). Preparation and reactivity of hierarchical catalysts in catalytic cracking, *Fuel Process. Technol.*, 194, 106116.

Jain, A., Jayaraman, S., Balasubramanian, S., Srinivasan, M. P. (2014). Hydrothermal pre-treatment for mesoporous carbon synthesis: Enhancement of chemical activation, *J. Mater. Chem. A*, 2, 520–528.

Jeong, Y., Cui, M., Choi, J., Lee, Y., Kim, J., Son, Y., Khim, J. (2020). Development of modified mesoporous carbon (CMK-3) for improved adsorption of bisphenol-A, *Chemosphere*, 238, 124559.

Koyuncu, D. D. E., Okur, M. (2021). Removal of AV 90 dye using ordered mesoporous carbon materials prepared via nanocasting of KIT-6: Adsorption isotherms, kinetics and thermodynamic analysis, *Sep. Purif. Technol*, 257, 117657.

Lian, Q., Yao, L., Ahmad, Z. U., Konggidinata, M. I., Zappi, M. E., Gang, D. D. (2020). Modeling mass transfer for adsorptive removal of Pb(II) onto phosphate modified ordered mesoporous carbon (OMC), *J. Contam. Hydrol.*, 228, 103562.

Lian, Q., Yao, L., Uddin Ahmad, Z., Gang, D. D., Konggidinata, M. I., Gallo, A. A., Zappi, M. E. (2020). Enhanced Pb(II) adsorption onto functionalized ordered mesoporous carbon (OMC) from aqueous solutions: The important role of surface property and adsorption mechanism, *Environ. Sci. Pollut. Res.*, 20, 23616–23630.

Liang, C., Hong, K., Guiochon, G. A., Mays, J. W., Dai, S. (2004). Synthesis of a large-scale highly ordered porous carbon film by self-assembly of block copolymers, *Angew. Chem. Int. Ed.*, 43, 5785–5789.

Liu, H. L., Cui, W. J., Jin, L. H., Wang, C. X., Xia, Y. Y. (2009). Preparation of three-dimensional ordered mesoporous carbon sphere arrays by a two-step templating route and their application for supercapacitors, *J. Mater. Chem.*, 19, 3661–3667.

Liu, R., Wang, X., Zhao, X., Pingyun, F. (2008). Sulfonated ordered mesoporous carbon for catalytic preparation of biodiesel, *Carbon*, 46, 1664–1669.

Liu, Y., Xiong, Y., Xu, P., Pang, Y., Du, C. (2020). Enhancement of Pb (II) adsorption by boron doped ordered mesoporous carbon: Isotherm and kinetics modeling, *Sci. Total Environ.*, 708, 134918.

Luo, W., Zhao, T., Li, Y., Wei, J., Xu, P., Li, X. (2016). A micelle fusion-aggregation assembly approach to mesoporous carbon materials with rich active sites for ultrasensitive ammonia sensing, *Am. Chem. Soc.*, 138, 12586–12595.

Meng, Y., Gu, D., Zhang, F., Shi, Y., Yang, H., Li, Z., Yu, C., Tu, B., Zhao, D. (2005). Ordered mesoporous polymers and homologous carbon frameworks: Amphiphilic surfactant templating and direct transformation, *Angew. Chem. Int. Ed.*, 117, 7215–7221.

Nanaji, K., Jyothirmayi, A., Varadaraju, U., Rao, T. N., Anandan, S. (2017). Facile synthesis of mesoporous carbon from furfuryl alcohol-butanol system by EISA process for supercapacitors with enhanced rate capability, *J. Alloys Compd.*, 723, 488–497.

Okamura, M., Takagaki, A., Toda, M., Kondo, JN., Domen, K., Tatsumi, T., Hara, M., Hayashi, S. (2006). Acid-catalyzed reactions on flexible polycyclic aromatic carbon in amorphous carbon, *Chem. Mater.*, 18, 3039–3045.

Rouquerol, J., Avnir, D., Fairbridge, C. W., Everett, D. H., Haynes, J. M., Pernicone, N., Ramsay, J. D. F., Sing, K. S. W., Unger, K. K. (1994). Recommendations for the characterization of porous solids (technical report), *Pure Appl. Chem.*, 66, 1739–1758.

Ryoo, R., Joo, S. H., Jun, S. (1999). Synthesis of highly ordered carbon molecular sieves via template-mediated structural transformation, *J. Phys. Chem. B*, 103, 7743–7746.

Shuangzhu Jia, S., Pan, H., Lin, Q., Wang, X., Li, C., Wang, M., Shi, Y. (2020). Study on the preparation and mechanism of chitosan-based nano-mesoporous carbons by hydrothermal method, *Nanotechnology*, 31, 365604.

Stein, A. (2020). In: V. Gitis and R. Gadi (eds.), *Handbook of Porous Materials*, Vol. 4. Singapore: World Scientific. doi:10.1142/11909.

Sudipta, De., Balu, A. M., Van der Waal, J. C., Luque, R. (2015). Biomass-derived porous carbon materials: Synthesis and catalytic applications, *Chemcatchem*, 7, 1608–1629.

Suganuma, S., Nakajima, K., Masaaki, K., M., Kato, H., Tamura, A., Kondo, H., Yanagawa, S., Hayashi, S., Hara., M. (2011). *Micropor. Mesopor. Mater.*, 143, 443–450.

Tan, F., Zhao, Q., Teng, F., Sun, D., Gao, J., Quan, X., Chen, J. (2014). Molecularly imprinted polymer/mesoporous carbon nanoparticles as electrode sensing material for selective detection of ofloxacin, *Mater. Lett.*, 129, 95–97.

Trewyn, B. G., Slowing, I. I., Giri, S., Chen, H. T., Lin, V. S.-Y. (2007). Synthesis and functionalization of a mesoporous silica nanoparticle based on the sol-gel process and applications in controlled release, *Acco. Chem. Res.*, 40, 846–53.

Vinu, A., Miyahara, M., Mori, T., Ariga, K. (2006). Carbon nanocage: A large-pore cage-type mesoporous carbon material as an adsorbent for biomolecules, *J. Porous Mater.*, 13, 379–383.

Wan, G., Weidong, A., Zhan, Y., Li, Z., Yan, S., Ji, P., Din, D. (2022). Mesoporous carbon framework supported Cu-Fe oxides as efficient peroxymonosulfate catalyst for sustained water remediation, *Chem. Eng. J.*, 430, 133060.

Wang, F., Shi, X., Zhang, J., He, T., Yang, L., Zhang., T. Ran., F. (2022). Bacterial cellulose-derived micro/mesoporous carbon anode materials controlled by poly(methyl methacrylate) for fast sodium ion transport, *Nanoscale*, 14, 3609–3617

Wang, G., Gao, G., Yang, S., Wang, Z., Jin, P., Wei, J. (2021). Magnetic mesoporous carbon nanospheres from renewable plant phenol for efficient hexavalent chromium remova, *Micropor. Mesopor. Mater.*, 310, 110623.

Wang, T., Zhang, L., Zhang, J., Hua, G. (2012). Synthesis and characterization of mesoporous CeO_2 nanotube arrays, *Micropor. Mesopor. Mater.*, 171, 196–200.

Xin, W., Song, Y. (2013). Mesoporous carbons: Recent advances in synthesis and typical applications, *RSC Adv.*, 5, 1–3.

Xu, J., Wu, F., Wu, H. T., Xue, B., Li, Y. X., Cao, Y. (2014). Three-dimensional ordered mesoporous carbon nitride with large mesopores: Synthesis and application towards base catalysis, *Micropor. Mesopor. Mater.*, 198, 223–229.

Yan, J., Guo, C., Guo, X., Tong, X. (2022). Highly active nitrogen-doped mesoporous carbon materials for supercapacitors, *J. Electron. Mater.*, 51, 1021–1028.

Zbair, M., Bottlinger, M., Ainassaari, K., Ojala, S., Stein, O., Keiski, R. L., Bensitel, M., Brahmi, R. (2020). Hydrothermal carbonization of argan nut shell: Functional mesoporous carbon with excellent performance in the adsorption of bisphenol a and diuron, *Waste Biomass Valoriz.*, 11, 1565–1584.

Zhang, D., Zheng, L., Ma, Y., Lei, L., Li, Q., Li, Y., Luo, H., Feng, H., Hao, Y. (2014). *ACS Appl. Mater. Interfaces*, 6, 2657–2665.

Zhang, F., Zong, S.,Zhang, Y., Lv, H., Liu, X., Du, J., Chen, A. (2021). Preparation of hollow mesoporous carbon spheres by pyrolysis-deposition using surfactant as carbon precursor, *J. Power Sour.*, 484, 229274.

9 Metal Oxide-based Nanocomposites for High-Energy Applications

Shahid Bashir, Pershaanaa Manogran, Z. L. Goh, N. R. Elizer, K. Ramesh, and S. Ramesh

9.1 INTRODUCTION

Top-down approach of materials has been highly appreciated with the discovery of existence of materials in nanoscale. Nanoscale materials got striking attention in both fundamental academic research and potential industrial sectors. This significant attention is because of the prominent features such as large surface area, enhanced porosity, rich surface chemistry, and enormous binding sites (Maduraiveeran et al., 2019). These features could play a major part in the adsorption and chemical reactions that occur at the surface and interface in various applications, specifically energy conversion and storage technologies. The smaller size and larger surface area of nanomaterials support the thermodynamics of chemical reactions and transportation kinetics taking place at the surface of the nanomaterials in terms of heat and charge transfer, mass and phase transitions, and dimensional changes owing to the chemical reactions (Nanayakkara et al., 2015; Chen and Ostrom, 2015).

Numerous materials exist in nanostructure. Among the nanomaterials, transition metal oxide is regarded as a promising aspirant that can be utilized in a vast range of applications (Islam et al., 2022). The applications of the transition metal oxide as electrode in energy conversion and storage possess salient features such as environmental friendliness, easy processability, ideal morphology, greater surface area, porous surface, higher catalytic and electrochemical conductivity, significant thermal stability, and outstanding electronic conductivity (Yu et al., 2011). Another intriguing feature is the higher theoretical capacitance which plays a central part in their exploitation in energy conversion and storage devices and offers a conspicuous capacitance improvement by adjusting and controlling their defects and surface/interfaces under a certain nanoscale (Lu et al., 2017). For energy applications, transition metal oxide nanostructures attained significant interest because of the ability to conduct charge through the electrical devices (Mahmood et al., 2019).

Transition metal oxides are found to be ubiquitous throughout energy applications involving solar-to-electrical conversion, solar-to-fuel conversion, and energy storage in supercapacitors and batteries (Gogoi et al., 2022). The band gap of the transition metal oxide depends on the nature of the transition metal, whereby some metal oxides possess wide band gap while others have narrow band gap. The performance

 DOI: 10.1201/9781003481157-9

of the transition metal oxides relies on the band gap. Narrow band gap transition oxides have more metallic character and are heavily utilized as cathodes in batteries and supercapacitors (Bredar et al., 2020; Carey et al., 2015; Kim and Ma, 2015).

A series of transition metal oxide nanostructures and their composites with other materials such as polymers and carbonaceous materials (carbon nanotubes, graphene, etc.) have been reported, including titanium oxide (TiO_2), zinc oxide (ZnO), nickel oxide (NiO), manganese oxide (MnO_2), and copper oxide nanostructures. These nanostructures and their composites demonstrate diverse morphologies such as spherical, nanowires, nanotubes, nanorods, stars, and triangular (Abdah et al., 2020; Liu et al., 2018). These metal oxide nanostructures either alone or in combination with polymers revealed exciting results in terms of physical and chemical properties; therefore, it is highly desirable to understand their various aspects in terms of synthesis, properties, and applications. One problem is the dispersion of nanoparticles in the matrices like polymers. To prepare homogeneous phase composites with non-separated structures, metal oxide nanostructures are dispersed in the polymers homogeneously. The homogeneous phase composites and aspect ratio of polymer to metal oxide control the physical and chemical properties (An et al., 2019). Metal oxide improved the thermal stability, toughness, optoelectronic, and mechanical strength of composites. These composites are extensively used in energy applications (Abdah et al., 2019). This chapter has been prepared to provide an insight of the transition metal composites for energy applications.

9.2 MANGANESE OXIDE-BASED NANOCOMPOSITES

Manganese (Mn)-based metal oxides, especially Mn (II) and Mn (IV) oxides, are one of the most popular *d*-block metal oxides used in energy storage devices such as supercapacitors, supercapattery, lithium-ion batteries, and solar cells due to their lower toxicity, good ion diffusion, excellent theoretical specific capacitance (1,370 F/g), and outstanding chemical stability (Liu et al., 2018; Ramesh et al., 2019). However, MnO_x often suffers from low electrical conductivity and drastic volume change during charging and discharging process which leads to significant drop in the resultant specific capacitance and other electrochemical performances of an electrochemical energy storage (EES) device (Liu et al., 2018; Xiao et al., 2018). Thus, modification of MnO_x's morphological structure is done to fine-tune according to the desired electrochemical performance (Mahdi et al., 2022). For instance, Mn-O is frequently reduced to micro- or nanoscale as it promotes the surface wettability of the material which enhances the charge transfer processes at the electrode/electrolyte interface. As a direct consequence, the pseudocapacitance of the electrode materials improves significantly. By the same token, fabrication of MnO_x nanostructures (Figure 9.1) of tunable pore and particle size is explored by researchers to increase the effective surface area and electrical conductivity of the MnO_x to strikingly raise the capacitance of the electrode material (Julien and Mauger, 2017). As a result, the energy density of the EES device is escalated together with its cyclic stability (Hu et al., 2018). Figure 9.1 shows commonly explored types of nanostructured MnO_2 for EES devices.

FIGURE 9.1 SEM images show the common types of MnO_2 nanostructures (NS) studied by researchers for EES devices (CC BY 4.0, Nanomaterials, MDPI) (Julien and Mauger, 2017).

To further strengthen the electrochemical properties and counter the drawbacks of MnO_x nanostructures (NSs), NSs are integrated to highly conductive materials like carbon nanotubes, carbon cloth, graphene, and carbon nanosheets. For instance, in a study, Xiao et al. (2018) encapsulated MnO nanoparticles into a foam-like carbon nanosheet for lithium-ion battery and revealed the nanocomposite has splendid BET surface area about 330.2 $m^2 g^{-1}$. On top of that, this MnO/C nanocomposite also showed outstanding reversibility property as the capacity of the nanocomposite rose from 954 to 1,212 mAh/g over 1,000 charge/discharge cycles (capacity retention is 127%). By the same token, Rosaiah et al. (2017) incorporated Mn_3O_4 into reduced graphene and reported good specific capacitance (194.8 F/g at 1 A/g) with excellent capacity retention over 1,000 cycles (90%). Besides, introduction of heteroatoms like nitrogen into the carbon matrix is encouraged before incorporating MnO_x because the presence of heteroatoms in carbon materials jacks up the charge transfer at the interface. Ahamad et al. (2020) doped nitrogen into carbon matrix (NDC) and integrated it with MnO_2 to produce a mesoporous structure. In the work, Ahamad et al. (2020) discovered that BET surface area of MnO_2/NDC (645 $m^2 g^{-1}$) was slightly lower compared to pure NDC (685 $m^2 g^{-1}$) due to blockage of the MnO_2 nanoparticles on the porous carbon. However, when asymmetric supercapacitor was constructed using NDC//MnO_2/NDC, it contributed to excellent specific capacitance (895 F/g at 1 A/g) with magnificent energy density and power density, 402.5 Wh kg^{-1} and 2,250 W kg^{-1}, respectively. According to Ahamad et al. (2020), the remarkable electrochemical performance was due to its highly interactive structure, presence of nitrogen atom in the carbon, and lastly, the synergistic effect between the elements in the nanocomposite. In another study, Ramesh et al. (2019) fabricated a hybrid composite by integrating MnO_2 with nitrogen-doped graphene oxide (NGO) and polypyrrole (PPy). It was reported that the addition of PPy to the MnO_2/NGO had boosted the specific capacitance (from 360 to 480 F/g) and capacity retention (from 90% to 95%) over 6,000 cycles of the hybrid composite due to the synergetic effect between PPy, MnO_2, and NGO. Apart from that, the properties of MnO_x nanostructures are improved by combining two transition metal oxides because integration of two or more transition metal oxides amplifies the electrochemical activity because of their synergetic effect. As an example, Ray et al. (2019) and X. Liu et al. (2018) studied the combination of nickel (Ni) with Mn oxide nanocomposite to synthesize asymmetric supercapacitors with different negative electrodes, activated carbon (AC) and iron (III) oxide (Fe_2O_3), respectively. Both the studies showed outstanding cyclic stability (>90%) and enhancement in other electrochemical performances, even though its morphology and the electrode material used to construct asymmetric supercapacitors differed according to their desired properties. Lately, researchers have fabricated ternary metal oxide as it has improved electrochemical properties compared to binary and monometal oxide. As an example, Sanchez et al. (2018) constructed ternary metal oxides/graphene nanocomposite comprised of Mn, Ni, and Co and reported high surface area around 200 $m^2 g^{-1}$ with a higher energy density (27 Wh/kg) with impressive capacity retention, which is 96% over 2,000 charge/discharge cycles when integrated with reduced graphene oxide as negative electrode.

9.3 NICKEL OXIDE-BASED NANOCOMPOSITES

Similar to MnO_x, nickel oxide (NiO) is one of the most frequently selected transition metal oxides due to its striking theoretical specific capacitance, which is higher than MnO_x, 2,573 F/g (Kate et al., 2018), excellent redox reaction, good thermal and chemical stability, and pollution free. Howbeit, it suffers from low electrical conductivity like other metal oxides which makes it unable to cope in extremely fast electron transfer processes in EES devices. To add on, NiO also has restricted potential window (around 0.5 V). As a direct consequence of the condition, the overall specific capacitance does not meet its theoretical specific capacitance value. Therefore, NiO's morphology, pore size distribution, and surface area are modified to increase the specific capacitance and energy density of the electrode material. Usually, NiO is reduced to nanostructure because reduction in the size jacks up the number of active sites for redox reaction; besides, it also increases the surface-to-volume ratio and reduces the diffusion path for the ions to transfer. Subsequently, the electrochemical activity of the nanostructured NiO spikes. NiO nanostructures are then subjected to morphological and structural changes by doping heteroatoms, hybridization, and formation of binary or ternary metal oxide structures or by changing the shape of the NiO nanostructure itself. For instance, the shape and size of the NiO nanostructures are often controlled via changing the fabrication method as shown in Figure 9.2 (Kate et al., 2018). It is perceived that different morphologies contribute to different specific capacitances as they result in different surface-to-volume ratios, effective surface areas, active sites, and charge transfer speeds at the electrode/electrolyte interface during charge/discharge processes. In a study, Chuminjak et al. (2017) synthesized NiO nanoparticles and coated them on a nickel foam (substrate) using sparking method to produce a highly porous NiO nanostructure film where the fabricated nanoparticles were sparked for 45 min and more. As a consequence, the NiO nanoparticles deposited on the nickel foam aggregated and formed a network with smaller surface area and bigger pores. The highly porous NiO nanofilms revealed a terrific electrochemical performance and specific capacitance of 1,840 F/g at a current density of 1 A/g with a capacity retention of 96% over 1,000 charge/discharge cycles.

Aside from that, NiO nanocrystals are also embedded into highly conductive materials like MnO_x to elevate their specific capacitance and other electrochemical performances by improving the electrical conductivity and widening the potential window of the EES devices. For instance, Vellakkat and his co-worker Devendrappa constructed a honeycomb-shaped nanocomposite for supercapacitor application by integrating polyaniline, chitosan, and NiO nanoparticles and reported distinctive specific capacitance, energy density, and power density of 554 F/g, 76.9 Wh/kg, and 2 kW/kg, respectively. In another study, Golkhatmi et al. (2021) fabricated a nickel oxide/graphene/polypyrrole hybrid nanocomposite and disclosed that combination of polypyrrole and graphene can give impressive synergetic effect which can enhance the pseudocapacitive behaviour of the nickel oxide via dynamic electron transportation. Not only that, when the ternary hybrid nanocomposite was immersed

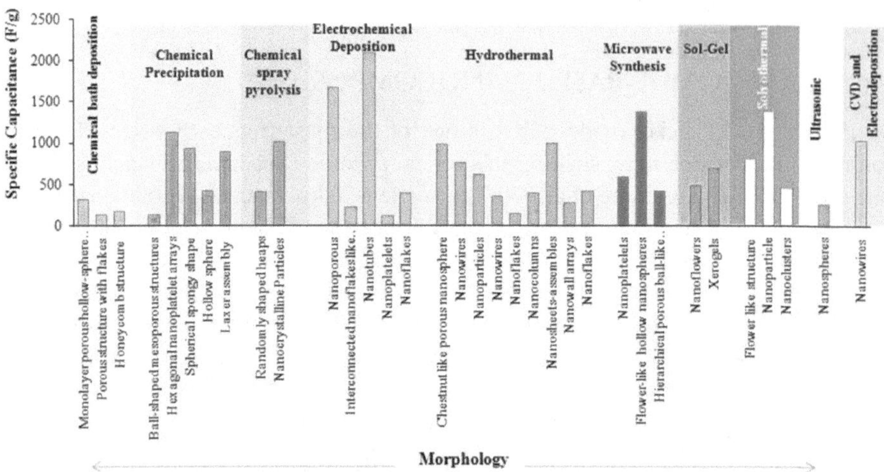

FIGURE 9.2 Graphical representation on how change in fabrication method varies the morphology of the NiO nanostructures against the specific capacitance of the synthesized nanostructures. Reprinted with permission from Kate et al. (2018).

into aqueous electrolyte (KOH) and organic electrolyte (TEA-BF_4 in acetonitrile), it was observed that the nanocomposite in organic electrolyte (36.76 Wh/ kg) showed slightly higher energy density than aqueous electrolyte (33.71 Wh/ kg) and specific capacitance is greater in aqueous KOH (970.85 F/g) compared to TEA-BF_4 salt in acetonitrile (66.71 F/g). This is mainly due to the cation and anion involved. Since the size of BF_4^- (anion of the organic electrolyte) is much bigger than OH^-, the diffusion rate is restricted as BF_4^- can only diffuse into pore size equal to or greater than the size of BF_4^-. Consequently, the ion diffusion rate drops significantly which then affects the specific capacitance of the supercapacitor cell. Along with it, formation of binary or ternary NiO nanocomposites has recently been explored by the researchers as mixing different metal oxide nanoparticles can boost the electrochemical activity, yet the electrode materials' rate capability and performance of device are poor. Thus, to counter the issue, Harilal et al. (2017) constructed hybrid nanobelts comprised of continuous NiO-Co_3O_4 nanobelts and reported that the surface area of the fabricated nanobelt was five times larger than the ordinary binary metal which opens space for better ion intercalation and allows a wide range of ions to diffuse in the nanobelts (due to bigger pore size). At the same time, it was perceived that the nanobelt synthesized had higher rate capability (capacity retention: $\sim$99% over 5,000 cycles) because the pore size of the nanobelt (approximately 21 nm) was twice as big as the thickness of the nanobelt (approximately 10 nm). Lastly, researchers are exploring the incorporation of transition metal oxide nanocrystals into a new 2D metal carbide family (MXene) due to its unique metallic conductivity and hydrophilicity that help to improve the electrochemical activity of the EES devices. For example, Li et al. (2020) embedded NiO nanoparticles into a highly conductive MXene for Li-O_2 battery application and reported capacity as high as 13,350 mAh/ g and maintained outstanding cycle performance over 90 cycles. It was revealed that the addition of the NiO nanoparticles in MXene helps to prevent restacking of the MXene, and at the same time, it revamps the ion transfer due to an increase in active sites and shortens the path for intercalation of lithium ions into the electrode material.

9.4 COPPER OXIDE-BASED NANOCOMPOSITES

Along the lines with other transition metal oxides, copper oxide (CuO) has excellent theoretical specific capacitance (1,800 F/g) (Pawar et al., 2016), low toxicity, abundant and simple preparation methods that are required to fabricate and modify nanosized CuO's morphology (Majumdar and Ghosh, 2021). Together with the pros, CuO also shares the same cons as other transition metal oxides which have low electricity conductivity and unsatisfactory cyclic stability due to structural breakdown during charge/discharge or ion insertion-extraction process (Viswanathan and Shetty, 2018; Wang et al., 2019; Viswanathan and Shetty, 2017). However, its nanostructures have better electrochemical performances due to enhanced quantum confinement effect and surface-to-volume ratio (Ameri et al., 2017). As we know, physiochemical properties of nanostructures are affected by their hydrous state, crystallinity, and morphology. For instance, Ameri et al. (2017) fabricated different types of CuO nanostructures, namely nanoparticles, nanorods, and nanowires (Figure 9.3) to study how hydrous state, crystallinity, and morphology of CuO nanostructures change the

FIGURE 9.3 SEM images show different types of nanostructures: (a) CuO nanorods, (b) CuO nanoparticles, (c) CuO nanoparticles (larger particle size than (b)), and (d) CuO nanowires. Reprinted with permission from Ameri et al. (2017).

electrochemical performance and revealed that the morphology of CuO nanostructure contributed to the highest specific capacitance 113.5 F/g with distinctive capacity retention (92%) over 1,300 cycles. To add on, it was predicted that CuO nanostructures' morphology influences its charge storage ability the most, while its crystallinity influences its cyclic stability. Not only that, but it was also revealed from his study that there is no significant difference between the crystallinity of the CuO nanostructures and its hydrous state. Aside from the above nanostructures, CuO nanoflakes exhibit higher specific capacitance as the tiny flake shapes form highly porous surfaces that ease the ion transportation at the electrode/electrolyte interfaces. For example, Shinde et al. (2019) constructed CuO hybrid nanostructure (nanoflakes interconnected with nanowires) for supercapacitor application and disclosed that CuO nanoflakes can deliver specific capacitance as high as 464 F/g at scan rate of 5 mV/s.

Next, to promote the physiochemical properties of the CuO-based electrode material, CuO nanoparticles are often embedded into highly porous and conductive materials like carbonaceous material to construct CuO-based nanocomposites as they act as a smashing base material besides boosting its overall conductivity. For example, Dipanwita et al. fabricated reduced graphene oxide@ copper (II) oxide (rGO@CuO) nanocomposite using the ultrasound-assisted preparation method and reported better specific capacitance, 277 F/g at 1 A/g than pure CuO (30.1 F/g) and pure rGO (136.4 F/g). Besides, to further improve the CuO-based nanocomposites, conductive polymers such as polyaniline (PANI), polypyrrole (PPy), and polythiophene (PTh) were added into the CuO @carbonaceous nanocomposite to jack up the pseudocapacitance of the nanocomposite and, at the same time, to act as a spacer to avoid the carbonaceous like rGO from restacking. Viswanathan and Shetty (2017) constructed a rGO/CuO/PANI nanocomposite with weight ratios of 6.6%, 13.40%, and 80% for rGO, CuO, and PANI, respectively, and reported good overall electrochemical performance, specific capacitance of 213.20 F/g with energy density of 18.95 Wh/kg, and power density of 545.79 W/kg maintaining up to 97.6% of its initial capacitance over 5,000 charge/discharge cycles. The resultant electrochemical performance was then significantly improved (specific capacitance of 684.93 F/g with energy density of 136.98 Wh/ kg and power density of 1315.76 W/ kg with a capacity retention of 84% over 5,000 charge/discharge cycles at a scan rate as high as 700 mV/s by changing the weight ratio of the nanocomposite to 12%, 40%, and 48% for rGO, CuO, and PANI, respectively (Viswanathan and Shetty, 2018). It was reported that change in the composition produced a honeycomb-shaped nanocomposite which allows the electrolyte to pass through the electrode material, resulting in better ion intercalation process. On top of that, above composition contributed to a flexible ternary nanocomposite that staves off the nanocomposite from excessive strain when exposed to high current. Lastly, a new approach is explored by researchers to fabricate a Cu-based coordination supramolecular network (CSN) binder-free electrode for supercapacitors or batteries deriving from CuO nanosheets to boost the chemical stability of electrode together with its electrochemical properties. Yao et al. (2019) fabricated a novel copper hexacyanoferrate (CuHCF) nanosheet array on a carbon cloth and reported splendid areal capacitance up to 1,441.4 mF cm^{-2}, retaining 87% of the initial capacitance over 2,000 charge/discharge cycles. It was disclosed that the CuHCF nanosheet had interconnected nanoparticles that were uniformly dispersed to ease the ion interaction with the active sites. As a direct consequence, the electrochemical behaviour of the CuHCF nanosheet was uplifted.

9.5 TITANIUM DIOXIDE-BASED NANOCOMPOSITES

Titanium dioxide, commonly known as titania or titanium (IV) oxide, is a naturally occurring oxide of titanium, with the chemical formula TiO_2. It is a prospective semiconductor with a variety of applications, including fuel cells, supercapacitors, and solar cells. It has several advantages, such as abundance, high chemical stability, low cost, effective oxidizing capability, and low toxicity. It also possesses a 3.0–3.2 eV intrinsic bandgap (Nowotny, 2019). Although TiO_2 has traditionally been utilized as the electrode material for supercapacitors, its electronic conductivity is often poor, resulting in a decrease in specific capacitance when the scan rate is increased. Therefore, many research works have been done on the development of TiO_2-based nanocomposites with other conductive materials that can facilitate faster charge transfer. Vishal et al. demonstrated the utilization of TiO_2-carbon nanocomposite as a supercapacitor electrode material (Shrivastav et al., 2020). The developed TiO_2/C electrodes have a large surface area, which is crucial for effective charging and discharging. The device, which consists of two TiO_2/C electrodes with equal weights and a gel electrolyte, produced an energy density of 43.5 Wh kg^{-1} and a power output of 0.865 kW kg^{-1}. Additionally, TiO_2 is also frequently employed in the preparation of working electrode for solar cells, such as dye- or quantum dot (QD)-sensitized solar cells and perovskite solar cells. The working electrode serves both as a support for the loading of the sensitizer and as a transporter of photo-excited electrons from the sensitizer to the external circuit. To ensure high dye loading and high charge transport rate, working electrode materials must have a large surface area to provide high electron collection efficiency, which is why TiO_2 was chosen. TiO_2 nanoparticles are normally coated on transparent conductive oxide (TCO) glass. However, the TiO_2 electrode is prone to problems such as dark current. Due to the disordered networks of TiO_2 nanoparticles, electron recombination occurs in the conduction band of TiO_2, reducing device efficiency. To overcome this problem, double-layered TiO_2 nanoparticles with different sizes are usually being practiced in preparation of the working electrode. A compact layer of TiO_2 P90 with a size of 14 nm is first coated on the TCO followed by a mesoporous layer of TiO_2 P25 with a size of 21 nm coated on top of it. The crack-free, homogeneous P90 compact layer acts as a semiconducting shield for back electron transport across the P90/P25 interface. Senadeera et al. further developed a three-layered nanostructured TiO_2 working electrode (Senadeera et al., 2020). A hydrothermally synthesized TiO_2 nanotube (TNT) layer was deposited on the double-layered TiO_2 electrode. They demonstrated that adding TNT to dye-sensitized solar cells (DSSCs) enhanced short-circuit current density (from 14.00 to 15.61 mA cm^{-2}) and efficiency (from 6.52% to 7.08%). To fabricate a more effective working electrode for DSSC, Tsai et al. (2020) also added nickel oxide (NiO) nanoparticles into TiO_2 paste to form a nanocomposite electrode. The authors reported that the energy level difference between TiO_2 and NiO created a potential barrier that prevented the electrons in the TiO_2 conduction band from recombining with the I_3^- ions in the electrolyte. Besides, TiO_2 nanoparticles have also been utilized as an additive in DSSC application and introduced into the electrolyte. Seung et al. formulated a solid polymer electrolyte by incorporating hydrothermally synthesized one-dimensional (1D) TiO_2 nanofiller into a polyethylene glycol (PEG) polymer matrix, suggesting that the 1D TiO_2-PEG electrolyte promotes the electron transport and enhances efficiency of DSSC (Lim et al., 2021). While Yu et al. (2018) fabricated iodine-free nanocomposite gel electrolytes with three distinct metal oxide nanoparticles, with TiO_2 nanoparticles providing the highest performance, they also observed that DSSC-TiO_2 nanocomposite gel electrolyte had greater long-term at-rest stability than both DSSC-based iodine liquid and gel electrolyte. This solved typical iodine-based electrolyte issues such as volatility, corrosion of iodine electrolyte with counter electrode, and electrolyte leakage (Figure 9.4).

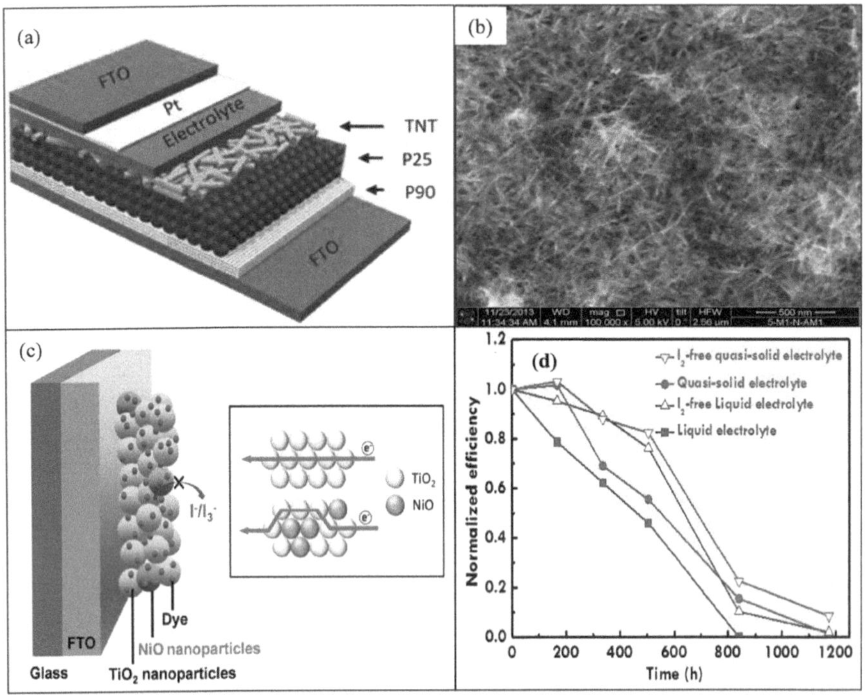

FIGURE 9.4 (a) Schematic diagram of DSSC with three-layered nanostructured TiO_2 working electrode, (b) SEM image of the three-layered nanostructured TiO_2 working electrode (CC BY 3.0, *Journal of Physics*, IOPScience) (Senadeera et al., 2020), (c) electron transport mechanism in TiO_2 and TiO_2–NiO working electrodes (CC BY 4.0, *Coatings*, MDPI) (Tsai et al., 2020) and (d) at-rest stability of DSSCs at 65°C with 65% relative humidity. Reprinted with permission from Yu et al. (2018).

9.6 COBALT OXIDE-BASED NANOCOMPOSITES

Cobalt (II) oxide (CoO), cobalt (III) oxide (Co_2O_3), and cobalt (II,III) oxide (Co_3O_4) are all known oxides of cobalt. Only Co_3O_4-based nanocomposite, which is generally employed in the form of nanoparticles, has potential applications in lithium-ion batteries, supercapacitors, and DSSCs. This is due to its stability and simplicity of production when compared to other forms of cobalt oxide. Co_3O_4 is a *p*-type semiconductor with a direct optical band gap of 3.95–2.13 eV (Sharma et al., 2015) and a theoretical capacitance of 3,560 F g^{-1} (Deori et al., 2013). For instance, the researchers fabricated a spray-deposited Co_3O_4 on a graphite plate that was utilized as an electrode in supercapacitor application. The designed device had a specific capacitance of 476 F g^{-1} at current density of 0.5 A g^{-1} (Deori et al., 2013). This showed that Co_3O_4 surpasses the performance of other metal oxides such as TiO_2, MnO_2, and others. Dong et al. have developed a supercapacitor-compatible 3D graphene/Co_3O_4 composite electrode. This composite electrode has a specific capacitance of 1,100 F g^{-1} at a current density of 10 A g^{-1} due to the high mechanical strength of graphene (Dong et al., 2012). On the other hand, the utilization of Co_3O_4 in the counter electrode of DSSCs is regarded uncommon. This is because platinum is widely used as a counter electrode due to its inertness. However, it is one of the rarest elements on the Earth. Sharma et al. began fabricating Co_3O_4 nanoparticles counter electrodes for DSSC application, but the cell efficiency was only 0.66%, which is extremely low when compared to platinum-based DSSC (Sharma et al., 2015). Later, another group of researchers found that utilizing a Co_3O_4-based nanocomposite counter electrode is more efficient than using Pt-based counter electrode, with the efficiency of 7.38% and 6.85%, respectively (Chen et al., 2018). In their work, they synthesized Co_3O_4-tungsten carbide (WC)-nitrogen-doped carbon (CN)/reduced graphene oxide (rGO) nanocomposite that combines strong electrochemical capacity of Co_3O_4, Pt-like properties from WC, and superior conductivity of rGO. Some researchers, like TiO_2, introduce Co_3O_4 nanoparticles to the electrolyte as an additive. The influence of nanoparticles on the characteristics of agarose polymer electrolyte was investigated by Ying et al. (2014). DSC measurements revealed a decrease in crystallinity, implying a cross-linking effect of the nanoparticles with the polymer matrix. The acidic surface of Co_3O_4 nanoparticles acts as a Lewis acid-base interaction centre with the ionic species in the electrolyte. For instance, triiodide ions, I_3^- will attract to the surface of Co_3O_4 nanoparticles, forming a bridge network in the polymer matrix and improving the ionic conductivity of electrolyte. Another group of researchers studies examined the effect of Co_3O_4 concentration in the electrolyte and discovered that electrolytes with Co_3O_4 are two times more efficient than electrolytes without Co_3O_4 (Saidi et al., 2019). However, the ionic conductivity was reduced above a critical concentration threshold due to the agglomeration of Co_3O_4 nanoparticles, which impeded charge transport in the electrolyte, according to electrochemical impedance studies (Figure 9.5).

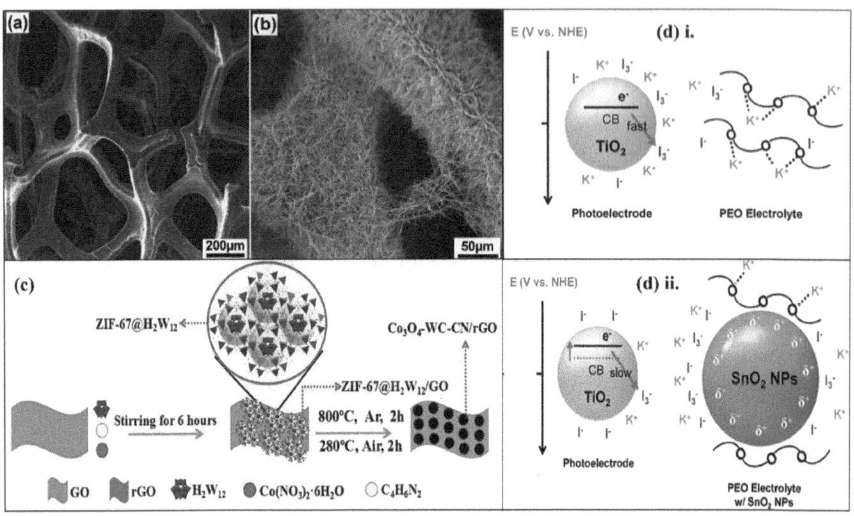

FIGURE 9.5 SEM images of (a) 3D graphene foam and (b) 3D graphene/Co_3O_4 nanowire composite. Reprinted with permission from Dong et al. (2012), (c) illustration of the Co_3O_4-WC-CN/rGO preparation process. Reprinted with permission from Chen et al. (2018) and (d) schematic diagrams of ionic species movement in (i.) polymer electrolyte and (ii.) polymer electrolyte with nanoparticles. Reprinted with permission from Chae et al. (2014).

9.7 ZINC OXIDE-BASED NANOCOMPOSITES

Zinc oxide, ZnO, is an *n*-type semiconductor with a direct band gap of 3.3 eV and a strong electron mobility of 100 $cm^2V^{-1}s^{-1}$. This makes it a good candidate for working electrode of DSSC application other than TiO_2. When compared to DSSC-based TiO_2, a DSSC-based ZnO is believed to show reduced recombination reactions. However, the greatest efficiency recorded for DSSC-based ZnO was 8.03% (He et al., 2015), whereas DSSC-based TiO_2 was 14.2% (Ji et al., 2020), indicating that ZnO still has a long way to go for replacing TiO_2 as the working electrode in DSSCs. This is because DSSC-based ZnO often encounters problems such as chemical instability of ZnO which causes ZnO to dissolve and create dye/Zn^{2+} complexes. Eventually, a thick insulating layer was formed between dye and ZnO, limiting the charge transfer from dye to the conduction band of ZnO. To overcome the problem, many core-shell structures have been developed using ZnO nanoarrays (NAs) by depositing a TiO_2 coating layer on the surface of ZnO-NAs. For instance, Feng et al. found that the core-shell structure ZnO@TiO_2 NA composite electrode was 2.6 times more efficient than pure ZnO-NAs electrode (Feng et al., 2012). This proved that the TiO_2 shell efficiently prevents ZnO and dye interaction. On the other hand, ZnO nanoparticles (NPs) can be incorporated into TiO_2 as a light scattering layer to improve the efficiency of DSSC applications. As previously discussed, grain boundaries of TiO_2 inhibit the transport of electron and result in the electron recombination with an electrolyte, leading to a dark current formation and lower device efficiency. ZnO NPs can reflect the incoming solar radiation and increase incident light travel distance in the working electrode. Using density functional theory (DFT), Ahmad et al. (2016) investigated the impact of ZnO on the conduction band of a composite working electrode (ZnO/TiO_2). The band gap of TiO_2 was reported as 3.10 eV, while the band gap of ZnO/TiO_2 composite was reported as 3.52 eV. This revealed that introduction of ZnO to the composite electrode causes the conduction band of the composite electrode to shift negatively, resulting in improved photovoltaic performance, particularly in terms of short-circuit current density and efficiency. ZnO-based nanocomposite can be used as an electrode in supercapacitor applications as well. Traditionally, porous carbon has been widely used as the electrode, and recently, researchers are focussed on the development of pseudocapacitive materials that can be combined with porous carbon materials. ZnO is a suitable material due to its inexpensive raw material cost, high electrochemical activity, and environmental friendliness; however, it was restricted by its low-rate capability and poor reusability during cycling. The double-layer capacitance of carbon materials with large specific surface area and the faradaic capacitance of ZnO can be coupled when ZnO is introduced into carbon materials. As a result, the capacitance and energy/power density are improved. For instance, Chang and Kim fabricated ZnO/activated carbon nanofiber (ACNF) composite to serve as the supercapacitor electrodes (Kim and Kim, 2015). Both the specific capacitance and energy density of the ZnO-based composite electrodes are greater than those of pure ACNF electrodes, indicating that ZnO-based composite electrodes have superior capacitive performance (Figure 9.6).

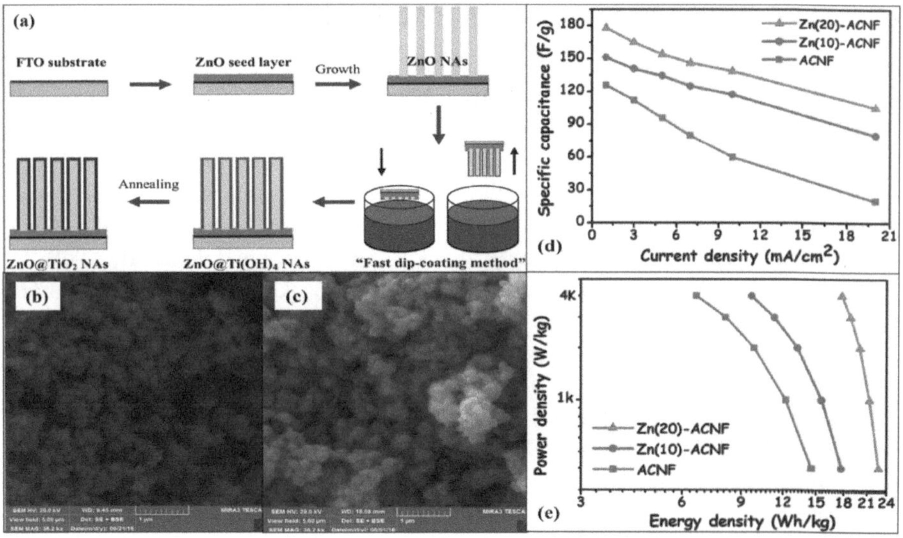

FIGURE 9.6 (a) Illustration of the synthesis process of ZnO@TiO_2 NAs. Reprinted with permission from Feng et al. (2012), SEM images of ZnO/TiO_2 composites: (b) bare TiO_2 and (c) ZnO/TiO_2 at 1 mm. Reprinted with permission from Ahmad et al. (2016), (d) specific capacitances as a function of various current densities and (e) Ragone plots of the ACNF and ZnO/ACNF composite. Reprinted with permission from Kim and Kim (2015).

9.8 CHROMIUM OXIDE-BASED NANOCOMPOSITES

Chromium oxide (Cr_2O_3) is a p-type transition metal oxide (TMO) semiconductor with a wide band gap of 3.4 eV and a theoretical capacity of 1,058 mAh/g, leading it to be utilized in many fields, ranging from catalysts, sensors, Li-ion batteries, and especially for supercapacitors due to its attractive properties such as cost-effectiveness, variable oxidation state, good cyclability, easy availability, and high thermal stability (Shafi et al., 2021). Chromium compounds consisting of Cr^{2+}/Cr^{3+} redox couples can be viewed as a promising candidate in supercapacitor application. There are, however, drawbacks of Cr_2O_3, which are the poor cycling performance caused by significant volume change during charge-discharge processes and poor electrical conductivity that restricts its practical application in supercapacitors. A strategy that can be implemented to enhance the cyclic performance of Cr_2O_3 is by forming a composite with carbon nanomaterials, for example graphene which is known for its excellent electrical conductivity. By doing so, the carbon nanomaterials can reduce the change in volume of Cr_2O_3 during the charge-discharge processes and improve the electrical conductivity of Cr_2O_3 (Asen and Shahrokhian, 2018). Ullah et al. had fabricated Cr_2O_3-carbon porous nanoribbons by carbonizing a mixture of polyfurfuryl alcohol and MIL-101(Cr) at 900°C and used as the electrode applied in supercapacitor. The nanocomposite produced showed excellent capacitive performance where the specific capacitance was greater than 290 F/g at a scan rate of 2 mV/s and a current density of 0.25 A/g. The produced sample also showed excellent long-term cyclability by maintaining 95.5% of its initial capacitance after 3,000 cycles.

The outstanding capacitor performance can be attributed to the combination of micro-mesopores with the polar functionalities of the carbon matrix which ensures maximum ionic wettability and storage during electrochemical assessments (Ullah et al., 2015). On a separate note, Shafi et al. investigated ultrafine Cr_2O_3 particles produced through the sol-gel method which possess excellent structural and electrochemical features as a promising capacitive energy candidate for supercapacitors. The theoretically estimated energy band gap calculated by the team was 2.65 eV, while the obtained energy band gap from the absorbance data from UV spectroscopy was 2.6 eV. A suitable energy band gap aids in the transfer of electrons between the conduction and valence bands during faradaic reactions and thus improves charge storage which leads to enhanced capacitance as well. The produced nanoparticles showed an impressive specific capacitance of 340 F/g at a specific current density of 0.5 A/g. The sample also showed an excellent cyclic retention of 85% after 3,000 cycles. Their study on the preparation method and nanostructure of the samples demonstrated the potential of implementing Cr_2O_3-based materials in energy storage devices (Figure 9.7) (Shafi et al., 2021). In a separate study conducted by Shafi and team, they have synthesized Cr_2O_3/reduced graphene oxide (rGO) via hydrothermal method and reported excellent electrochemical capacitive properties from the produced nanocomposite. It was discovered that the electrochemical properties of Cr_2O_3/rGO nanocomposite have strong synergistic effects due to the good structural compatibility between Cr_2O_3 nanoplates and rGO. The produced Cr_2O_3/rGO nanocomposite exhibited excellent cyclic retention by maintaining 92% of the initial

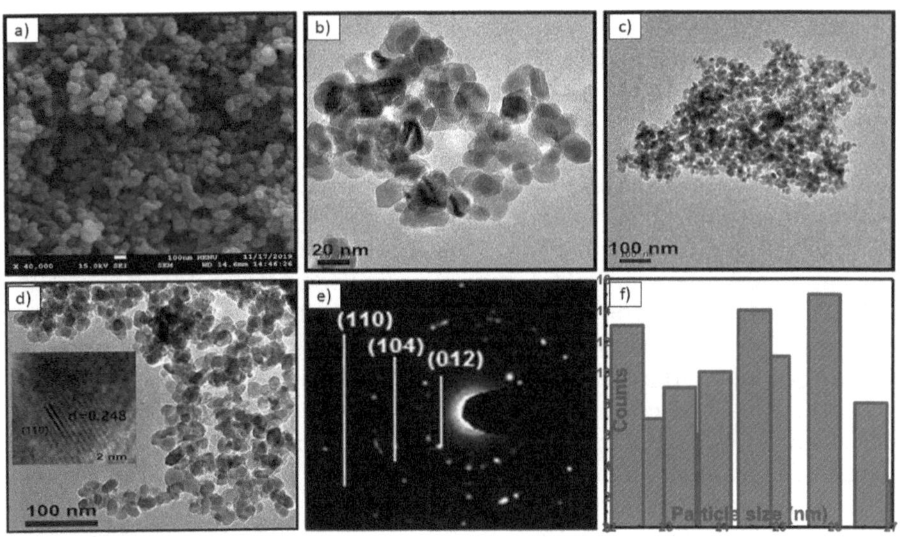

FIGURE 9.7 (a) SEM image of Cr_2O_3 nanoparticles at 200 nm, (b) TEM image of Cr_2O_3 nanoparticles at 20 nm, (c) TEM image of Cr_2O_3 nanoparticles at 100 nm, (d) TEM image of Cr_2O_3 nanoparticles at 100 nm (inset of Figure 1(d) shows the HRTEM image of Cr_2O_3 nanoparticles), (e) SAED of Cr_2O_3 nanoparticles, and (f) particle size distribution of Cr_2O_3 nanoparticles. Reprinted with permission from (Shafi et al., 2021).

value after 3,500 cycles. The specific capacitance recorded was 556 and 310 F/g at a specific current of 0.75 and 1.75 A/g, respectively. The CV analysis for the sample achieved 635 F/g at a scan rate of 1.1 mV/s. This study shows an example of the introduction of rGO for the electronic conductivity enhancement of the composite by adding rGO through the formation of heterojunctions (Shafi et al., 2020).

9.9 IRON OXIDE-BASED NANOCOMPOSITES

Iron oxide materials are inexpensive and can be found in abundance, and when compared to other TMOs, iron oxide materials are environmentally friendly and are less toxic. Iron oxide materials are capable of achieving high theoretical specific capacitance and wide operating potential which are key in developing supercapacitors with high energy densities. The various forms, such as iron(III) oxide (Fe_2O_3) and iron(II,III) (Fe_3O_4) with theoretical specific capacitance of 3,625, and 2,299 F/g, respectively, have become the centre of attention for electrode materials (Chen et al., 2015). Iron oxides suffer from two major drawbacks, namely poor electronic conductivity and capacitance decay, which result in a limitation in electrochemical performance. There are a few methods that can be used to overcome these drawbacks. The efficient method would be to design special nanostructures, such as nanotubes (NTs), nanorods (NRs), and nanoparticles (NPs), which will increase specific surface area, decrease the diffusion length of ions, and help accommodate strain during volume expansion (Zheng et al., 2018). Other than that, the combination with other materials such as carbon nanostructures, conducting polymers, and other metal oxides can also aid in elevating the limitations of iron oxide materials (Zhi et al., 2013). Among the various iron oxide materials, Fe_2O_3, also known as ferric oxide, has gained interest as an electrode for supercapacitors. Among the crystal structures for Fe_2O_3, α-Fe_2O_3 is the most thermodynamically stable phase. However, despite the advantages of Fe_2O_3 as an iron oxide in general, it still suffers from limitations in the form of electrical conductivity, cyclic stability, and rate capability. It was found that combining different materials to form Fe_2O_3 composites is an effective method to minimize the drawbacks of Fe_2O_3 (Xia et al., 2016). Guan and team reported a novel architecture of carbon nanotube (CNT) and graphite foam (GF) framework with Fe_2O_3 nanoparticles coated on the surface (Guan et al., 2015). The CNTs were grown on ultrathin GF after utilizing the chemical vapour deposition (CVD) method as shown in Figure 9.8a. It was found that the obtained CNT-GF had high surface area, high electrical conductivity, and good mechanical flexibility. The CNT-GF substrate was subsequently coated with crystalline Fe_2O_3 after atomic layer deposition (ALD) as shown in Figure 9.8b and c. Among the samples prepared from different ALD cycles (200, 400, and 600 cycles), the highest areal capacitance of 659.5 mF cm^{-2} at a current of 5 mA cm^{-2} was achieved by GF-CNT/600 Fe_2O_3 sample. However, the GF-CNT/400 Fe_2O_3 sample managed to maintain 111.2% of its initial capacitance of 470 mF cm^{-2} at 20 mA cm^{-2} after 50,000 cycles. The fabricated supercapacitor consisting of GF-$CoMoO_4$ as a cathode, as shown in Figure 9.8d, achieved a 115.5 F/g at 14 A/g, an energy density of 74.7 Wh/kg, and an excellent cyclic stability of 95.4% retention after 50,000 cycles. The superior performance of the fabricated supercapacitor makes Fe_2O_3 a promising

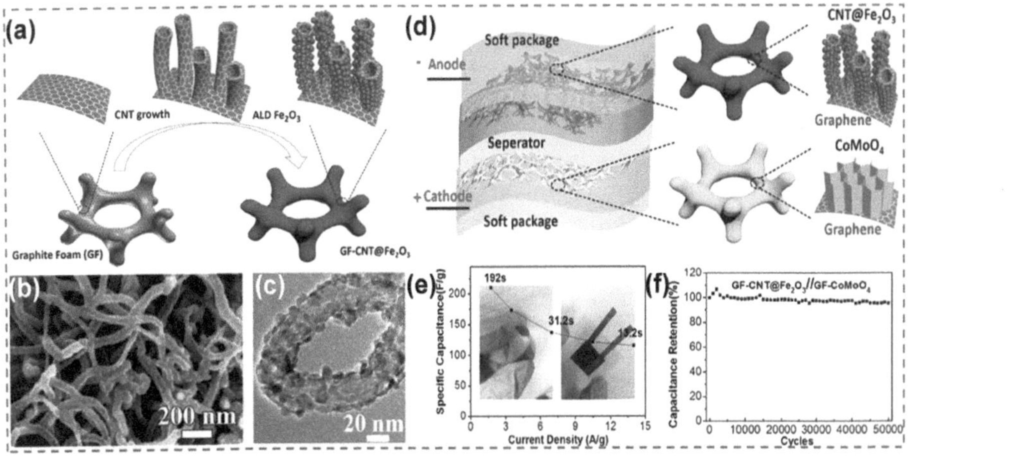

FIGURE 9.8 (a) Schematic of the fabrication of GF-CNT/Fe_2O_3 from GF. (b) SEM and (c) high-resolution transmission electron microscopy (HRTEM) images of GF-CNT/Fe_2O_3. (d) Schematic illustration of GF-CNT/Fe_2O_3//GF–$CoMoO_4$ ASC. (e) Rate capability and (f) cyclical stability of the GF-CNT/Fe_2O_3//GF–$CoMoO_4$. Reprinted with permission from Guan et al. (2015).

candidate for energy storage devices. The other suitable iron oxide material chosen to be used as an electrode for supercapacitors is Fe_3O_4, also known as ferrous ferric oxide, due to its high electrical conductivity of approximately 200 S/cm, wide operating potential window, low-cost, eco-friendly, and high theoretical specific capacitance as mentioned above (Kim and Kim, 2017). Unfortunately, similar to other metal oxides, it suffers from poor cyclic stability and inferior rate capability. To enhance the conductivity and cyclic stability of Fe_3O_4, researchers have combined this variation of iron oxide with various carbon materials such as CNTs and graphene to form nanocomposites. Graphene is often used given its large theoretical surface area and superb electroconductivity. On this note, Lin et al. had fabricated a novel configuration of Fe_3O_4 nanosheets *in situ* encapsulated by graphene layers (G/Fe_3O_4) (Lin et al., 2017). The G/Fe_3O_4 was grown vertically on conductive substrates through hydrothermal treatment followed by a plasma-enhanced CVD. Due to the nature of the vertical-standing nanosheet arrays, the effective surface area was dramatically enhanced. The graphene coating not only improved the conductivity but also provided enhanced structural stability. As a result, when G/Fe_3O_4 was utilized as an electrode, it exhibited a high specific capacitance of 732 F/g at 2 A/g and an impressive cyclic stability of ~90.4% after 10,000 cycles. A fabricated asymmetric supercapacitor using G/Fe_3O_4 as the anode and $CuCo_2O_4$ as the cathode produced an energy density of 82.8 Wh/kg at 2,047 W/kg. From this research, it can be seen that graphene coating was able to shield the active Fe_3O_4 from degradation during cycling to obtain excellent cycling performance. Other than that, formation of Fe-O-C bonds between the Fe_3O_4 and graphene led to an enhanced chemical connection that resulted in improved conductivity and even cycling performance of the hybrid electrode.

9.10 RUTHENIUM OXIDE-BASED NANOCOMPOSITES

Ruthenium oxide (RuO_2) is a transition metal oxide that has been regarded as a promising electrode material for electrochemical capacitors, Li-ion batteries, and DSSCs, in regards to its many attractive features such as high theoretical capacitance of 2,000 F/g, capability of carrying out fast faradaic redox reactions, good electrical conductivity, high chemical and thermal stability, wide potential window, and good cycling stability (Majumdar et al., 2019). Despite all these advantages, RuO_2 particles tend to cause aggregation which would decrease the surface area and disrupt ion and electron transfer. Besides that, one other limitation that severely affects the utilization of RuO_2 as an electrode material is its extremely high cost (Zhou et al., 2014). Carbon materials have been regarded as excellent electrode materials due to the many advantages they provide, such as good electrical conductivity, high specific surface area, excellent mechanical and structural properties, and their capability of fast electron transfer rate during charging/discharging. When used as a substrate, carbon materials are able to avoid agglomeration of metal oxides, which leads to an efficient utilization of valuable metal oxides such as RuO_2. To further enhance specific capacitance and reduce the cost of RuO_2 electrodes, researchers have been synthesizing composites of RuO_2 with various carbon materials such as graphene, CNT, and AC (Hossain et al., 2018). Ji and team designed and developed a high-energy aqueous asymmetric supercapacitor (SC) by combining hexagonal tungsten oxide (h-WO_3) with hydrous RuO_2 on three-dimensional conducting carbon cloth (CC) via the hydrothermal method (Ji et al., 2019). It was found that the produced composite exhibited excellent electrochemical features in both the positive and negative potential windows in the presence of an aqueous sulphuric acid (H_2SO_4) electrolyte. This was due to their high electrical conductivity, nano-porous surface morphology, and redox activity. The asymmetric SC was fabricated in aqueous H_2SO_4 electrolyte with hydrated RuO_2/CC as the positive electrode and h-WO_3/CC as the negative electrode. The RuO_2//h-WO_3 asymmetric SC was optimized at an operating voltage of 1.6 V at 47.59 F/g with a volumetric capacitance of 3.52 F cm^{-3} at 5 mA cm^{-2} (0.68 A/g). At a power density of 40 W cm^{-3}, the asymmetric SC showed an excellent energy density of 1.25 W h cm^{-3} (Figure 9.9).

Moreover, Dao et al. synthesized RuO_2 and rGO composites via facile and scalable liquid plasma-assisted method and utilized them as electrode materials in SC and counter electrode for DSSC. The advanced liquid plasma-assisted method allowed the fabrication of high-quality RuO_2 nanoparticles/rGO composites in ambient atmosphere without the usage of inert gases. The RuO_2 nanoparticles with a size range of 4–10 nm that had homogenously decorated the surface of the rGO resulted in a large specific surface area which enhanced electrochemical capacitance of the materials. The RuO_2/rGO nanocomposite possessed a high capacitance of ~ 136.7 F/g at a scan rate of 20 mV/s. When the developed nanocomposite was used as a counter electrode in the DSSC, the efficiency obtained was 6.78% as compared to an efficiency of 6.20% when platinum was used as the counter electrode (Dao et al., 2019). This shows that composites of RuO_2 have high potentials in being used as electrodes in various energy storage devices.

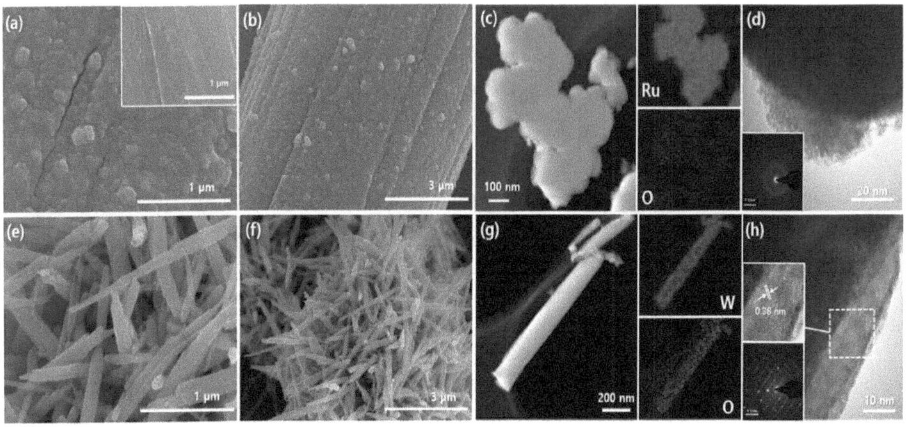

FIGURE 9.9 (a, b) Inset of (a) FESEM images of hydrous RuO_2 on carbon cloth at different magnifications; (c) TEM image of the hydrous RuO_2 and its EDS mapping; (d) HRTEM image of hydrous RuO_2 and its SAED pattern in the inset of (d); (e, f) FESEM images of h-WO_3 on carbon cloth at different magnifications; (g) TEM image of the h-WO_3 nanorod and its EDS mapping; and (h) HRTEM image of h-WO_3 nanorod. The inset shows the SAED pattern. Reprinted with permission from Ji et al. (2019).

9.11 CONCLUSION AND FUTURE PERSPECTIVES

This chapter mainly focussed on transition metal oxides and their nanocomposites with polymers and carbonaceous materials for energy generation and storage applications. Numerous transition metal oxides such as manganese dioxide, nickel oxide, copper oxide, titanium dioxide, zinc oxide, cobalt oxide, iron oxide, chromium oxide, ruthenium oxides, and their composites with polymers and carbonaceous materials were reviewed. These transition metal oxides and their nanocomposites can be used in energy generation and storage devices as electrode and electrolytes. The performance of metal oxide nanostructures and nanocomposites is quite satisfactory and highly accepted in the researchers' community. To fulfil the energy requirement and to get rid of global warming, energy generation and storage devices attained striking attention and are playing a vital part to achieve the sustainable development goals set by the United Nations (UN). In these devices, transition metal oxides and their nanocomposites are playing an outstanding role owing to the remarkable features of higher electronic conductivity, narrow band gap, greater thermal and electrochemical stability, and enhanced cycling efficiency. However, morphology, size, and shape of the transition metal oxides and their dispersion in the polymers and carbonaceous matrices remained a considerable challenge and need researchers' attention. In addition, these metal oxides and composite fabrication also seek inexpensive and green manufacturing technology. The physico-chemical, rheological, optoelectronic, conductive, and thermal properties of the composites need further exploration.

ACKNOWLEDGEMENTS

This work is financially supported by Technology Development Fund 1 (TeD1) from the Ministry of Science, Technology, and Innovation (MOSTI), Malaysia (MOSTI002-2021TED1).

REFERENCES

Abdah, Muhammad Amirul Aizat Mohd, Nur Hawa Nabilah Azman, Shalini Kulandaivalu, and Yusran Sulaiman. 2020. "Review of the use of transition-metal-oxide and conducting polymer-based fibres for high-performance supercapacitors." *Mater. Des.* 186: 108199.

Abdah, Muhammad Amirul Aizat Mohd, Norizah Abdul Rahman, and Yusran Sulaiman. 2019. "Ternary functionalised carbon nanofibers/polypyrrole/manganese oxide as high specific energy electrode for supercapacitor." *Ceram. Int.* 45 (7): 8433–8439.

Ahamad, Tansir, Mu Naushad, Mohd Ubaidullah, Yahya Alzaharani, and Saad M Alshehri. 2020. "Birnessite-type manganese dioxide nanoparticles embedded with nitrogen-doped carbon for high-performance supercapacitor." *J. Energy Storage* 32: 101952.

Ahmad, Waqar, Umer Mehmood, Amir Al-Ahmed, Fahad A Al-Sulaiman, M Zaheer Aslam, Muhammad Shahzad Kamal, and RA Shawabkeh. 2016. "Synthesis of zinc oxide/titanium dioxide (ZnO/TiO_2) nanocomposites by wet incipient wetness impregnation method and preparation of ZnO/TiO_2 paste using poly (vinylpyrrolidone) for efficient dye-sensitized solar cells." *Electrochim. Acta* 222: 473–480.

Ameri, Bahareh, Saied Saeed Hosseiny Davarani, Reza Roshani, Hamid Reza Moazami, and Azadeh Tadjarodi. 2017. "A flexible mechanochemical route for the synthesis of copper oxide nanorods/nanoparticles/nanowires for supercapacitor applications: The effect of morphology on the charge storage ability." *J. Alloys Compd.* 695: 114–123.

An, Cuihua, Yan Zhang, Huinan Guo, and Yijing Wang. 2019. "Metal oxide-based supercapacitors: Progress and prospectives." *Nanoscale Adv.* 1 (12): 4644–4658.

Asen, Parvin, and Saeed Shahrokhian. 2018. "Ternary nanostructures of Cr_2O_3/graphene oxide/conducting polymers for supercapacitor application." *J. Electroanal. Chem.* 823: 505–516.

Bredar, Alexandria RC, Amanda L Chown, Andricus R Burton, and Byron H Farnum. 2020. "Electrochemical impedance spectroscopy of metal oxide electrodes for energy applications." *ACS Appl. Energy Mater.* 3 (1): 66–98.

Carey, Graham H, Ahmed L Abdelhady, Zhijun Ning, Susanna M Thon, Osman M Bakr, and Edward H Sargent. 2015. "Colloidal quantum dot solar cells." *Chem. Rev.* 115 (23): 12732–12763.

Chae, Hwaseok, Donghoon Song, Yong-Gun Lee, Taewook Son, Woohyung Cho, Yong Bum Pyun, Tea-Yon Kim, Jung Hyun Lee, Francisco Fabregat-Santiago, and Juan Bisquert. 2014. "Chemical effects of tin oxide nanoparticles in polymer electrolytes-based dye-sensitized solar cells." *J. Phys. Chem.* 118 (30): 16510–16517.

Chen, Aicheng, and Cassandra Ostrom. 2015. "Palladium-based nanomaterials: Synthesis and electrochemical applications." *Chem. Rev.* 115 (21): 11999–12044.

Chen, Li, Weilin Chen, and Enbo Wang. 2018. "Graphene with cobalt oxide and tungsten carbide as a low-cost counter electrode catalyst applied in Pt-free dye-sensitized solar cells." *J. Power Sources* 380: 18–25.

Chen, Xu, Kunfeng Chen, Hao Wang, and Dongfeng Xue. 2015. "Composition design upon iron element toward supercapacitor electrode materials." *Mater. Focus* 4 (1): 78–80.

Chuminjak, Yaowamarn, Suphaporn Daothong, Aekapong Kuntarug, Ditsayut Phokharatkul, Mati Horprathum, Anurat Wisitsoraat, Adisorn Tuantranont, Jaroon Jakmunee, and Pisith Singjai. 2017. "High-performance electrochemical energy storage electrodes based on nickel oxide-coated nickel foam prepared by sparking method." *Electrochim. Acta* 238: 298–309.

Dao, Van-Duong, Nguyen Duc Hoa, Ngoc Hung Vu, Dang Viet Quang, Nguyen Van Hieu, Tran Thi Ngoc Dung, Nguyen Xuan Viet, Chu Manh Hung, and Ho-Suk Choi. 2019. "A facile synthesis of ruthenium/reduced graphene oxide nanocomposite for effective electrochemical applications." *Sol. Energy* 191: 420–426.

Deori, Kalyanjyoti, Sanjeev Kumar Ujjain, Raj Kishore Sharma, and Sasanka Deka. 2013. "Morphology controlled synthesis of nanoporous Co_3O_4 nanostructures and their charge storage characteristics in supercapacitors." *ACS Appl. Mater. Interfaces* 5 (21): 10665–10672.

Dong, Xiao-Chen, Hang Xu, Xue-Wan Wang, Yin-Xi Huang, Mary B Chan-Park, Hua Zhang, Lian-Hui Wang, Wei Huang, and Peng Chen. 2012. "3D graphene-cobalt oxide electrode for high-performance supercapacitor and enzymeless glucose detection." *ACS Nano* 6 (4): 3206–3213.

Feng, Yamin, Xiaoxu Ji, Jinxia Duan, Jianhui Zhu, Jian Jiang, Hao Ding, Gaoxiang Meng, Ruimin Ding, Jinping Liu, and Anzheng Hu. 2012. "Synthesis of ZnO@ TiO_2 core-shell long nanowire arrays and their application on dye-sensitized solar cells." *J. Solid State Chem.* 190: 303–308.

Gogoi, Debika, Priyanka Makkar, Manash R Das, and Narendra Nath Ghosh. 2022. "$CoFe_2O_4$ Nanoparticle decorated hierarchical biomass derived porous carbon based nanocomposites for high-performance all-solid-state flexible asymmetric supercapacitor devices." *ACS Appl. Electron. Mater.* 4 (2): 795–806.

Golkhatmi, Sanaz Zarabi, Arman Sedghi, Hoda Nourmohammadi Miankushki, and Maryam Khalaj. 2021. "Structural properties and supercapacitive performance evaluation of the nickel oxide/graphene/polypyrrole hybrid ternary nanocomposite in aqueous and organic electrolytes." *Energy* 214: 118950.

Guan, Cao, Jilei Liu, Yadong Wang, Lu Mao, Zhanxi Fan, Zexiang Shen, Hua Zhang, and John Wang. 2015. "Iron oxide-decorated carbon for supercapacitor anodes with ultrahigh energy density and outstanding cycling stability." *ACS Nano* 9 (5): 5198–5207.

Harilal, Midhun, Syam G Krishnan, Bincy Lathakumary Vijayan, M Venkatashamy Reddy, Stefan Adams, Andrew R Barron, Mashitah M Yusoff, and Rajan Jose. 2017. "Continuous nanobelts of nickel oxide-cobalt oxide hybrid with improved capacitive charge storage properties." *Mater. Des.* 122: 376–384.

He, Yitao, Jing Hu, and Yahong Xie. 2015. "High-efficiency dye-sensitized solar cells of up to 8.03% by air plasma treatment of ZnO nanostructures." *Chem. Commun.* 51 (90): 16229–16232.

Hossain, M Nur, Shuai Chen, and Aicheng Chen. 2018. "Fabrication and electrochemical study of ruthenium-ruthenium oxide/activated carbon nanocomposites for enhanced energy storage." *J. Alloys Compd.* 751: 138–147.

Hu, Yating, Yue Wu, and John Wang. 2018. "Manganese-oxide-based electrode materials for energy storage applications: How close are we to the theoretical capacitance?" *Adv. Mater.* 30 (47): 1802569.

Islam, Mobasserul, Mostafizur Rahaman, Ali Aldalbahi, Bibhudatta Paikaray, Jayakrushna Moharana, Subhadip Mondal, Narayan Ch Das, Prashant Gupta, and Radhashyam Giri. 2022. "High density polyethylene and metal oxides based nanocomposites for high voltage cable application." *J. Appl. Polym. Sci.* 139 (11): 51787.

Ji, Jung-Min, Haoran Zhou, Yu Kyung Eom, Chul Hoon Kim, and Hwan Kyu Kim. 2020. "14.2% efficiency dye-sensitized solar cells by co-sensitizing novel thieno [3, 2-b] indole-based organic dyes with a promising porphyrin sensitizer." *Adv. Energy Mater.* 10 (15): 2000124.

Ji, Su-Hyeon, Nilesh R Chodankar, and Do-Heyoung Kim. 2019. "Aqueous asymmetric supercapacitor based on RuO_2-WO_3 electrodes." *Electrochim. Acta* 325: 134879.

Julien, Christian M, and Alain Mauger. 2017. "Nanostructured MnO_2 as electrode materials for energy storage." *Nanomaterials* 7 (11): 396.

Kate, Ranjit S, Suraj A Khalate, and Ramesh J Deokate. 2018. "Overview of nanostructured metal oxides and pure nickel oxide (NiO) electrodes for supercapacitors: A review." *J. Alloys Compd.* 734: 89–111.

Kim, Chang Hyo, and Bo-Hye Kim. 2015. "Zinc oxide/activated carbon nanofiber composites for high-performance supercapacitor electrodes." *J. Power Sources* 274: 512–520.

Kim, Mee Rahn, and Dongling Ma. 2015. "Quantum-dot-based solar cells: Recent advances, strategies, and challenges." *J. Phys. Chem. Lett.* 6 (1): 85–99.

Kim, Myeongjin, and Jooheon Kim. 2017. "Synergistic interaction between pseudocapacitive Fe_3O_4 nanoparticles and highly porous silicon carbide for high-performance electrodes as electrochemical supercapacitors." *Nanotechnology* 28 (19): 195401.

Li, Xingyu, Caiying Wen, Mengwei Yuan, Zemin Sun, Yingying Wei, Luyao Ma, Huifeng Li, and Genban Sun. 2020. "Nickel oxide nanoparticles decorated highly conductive Ti_3C_2 MXene as cathode catalyst for rechargeable Li-O_2 battery." *J. Alloys Compd.* 824: 153803.

Lim, Seung Man, Juyoung Moon, Uoon Chul Baek, Jae Yeon Lee, Youngjin Chae, and Jung Tae Park. 2021. "Shape-controlled TiO_2 nanomaterials-based hybrid solid-state electrolytes for solar energy conversion with a mesoporous carbon electrocatalyst." *Nanomaterials* 11 (4): 913.

Lin, Jinghuang, Haoyan Liang, Henan Jia, Shulin Chen, Jiale Guo, Junlei Qi, Chaoqun Qu, Jian Cao, Weidong Fei, and Jicai Feng. 2017. "In situ encapsulated Fe_3O_4 nanosheet arrays with graphene layers as an anode for high-performance asymmetric supercapacitors." *J. Mater. Chem. A* 5 (47): 24594–24601.

Liu, Qin, Jinghui Zhu, Liwen Zhang, and Yejun Qiu. 2018. "Recent advances in energy materials by electrospinning." *Renew. Sust. Energ. Rev.* 81: 1825–1858.

Liu, Xinyue, Jianxing Wang, and Guowei Yang. 2018. "Amorphous nickel oxide and crystalline manganese oxide nanocomposite electrode for transparent and flexible supercapacitor." *Chem. Eng. J.* 347: 101–110.

Lu, Xiaofeng, Ce Wang, Frédéric Favier, and Nicola Pinna. 2017. "Electrospun nanomaterials for supercapacitor electrodes: Designed architectures and electrochemical performance." *Adv. Energy Mater.* 7 (2): 1601301.

Maduraiveeran, Govindhan, Manickam Sasidharan, and Wei Jin. 2019. "Earth-abundant transition metal and metal oxide nanomaterials: Synthesis and electrochemical applications." *Prog. Mater. Sci.* 106: 100574.

Mahdi, Fatemeh, Mehran Javanbakht, and Saeed Shahrokhian. 2022. "In-site pulse electrodeposition of manganese dioxide/reduced graphene oxide nanocomposite for high-energy supercapacitors." *J. Energy Storage* 46: 103802.

Mahmood, Nasir, Isabela Alves De Castro, Kuppe Pramoda, Khashayar Khoshmanesh, Suresh K Bhargava, and Kourosh Kalantar-Zadeh. 2019. "Atomically thin two-dimensional metal oxide nanosheets and their heterostructures for energy storage." *Energy Storage Mater.* 16: 455–480.

Majumdar, Dipanwita, and Srabanti Ghosh. 2021. "Recent advancements of copper oxide based nanomaterials for supercapacitor applications." *J. Energy Storage* 34: 101995.

Majumdar, Dipanwita, Thandavarayan Maiyalagan, and Zhongqing Jiang. 2019. "Recent progress in ruthenium oxide-based composites for supercapacitor applications." *ChemElectroChem* 6 (17): 4343–4372.

Nanayakkara, Sanjini U, Jao van de Lagemaat, and Joseph M Luther. 2015. "Scanning probe characterization of heterostructured colloidal nanomaterials." *Chem. Rev.* 115 (16): 8157–8181.

Nowotny, Janusz. 2019. *Oxide Semiconductors for Solar Energy Conversion: Titanium Dioxide*. Boca Raton: CRC Press/Taylor & Francis.

Pawar, Sambhaji M, Jongmin Kim, Akbar I Inamdar, Hyeonseok Woo, Yongcheol Jo, Bharati S Pawar, Sangeun Cho, Hyungsang Kim, and Hyunsik Im. 2016. "Multi-functional reactively-sputtered copper oxide electrodes for supercapacitor and electro-catalyst in direct methanol fuel cell applications." *Sci. Rep.* 6 (1): 1–9.

Ramesh, Sivalingam, Hemraj M Yadav, K Karuppasamy, Dhanasekaran Vikraman, Hyun-Seok Kim, Joo-Hyung Kim, and Heung Soo Kim. 2019. "Fabrication of manganese oxide@ nitrogen doped graphene oxide/polypyrrole (MnO_2@ NGO/PPy) hybrid composite electrodes for energy storage devices." *J. Mater. Res. Technol.* 8 (5): 4227–4238.

Ray, Apurba, Atanu Roy, Samik Saha, Monalisa Ghosh, Sreya Roy Chowdhury, T Maiyalagan, Swapan Kumar Bhattacharya, and Sachindranath Das. 2019. "Electrochemical energy storage properties of Ni-Mn-oxide electrodes for advance asymmetric supercapacitor application." *Langmuir* 35 (25): 8257–8267.

Rosaiah, P, Jinghui Zhu, Dadamiah PMD Shaik, OM Hussain, Yejun Qiu, and Lei Zhao. 2017. "Reduced graphene oxide/Mn_3O_4 nanocomposite electrodes with enhanced electrochemical performance for energy storage applications." *J. Electroanal. Chem.* 794: 78–85.

Saidi, Norshahirah M, Fatin Saiha Omar, Arshid Numan, David C Apperley, Mohammed M Algaradah, Ramesh Kasi, Alyssa-Jennifer Avestro, and Ramesh T Subramaniam. 2019. "Enhancing the efficiency of a dye-sensitized solar cell based on a metal oxide nanocomposite gel polymer electrolyte." *ACS Appl. Mater. Interfaces* 11 (33): 30185–30196.

Sanchez, Jaime S, Afshin Pendashteh, Jesus Palma, Marc Anderson, and Rebeca Marcilla. 2018. "Porous NiCoMn ternary metal oxide/graphene nanocomposites for high performance hybrid energy storage devices." *Electrochim. Acta* 279: 44–56.

Senadeera, GKR, CA Thotawatthage, and MAKL Dissanayake. 2020. "Efficiency enhancement in dye sensitized solar cells by light scattering in photoanode with TiO_2 nanotubes." *J. Phys.: Conf. Ser.* 1552: 012002.

Shafi, Imran, Erjun Liang, and Baojun Li. 2021. "Ultrafine chromium oxide (Cr_2O_3) nanoparticles as a pseudocapacitive electrode material for supercapacitors." *J. Alloys Compd.* 851: 156046.

Shafi, Imran, Yuanyue Liu, Gaojie Zeng, Zijiong Li, Baojun Li, and Erjun Liang. 2020. "Cr_2O_3/rGO nanocomposite with excellent electrochemical capacitive properties." *SN Appl. Sci.* 2 (11): 1–10.

Sharma, JK, Pratibha Srivastava, Gurdip Singh, M Shaheer Akhtar, and SJMS Ameen. 2015. "Green synthesis of Co_3O_4 nanoparticles and their applications in thermal decomposition of ammonium perchlorate and dye-sensitized solar cells." *Mater. Sci. Eng. B* 193: 181–188.

Shinde, SK, HM Yadav, GS Ghodake, AA Kadam, VS Kumbhar, Jiwook Yang, Kyojung Hwang, AD Jagadale, Sunil Kumar, and DY Kim. 2019. "Using chemical bath deposition to create nanosheet-like CuO electrodes for supercapacitor applications." *Colloids Surf. B* 181: 1004–1011.

Shrivastav, Vishal, Shashank Sundriyal, Ki-Hyun Kim, Ravindra K Sinha, Umesh K Tiwari, and Akash Deep. 2020. "Metal-organic frameworks-derived titanium dioxide-carbon nanocomposite for supercapacitor applications." *Int. J. Energy Res.* 44 (8): 6269–6284.

Tsai, Chih-Hung, Chia-Ming Lin, and Yen-Cheng Liu. 2020. "Increasing the efficiency of dye-sensitized solar cells by adding nickel oxide nanoparticles to titanium dioxide working electrodes." *Coatings* 10 (2): 195.

Ullah, Shaheed, Inayat Ali Khan, Mohammad Choucair, Amin Badshah, Ishtiaq Khan, and Muhammad Arif Nadeem. 2015. "A novel Cr_2O_3-carbon composite as a high performance pseudo-capacitor electrode material." *Electrochim. Acta* 171: 142–149.

Viswanathan, Aranganathan, and Adka Nityananda Shetty. 2017. "Facile in-situ single step chemical synthesis of reduced graphene oxide-copper oxide-polyaniline nanocomposite and its electrochemical performance for supercapacitor application." *Electrochim. Acta* 257: 483–493.

Viswanathan, Aranganathan, and Adka Nityananda Shetty. 2018. "Single step synthesis of rGO, copper oxide and polyaniline nanocomposites for high energy supercapacitors." *Electrochim. Acta* 289: 204–217.

Wang, Qiushi, Yifu Zhang, Jinqiu Xiao, Hanmei Jiang, Tao Hu, and Changgong Meng. 2019. "Copper oxide/cuprous oxide/hierarchical porous biomass-derived carbon hybrid composites for high-performance supercapacitor electrode." *J. Alloys Compd.* 782: 1103–1113.

Xia, Qiuying, Meng Xu, Hui Xia, and Jianping Xie. 2016. "Nanostructured Iron oxide/hydroxide-based electrode materials for supercapacitors." *ChemNanoMat* 2 (7): 588–600.

Xiao, Yu-Chen, Cheng-Yan Xu, Pan-Pan Wang, Hai-Tao Fang, Xue-Yin Sun, Fei-Xiang Ma, Yi Pei, and Liang Zhen. 2018. "Encapsulating MnO nanoparticles within foam-like carbon nanosheet matrix for fast and durable lithium storage." *Nano Energy* 50: 675–684.

Yang, Ying, Jiarui Cui, Pengfei Yi, Xiaolu Zheng, Xueyi Guo, and Wenyong Wang. 2014. "Effects of nanoparticle additives on the properties of agarose polymer electrolytes." *J. Power Sources* 248: 988–993.

Yao, Hua, Feng Zhang, Gaowei Zhang, and Yangyi Yang. 2019. "A new hexacyanoferrate nanosheet array converted from copper oxide as a high-performance binder-free energy storage electrode." *Electrochim. Acta* 294: 286–296.

Yu, Guihua, Liangbing Hu, Michael Vosgueritchian, Huiliang Wang, Xing Xie, James R McDonough, Xu Cui, Yi Cui, and Zhenan Bao. 2011. "Solution-processed graphene/MnO_2 nanostructured textiles for high-performance electrochemical capacitors." *Nano Lett.* 11 (7): 2905–2911.

Yu, Wan-Chin, Lu-Yin Lin, Wei-Cheng Chang, Si-Hua Zhong, and Chao-Chin Su. 2018. "Iodine-free nanocomposite gel electrolytes for quasi-solid-state dye-sensitized solar cells." *J. Power Sources* 403: 157–166.

Zheng, Shasha, Huaiguo Xue, and Huan Pang. 2018. "Supercapacitors based on metal coordination materials." *Coord. Chem. Rev.* 373: 2–21.

Zhi, Mingjia, Chengcheng Xiang, Jiangtian Li, Ming Li, and Nianqiang Wu. 2013. "Nanostructured carbon-metal oxide composite electrodes for supercapacitors: A review." *Nanoscale* 5 (1): 72–88.

Zhou, Zhou, Yirong Zhu, Zhibin Wu, Fang Lu, Mingjun Jing, and Xiaobo Ji. 2014. "Amorphous RuO_2 coated on carbon spheres as excellent electrode materials for supercapacitors." *RSC Adv.* 4 (14): 6927–6932.

10 Poly(amidoamine) Dendrimer-Encapsulated Metal Nanoparticles

Synthesis and Applications

Ravendra Kumar and Divya Kushwaha

10.1 INTRODUCTION

Dendrimers are a family of well-defined, regularly branched, nanosized, globular macromolecules, which can be synthesized in various shapes and sizes in a controlled manner. This 3D architecture is comprised of the following regions: (a) a central core, (b) repetitive units of monomers, called generations, (c) surface groups, and (d) void spaces, where molecules/ions can be entrapped. With an increase in the number of generations on the core, the diameter of the dendrimer increases linearly, while the number of surface groups increases exponentially. Typically, low-generation dendrimers are flexible; however, with each increasing generation, the macromolecule becomes more rigid and dense (Fréchet and Tomalia, 2001; Newkome et al., 2001). To date, a diverse class of dendrimer structures comprising distinct chemical features have been designed and synthesized by carefully tailoring the architectures, sizes, and peripheral surface groups. The potential application of these macromolecules is extensively explored in the field of biomedicine (for drug delivery, diagnostics, and imaging purposes) (Markowicz-Piasecka et al., 2016), supramolecular chemistry (host-guest reactions and self-assembly processes) (Zeng and Zimmerman, 1997; Baars and Meijer, 2000), catalysis (Astruc and Chardac, 2001), and energy transfer (Adronov and Fréchet, 2000). Among various types of dendrimers, poly(amidoamine) (PAMAM) dendrimer is one of the most widely explored and commercially viable classes. The discovery and synthesis of PAMAM dendrimer, which is also known as starburst dendrimer, are credited to Tomalia et al. in 1985. In general, PAMAM dendrimers are composed of a central core of ethylenediamine and the repeating units of methyl acrylate. As the number of repeating units in dendrimers increases, the number of generations increases (like G0, G1, G2, and so on) and so does the size (diameter), molecular weight, and the number of peripheral surface groups. PAMAM dendrimers are characterized by the number of functionalities present, such as the amide group, the interior tertiary amine groups, and the surface groups (-OH, -NH_2, or -COOH group) (Tomalia et al., 1986). The use of dendrimer to prepare metal-encapsulated nanoparticles by complexing the metal ions onto the cavities of dendrimer was pioneered by Zhao et al. (1998) and Tomalia

DOI: 10.1201/9781003481157-10

and Balogh (1998). The most widely used synthetic strategy for metal-dendrimer nanoparticles (Den-NPs) includes the complexation of metal ions within the interior cavity of dendrimer and the subsequent reduction of metal ions to elemental metal NPs. To date, various metal ions complexed with diverse generations of PAMAM dendrimers have been used to prepare the dendrimer-encapsulated metal nanoparticles in a controlled manner, and their most important applications have been found in the area of catalysis and energy storage (Devadas et al., 2021). This chapter will give an overview of the PAMAM-encapsulated monometallic metal NPs prepared for electrochemical energy applications.

10.2 DENDRIMER-ENCAPSULATED METAL NPs

The synthesis of zerovalent metal nanoparticles is challenging as the composites are prone to air oxidation and precipitation. Thus, during the preparation of metal NPs in the reduction step of metal ions, the use of stabilizers and templates such as polymers and surfactants is recommended. Though the use of stabilizers and templates may sometimes lead to an overall decrease in the activity of metal nanoparticles, dendrimers have been proven to be an excellent system in this regard, as they do not passivate the activity of metal nanoparticles. Moreover, they behave as a template over which monodisperse metal nanoclusters can be constructed. Additionally, they stabilize the metal nanoparticles and prevent agglomeration of the nanoparticles by confining them within their interiors (Yamamoto et al., 2020).

10.3 PAMAM-ENCAPSULATED METAL NPs

PAMAM dendrimer offers an excellent opportunity to prepare dendrimer-complexed nanoparticles because it allows control over the particle size, behaves as a suitable template, and prevents the agglomeration of nanocomposites. Many reports are available where diverse transition metal ions, such as Cu^{2+}, Pt^{2+}, Pd^{2+}, Ag^{+}, and Ni^{2+}, are conveniently complexed within the cavity of PAMAM dendrimers via interactions between the metal ion and lone pair electrons of the interior tertiary amine groups. Selective binding of the metal ions with interior amine groups can be achieved by decorating the surface with non-complexing groups, for instance, hydroxyl groups, or by converting the complexing surface groups, e.g., amine, into quaternary amine groups via adjusting the pH of the solution (Figure 10.1). This way, the bonded metal ions within the nanosized dendrimers remain nearly monodisperse and are later reduced in situ to form nanoclusters within the dendrimer (Devadas et al., 2021). The metal loading capacity of a PAMAM dendrimer depends on the size of the cavity and the functional groups present in the interior and exterior of the dendrimer (Scott et al., 2003). Higher-generation dendrimers with densely packed functionalities have shown more promising results in this regard. Interestingly, the surface functional groups present at the periphery of dendrimers provide a useful chemical handle, which enables surface immobilization (Zhao et al., 1998). Thus, the prepared metal-bound dendrimer nanocluster can be suitably used for the electrochemical reactions.

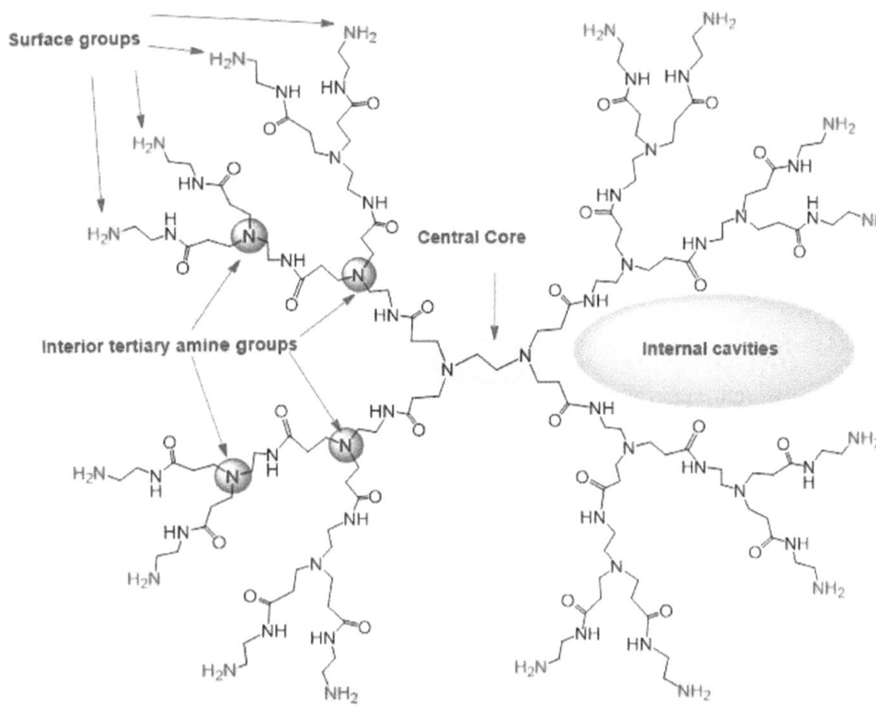

FIGURE 10.1 Second-generation (G2) PAMAM dendrimer.

10.3.1 PAMAM-Encapsulated Transition Metal NPs

The synthesis of nearly monodisperse nanosized zerovalent metal composites with controlled size growth remained a challenging task. Dendrimers are attractive nanoreactors in this regard, which can be used as a template that allows metal-binding sites for the preorganization of metal ions within their interiors separated by branch cells. Since the oxidation behavior, spectral properties, and complexation properties of copper metal have been extensively studied, the first report on the study of the structural aspect of PAMAM dendrimer-encapsulated metal was with Cu (II) ions by the research group of Ottaviani et al. (1994). Furthermore, the research groups of Tomalia and Crooks individually reported the synthesis and structural characterization of PAMAM dendrimer-encapsulated zerovalent Cu nanoparticles in 1998. Template synthesis strategy was introduced, where Cu^{2+} ions were extracted within the interior of a dendrimer and subsequently reduced to obtain zerovalent metal-Den-NPs (Figure 10.2) (Zhao et al., 1998; Tomalia and Balogh, 1998). Nanoparticles containing 4–64 Cu atoms were prepared by partitioning of Cu^{2+} within the PAMAM dendrimers. Spectrophotometric titration showed that G-4 PAMAM dendrimer, which possesses 62 interior amine groups, sorbed a total of 16 Cu^{2+} ions. Therefore, each Cu^{2+} ion was co-ordinated with about four amine groups within the dendrimer interior producing NPs of particle size 1.8 nm. Importantly,

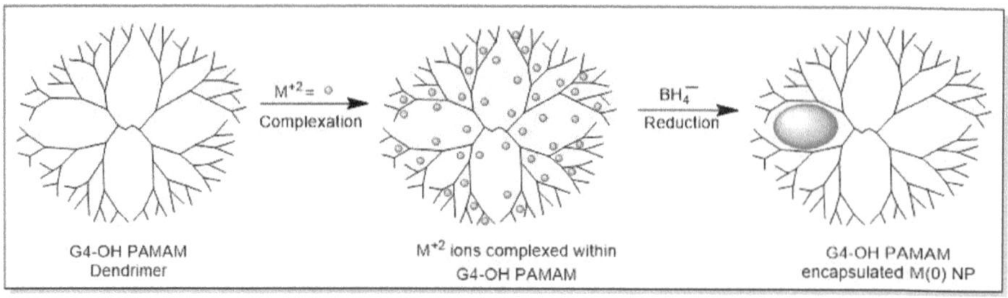

FIGURE 10.2 Schematic representation of zerovalent metal nanoparticle synthesis over PAMAM dendrimer template.

the use of different generations (sizes) of dendrimers allowed control over nanoparticle size (Zhao et al., 1998). Thus prepared [Cu(0)n-PAMAM] dendrimers remained stable in solution form from 7 to 90 days under oxygen-free conditions.

Later in 1999, Crooks et al. presented a new alternative "displacement method" of dendrimer-complexed metal NP synthesis, which was shown to be suitable for the preparation of NPs of noble metals that do not strongly complex with the amine groups of PAMAM dendrimers. The authors prepared sixth-generation PAMAM-encapsulated Cu (55 atoms) NPs using the template strategy and showed that Cu metal gets replaced when exposed to a solution of metal ions more noble than Cu (Figure 10.3) (Zhao and Crooks, 1999). Meijboom et al. investigated the maximum loading capacities of hydroxyl-terminated G4, G5, and G6 PAMAM dendrimers with Cu^{2+} ions using spectrophotometric titrations and Vivaspin filtration techniques, which showed Cu^{2+} loading of about 16, 32, and 57 for G4-OH, G5-OH, and G6-OH, respectively. However, nanoclusters had poor stability under atmospheric conditions or even under a nitrogen atmosphere (Patala et al., 2018).

Notably, Crooks et al. synthesized the small-sized, highly monodisperse Ni NPs (bearing 55 and 147 atoms) by the use of G-6 PAMAM dendrimer modified with alkyl groups at the periphery as a template and were able to attain the particle size <2 nm diameter. With the use of hydrophobic dendrimers and organic

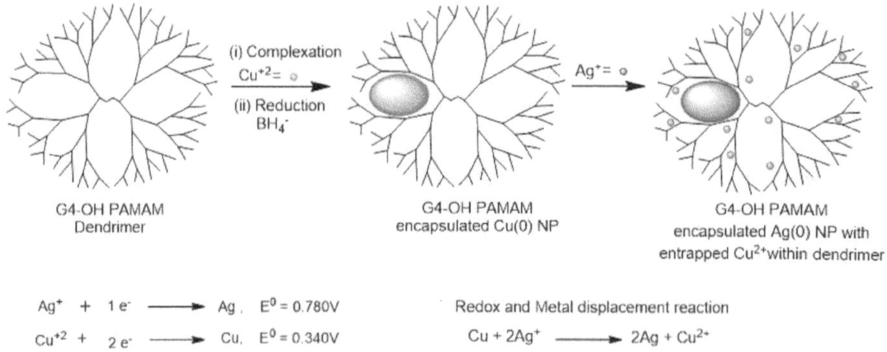

FIGURE 10.3 Schematic illustration of Ag Den-NP synthesis via metal displacement reaction of Cu Den-NPs with Ag^+ ion.

solvent (toluene), the authors successfully avoided the oxidation of Den-Ni (0) NPs (Knecht et al., 2006). Nevertheless, NPs were found stable under a nitrogen atmosphere only, and precipitation was observed upon exposure to air. Furthermore, under similar reaction conditions, the synthesis of Fe (0) NPs having 55 and 147 atoms was achieved by the use of G-6 PAMAM dendrimers decorated with peripheral dodecyl groups (Knecht and Crooks, 2007). Dendrimer-encapsulated Ni and Co NPs didn't show significant O_2 reduction reaction (ORR) activity (Goubert-Renaudin et al., 2011). However, their use as an antimicrobial agent (Mazumder et al., 2013) and as a catalyst in organic reactions has been reported (Wu et al., 2012; Kalbasi et al., 2014).

10.3.2 PAMAM-Encapsulated Platinum Nanoparticles

Pt metal efficiently catalyzes the O_2 reduction reaction (ORR), which is the most crucial cathode reaction in hydrogen-based fuel cells. To obtain high-performance Pt-derived electrocatalysts, several groups have worked toward the development of reduced particle size Pt nanocomposites with high stability and good O_2 accessibility (Shao et al., 2016). Crooks and co-workers reported the synthesis of G-4 PAMAM dendrimer (G4-OH)-encapsulated Pt (0) in two steps, which included sorption of a predetermined amount of Pt^{2+} into dendrimer followed by reduction of the encapsulated ions by the use of BH_4^-. A series of G4-OH NPs loaded with 12, 40, and 60 Pt^{2+} ions were prepared, where the Pt^{2+} ions were shown to complex with the interior tertiary amine groups rather than the exterior -OH groups. In the acidic solution of G4-OH dendrimer (at pH = 1 or less), no Pt^{2+} complexation was observed due to the protonation of amine groups. Apart from G4-OH PAMAM, when the NH_2-functionalized PAMAM dendrimer was used as a template, a Pt^{2+}-induced cross-linking was observed that led to the precipitation of dendrimer solution. Interestingly, G4-OH(Pt_{60}) NPs were immobilized onto the Au surface and a significant O_2 reduction catalytic activity was observed in them. This crucial finding led to the foundation for the preparation of new nanocomposites by varying surface functionalization (Zhao et al., 1998). In addition, G4-OH(Pt_{40}) Den-NPs were tested for their catalytic activity toward the hydrogenation reaction of alkenes in the aqueous medium (Zhao and Crooks, 1999).

Esumi et al. utilized G3- to G5-NH_2 PAMAM dendrimers to synthesize NPs in the particle size of 2.4–3.0 nm. Their study revealed that the particle size of NPs didn't vary with the change in concentrations and generations of the dendrimer. Moreover, stable Pt NPs were formed only when the ratio of surface groups of G3-NH_2 PAMAM/Pt^{4+} was above 40/1, attributed to the weak interactions between Pt colloids and dendrimers with amine groups (Esumi et al., 2000). Furthermore, their group prepared Pt NPs by the reduction of $PtCl_6^{2-}$ with dimethylamine borane (reducing agent) and ethyl acetate in the presence of methyl ester-terminated G1.5, G3.5, and G5.5 PAMAM dendrimers. The platinum NPs are obtained by stirring dendrimer/ Pt^{4+} solution for different time periods ranging from 5 min to three days. With the G3.3 and G5.5 dendrimers, a prolonged stirring of solutions led to the smaller-sized Pt NPs; however, a shorter mixing time resulted in Pt NPs at the exterior of the dendrimer (Esumi et al., 2000; Table 10.1).

TABLE 10.1
Monometallic Platinum (Pt)-Encapsulated Dend-NPs for Their Electrochemical Applications

Metal	Dendrimer-End Group	Generation	Applications	References
Pt^{2+}	PAMAM-OH, PAMAM-NH_2	G4	G4-OH(Pt_{60}) NP immobilized onto Au surface: electrocatalyst for O_2 reduction reaction ORR	Zhao and Crooks (1999)
Pt^{+2}/ Pd^{+2}	PAMAM-OH	G-4, G-6, G-8	G4-OH(Pd_{40}) NP: catalyst for hydrogenation of alkenes in aqueous solution	Zhao and Crooks (1999)
Pt^{+2}	PAMAM-OH	G-6	G6-OH(Pt_n) immobilized onto GCE: activity and kinetics of ORR are measured	Ye et al. (2007)
Pt^{+2}	PAMAM-OH	G-4	G4-OH(Pt_n) immobilized onto carbon fiber surfaces to study electrocatalytic activity for ORR	Ledesma-Garcia et al. (2008)
Pt^{+2}	PAMAM-OH	G-4	Carbon electrode modified with G4-OH(Pt_n) using electrochemical and physical impregnation methods was studied for fuel cell performance	Ledesma-Garcia, et al. (2009)
Pt^{+2}	PAMAM-OH	G-4	G4-OH(Pd_{40}) anchored onto carbon studied for ORR to develop cathode catalyst for polymer electrolyte fuel cells	Kottakkat et al. (2011)
Pt^{+2}	PAMAM-OH	G-4	G4-OH(Pt_n) as catalyst for ORR in direct methanol fuel cells	Ledesma-Garcia, et al. (2010)
Pt^{2+}	PAMAM-OH	G6	G6-OH Dend-Pt_{147} and Pt_{55} NPs were studied for the effect of mass transfer on the ORR	Dumitrescu et al. (2012)
Pt^{2+}	PAMAM-OH	G6	Study of G6-OH(Pt55) NPs/PPF electrode toward ORR in the presence and absence of dendrimer	Ostojic and Crooks et al. (2016)
Pt^{2+}	PAMAM-OH terminated with triple hydroxyl groups (-t) and single (-s) hydroxyl group	t-G-6 s-G-6	Catalytic potential of t-G6-OH/Pt_n and s-G6-OH/Pt_n examined for H_2 production via reduction of H^+	Yu et al. (2012)
Pt^{2+}	PAMAM-OH	G-4	Pt NPs immobilized on graphene oxides, carbon nanohorns (CNHs), and carbon nanotubes (CNTs) tested for hydrogen evolution reaction (HER)	Devadas et al. (2016)
Pt^{2+}	PAMAM-OH	G-4	Pt NPs fabricated with $PW_{12}O_{40}^{3-}$ evaluated for HER	Sun et al. (2005)

Moreover, Crooks et al. used G6-OH PAMAM dendrimers to prepare Den-NPs of size range 1–2 nm by complexing an average of 55, 100, 147, 200, and 240 Pt atoms. The effect of particle size on the activity of Pt catalyst and kinetics (quantitative) of ORR was examined using rotating disk voltammetry after immobilizing these Den-NPs on a glassy carbon electrode (GCE). The specific activity measurement showed the highest activity for the largest nanoparticle. For instance, G6-OH(Pt_{240}) NPs (size: 1.90±0.29 nm) reportedly had nine- to sixfold higher specific activity than the small-sized (1.44 ± 0.26 nm) G6-OH-(Pt_{55}) NPs. The surface area of NPs determined by CO-oxidation and H-desorption methods revealed a decrease in total Pt surface area with decreasing dendrimer size (Ye et al., 2007).

In an attempt to generate electrodes with enhanced electrocatalytic activity, the G4-OH-encapsulated Pt NPs were conjugated to the carbon fiber electrodes, which might behave as a fuel cell cathode for ORR. To achieve this, initially, the surfaces of carbon fiber were functionalized using three alternative electrochemical anodic pre-treatments in acidic media in such a way that gave activated surfaces bearing oxidized groups. Next, platinum-dendrimer nanocomposites were loaded by scanning of potential in a solution of previously synthesized dendrimer-encapsulated nanoparticles. Furthermore, an evaluation of their electrocatalytic activity in ORR suggested a good exchange current density at low platinum loading (Ledesma-Garcia et al., 2008).

G4-OH Dend-Pt NPs have been immobilized on porous carbon substrates (Kottakkat et al., 2011) to construct an H_2-O_2 fuel cell using electrochemical and physical impregnation. Three different membrane electrode assemblies (MEAs) were prepared from these modified surfaces that acted as electrodes and gas diffusion layers. The fuel cell performance was found to be better with MEA prepared by the physical impregnation method; however, a comparable performance was achieved with the MEA based on electrochemical immobilization by reducing the quantities of Pt, i.e., the electrocatalyst (Ledesma-Garcia et al., 2009).

The use of dendrimer-encapsulated Pt nanoparticles as cathodic catalysts in direct methanol fuel cells (DMFCs) has also been explored. The study of their electrocatalytic behavior toward ORR was conducted using rotating disk electrode configuration in an acidic medium with and without the presence of methanol (0.01, 0.1, and 1 M). The kinetic data recorded on dendrimer-encapsulated Pt nanoparticles displayed a higher selectivity for ORR in the presence of methanol when compared with the electrodes based on commercial Pt black catalysts. It was observed that in the presence of methanol, the dendrimer confers a protective effect on the Pt that prevents the interference of methanol and allows its use as a cathode in DFMCs working at various temperatures (Ledesma-Garcia et al., 2010).

With the use of a dual-electrode microelectrochemical device, Dumitrescu et al. studied the effect of mass transfer (k_t) on ORR, catalyzed by G6-OH Dend-Pt NPs of two different particle sizes each having 147 and 55 Pt atoms, respectively. ORR function was examined under high-k_t conditions with simultaneous detection of H_2O_2. At low values of k_t (0.02 cm s^{-1}), the effective number of electrons (n_{eff}) involved in the ORR, at Pt_{147} and Pt_{55} electrodes, was determined to be 3.7 and 3.4, respectively. The difference in the effective number of electrons involved in the ORR (n_{eff}) suggested the role of nanoparticle size on the ORR mechanism. At an increased k_t from 0.02 to 0.12 cm s^{-1}, a decreased ORR mass-transfer-limited current was observed for both Pt_{147} and Pt_{55}, which was concluded as a result of kinetic effects and not due to the change in the ORR mechanism (Dumitrescu and Crooks, 2012).

Recently, Ostojic et al. described the electrocatalytic properties of hydroxyl-terminated G6 PAMAM-encapsulated Pt nanoparticles (Pt DENs) that were immobilized onto a pyrolyzed photoresist film (PPF) electrode. Furthermore, they utilized ultraviolet/ozone (UV/O_3) treatment for the removal of PAMAM from the encapsulated Pt nanoparticles (Pt DENs) and have reported no change in the size, shape, and electrocatalytic property of Pt nanoparticles after dendrimer removal (Ostojic and Crooks, 2016).

Kim and co-workers reported a controllable synthesis of dendrimer-encapsulated Pt NPs (G6-OH, Pt atom no. = 220, 550, 880, and 1,320) by the use of repetitively coupled chemical reduction and galvanic exchange reactions. To synthesize Pt DENs, Cu^{2+} and Pt^{2+} precursors were co-added into a dendrimer solution that resulted in the selective complexation of Cu^{2+}, whereas Pt^{2+} remained uncomplexed and oxidized. The subsequent reduction reaction of Cu^{2+} with BH_4^- to generate Cu nanoparticles gets coupled with the galvanic exchange of Cu NPs with neighboring Pt^{2+} and proceeds continuously until all Pt^{2+} ions form Dend-Pt NPs. This synthesis strategy permitted the synthesis of large and homogenous Pt DENs having more than 1,000 Pt atoms encapsulated inside dendrimers within a short time of ~10 min (Cho et al., 2018).

The production of hydrogen via the electrocatalytic hydrogen evolution reaction (HER) is a promising alternative fuel and energy carrier. Platinum is an effective electrocatalyst in HER, and the reports are available for the use of dendrimer-encapsulated Pt NPs in catalyzing HER. Li et al. developed PAMAM G6-OH-encapsulated Pt NPs for their application as artificial hydrogenases. The prepared Dend-Pt NPs successfully produced hydrogen by the reduction of H^+ in the presence of visible light upon excitation with a photosensitizer. Excellent stability and effective catalytic activity were observed at a pH value of 9 and with the size of Pt clusters bearing 200 Pt atoms (Yu et al., 2012). Imae et al. prepared platinum nanoparticles (Pt NPs) anchored to graphene oxides (GOs), carbon nanohorns (CNHs), and carbon nanotubes (CNTs) to obtain an efficient electrocatalyst for HER. Nanocomposites of GO/ Dend-Pt NPs, CNH/ Dend-Pt NPs, and CNT/ Dend-Pt NPs loaded with ~1 wt% of Pt displayed low onset potential and high cathodic current density for HER, which was comparable to 40 times more loaded commercial 40 wt% Pt/C- and Pt/MoS_2-based electrocatalysts consisting of 10 and 2.03 wt% Pt (Devadas et al., 2016).

Sun et al. evaluated the HER efficiency of a nanocomposite of G4-OH (Pt NPs) and Keggin-type phosphotungstic acid ($PW_{12}O_{40}^{3-}$) prepared by immersing the electrode in indium-doped tin oxide (ITO)/3-aminopropyl triethoxysilane (3-APTES) solution. The electrocatalytic activity of PtNP-PAMAM was significantly improved in the presence of $PW_{12}O_{40}^{3-}$, and the HER was efficiently controlled by the composition of PtNP-PAMAM/$PW_{12}O_{40}^{3-}$ deposited on the electrodes. These results suggest a synergistic interaction between Pt NPs and $PW_{12}O_{40}^{3-}$ toward the enhancement of HER (Sun et al., 2005).

10.3.3 PAMAM-Encapsulated Palladium Nanoparticles

The first report of PAMAM dendrimer-encapsulated Pd nanoparticles was published by the research group of Crooks, in which they used template-based strategy to synthesize G4-OH(Pd_{40}) and G4-OH(Pd_{60}) NPs. Nanoparticles showed significant stability for several months under solvated conditions and displayed high catalytic activity toward the hydrogenation of alkenes in an aqueous solution, which was shown to be modulated by the use of variable generation of nanoparticles (Zhao and Crooks, 1999). Pd-based Den-NPs were also demonstrated to have significant catalytic potential toward the Heck organic reaction. However, the poor stability and Pd leakage during catalyst reuse are the major drawbacks that limit their application (Rahim et al., 2001).

Li et al. studied the catalytic potential of G3-OH- and G4-OH-encapsulated Pd NPs for the Suzuki reaction in aqueous conditions. On the one hand, G3 Den-NPs were found to be more efficient catalysts for the Suzuki reactions. On the other hand, G4 Den-NPs showed better stability than G3 Den-NPs owing to a strong complexation of Pd within the dendrimer cavity, however, which led to a loss of catalytic activity (Li and El-Sayed, 2001).

The chemical synthesis, properties, and stability of G4-OH and G4-NH_2 dendrimer-encapsulated Pd NPs were studied in detail by the Crooks group following the concerns regarding poor stability of the same. According to their report, Pd(0) DENs undergo oxidation to generate Pd^{2+} upon exposure to air and not to PdO. The rate of Pd oxidation didn't change with the dendrimer generation. Additionally, the oxidation of Pd DENs was found to be a reversible process, where oxidized Pd^{2+} was successfully reduced to zerovalent Pd nanoparticles by the use of H_2 gas without changing its particle size (Scott et al., 2003). Furthermore, their group synthesized Pd(0) NPs encapsulated with a series of G4-NH_2 dendrimers having surface -NH_2 groups functionalized with linkers glycidol and 2-methyl glycidol and examined their potential toward hydrogenation reaction (Oh et al., 2005).

Deraedt and colleagues described G4-OH PAMAM encapsulated metal nanoparticles (Pd and Pt) as an efficient and reusable catalyst for the dehydrogenation/hydrogenation of tetrahydroquinoline/indoline derivatives (Deraedt et al., 2017).

10.3.4 PAMAM-Encapsulated Silver Nanoparticles

There have been several reports on the synthesis of metal-Den-NPs (Cu^{2+}, Pd^{2+}, and Pt^{2+}) in a two-step protocol that includes the partition of metal into the interior of dendrimer followed by its reduction with BH_4^-. However, many metals (like Ag and Au) do not interact strongly with the interior amine groups of dendrimers to form complexes. Thus, the aforesaid method of nanocomposite formation doesn't work suitably with these metals. Crooks et al. developed a displacement methodology in this regard, in which dendrimer-encapsulated nanoclusters are prepared by the metals, which can complex strongly within the dendrimer, and next, in situ exchange of metal is performed to displace dendrimer-encapsulated metal with another noble metal. For instance, authors initially prepared G6-OH PAMAM-encapsulated Cu NPs, which upon exposure to a solution containing Ag^+ ions displaced Cu, whereas Ag^+ itself gets reduced to Ag(0) (Figure 10.3). Using the same protocol, G6-OH(Ag_{110})

NP (particle size <3 nm) was synthesized and immobilized onto Au electrode, which didn't exhibit any substantial catalytic activity toward the ORR process. Notably, Cu clusters can be displaced by Ag, Au, Pt, or Pd, and Ag clusters can be replaced by Au, Pt, or Pd (but not Cu) (Zhao and Crooks, 1999). Although the Ag^+ ions poorly complex within dendrimer, there are reports available in the literature for the synthesis of dendrimer-encapsulated Ag NPs by the use of reducing agent BH_4^-. Aggregation (Esumi et al., 2001) and large particle size NPs (size between 4 and 20 nm) have been observed in such composites, suggesting that NPs are stabilized at the dendrimer periphery rather than the dendrimer interior (Manna et al., 2001; Esumi et al., 2004; Fan et al., 2005; Xu et al., 2007; Ilunga and Meijboom, 2016; Štofik et al., 2009).

10.3.5 PAMAM-Encapsulated Gold (Au) Nanoparticles

Crooks et al. prepared Au colloids in the size range of 2–3 nm by the in-situ reduction of $HAuCl_4$ in the presence of PAMAM dendrimers. The colloidal solution was stabilized by the multiple dendrimers through their exterior amine groups. These solutions remained stable for one week (Gracia et al, 1999). Similarly, PAMAM-based Au NPs of size range 1–2 nm was synthesized with the use of 55 and 140 equiv, respectively, of $HAuCl_4$ per dendrimer. PAMAM dendrimers used were having positively charged quaternized amine terminal groups (Gn-Qp) (Kim et al., 2004). Furthermore, Au NPs are produced by performing the reduction of $HAuCl_4$ in formamide solvent within the PAMAM dendrimer (G1.5, G3.5, and G5.5) having methyl ester end groups. The size of NPs was found to be decreased with the increasing concentration of dendrimer and with an increase in the generation. This study suggested that stable Au NPs were obtained by the stabilization of Au NP within cavities of higher-generation dendrimers. However, with low-generation dendrimers, Au NPs were stabilized by several surrounding dendrimers (Esumi et al., 2001) (Figure 10.4).

Lately, Camarada et al. performed MD simulation studies to find the binding interaction of small-sized (G0)- and intermediate-sized (G4)-NH_2-terminated

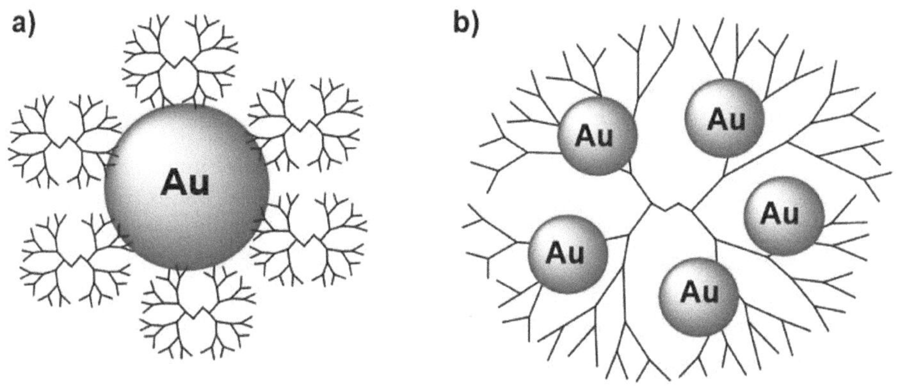

FIGURE 10.4 Schematic illustration of structures of Au Den-NPs: (a) Au NPs stabilized by the surrounding low-generation dendrimers; (b) Au NPs stabilized within the interior of high-generation dendrimers.

dendrimers with Au nanoparticles. Their study demonstrated that stabilization in low-generation dendrimers is done by the surrounding surface of the Au NPs, whereas high-generation dendrimers provide stabilization through internal cavities. On the one hand, three G0 PAMAM dendrimers were involved in the stabilization of one Au NP; on the other hand, the G4 PAMAM stabilized two Au NPs (Salas et al., 2020). Moreover, their group explored the interaction of PAMAM G0 dendrimer (terminal group -OH, -CHO, and -NH_3^+) with Au NPs via DFT calculations. The G0-NH_2 dendrimer displayed binding interactions with Au NP surface, whereas the surfaces of G0-OH and G0-CHO dendrimers were internally stabilized with Au NPs. Additionally, the DFT calculations of PAMAM G0 dendrimer (terminal group -OH, -CHO, and -NH_3^+) in Au NPs evidenced a strong interaction with Au NP surface, whereas the surfaces of G0-OH and G0-CHO dendrimers were internally stabilized with Au NPs (Camarada, 2017). An Au-electroactive nanostructured membrane (ENM) was fabricated via layer-by-layer (LbL) films comprising Au nanoparticle-G4-NH_2 PAMAM dendrimer to generate 3-bilayer PVS poly(vinylsulfonic acid)/PAMAM-Au electrode and successfully used for ORR (Crespilho et al., 2007). Recently, Strasser et al. prepared 1–2-nm-sized Au NPs (G6-NH_2(Au_{147}) and G6-NH_2(Au_{55})) to investigate the effect of electrochemical cleaning on their growth and on electrocatalysis. It was demonstrated that three electrochemical cleaning scans to modest positive potentials resulted in the growth of G6-NH_2(Au_{55}) DENs to a size similar to G6-NH_2(Au_{147}). Furthermore, the growth in NPs was correlated with the electrocatalytic ORR activity (Strasser et al., 2021).

Trindell et al. investigated the electrocatalytic reduction of CO_2 in PAMAM-OH stabilized Au NPs and reported that after 15 min of electrolysis at −0.8 V, the G6-OH encapsulated Au NPs of diameter ~2 nm increased to 6–7 nm. On the other hand, under similar conditions G8-OH-based Au NPs showed a size growth from 1.7 ± 0.3 nm to 2.2 ± 0.5 only, suggesting a substantial stabilization by the compact structure of higher generation dendrimer. Notably, the small-sized and relatively more stable G8-OH encapsulated Au NPs preferentially formed H_2 rather than CO in CO_2 reduction (Trindell et al., 2017).

10.4 CONCLUSION

In this chapter, we have discussed the synthesis of PAMAM-encapsulated transition and precious metal NPs, like (Cu), nickel (Ni), platinum (Pt), palladium (Pd), silver (Ag), and gold (Au), and also their properties, including their complexation behavior, stability, and solubility. PAMAM dendrimers are an attractive system that complexes with metal ions and upon reduction leads to the formation of zerovalent metal NPs. Dendrimers stabilize the metal nanoparticle through suitable binding interactions without complete passivation of the activity of metal ions. Therefore, the dendrimer-encapsulated metal NPs display the catalytic activities in various types of organic and electrocatalytic reactions. PAMAM-encapsulated Pt-, Pd-, and Au-based NPs have shown their potential in electrocatalysis of HER and oxygen reduction reaction (ORR), which finds their promising application in electrolyzers and fuel cells.

ACKNOWLEDGMENTS

DK gratefully acknowledges support from the University Grants Commission (UGC), New Delhi, for Start-up Research Grant (M-14-57), and IoE Seed Grant, BHU (Dev. Scheme No.-6031). RK thanks UGC, New Delhi, for Junior Research Fellowship.

REFERENCES

Adronov, A., & Fréchet, J. M. J., (2000). Light-harvesting dendrimers. *Chem. Commun.*, 18, 1701–1710. https://doi.org/10.1039/B005993P

Astruc, D., & Chardac, F., (2001). Dendritic catalysts and dendrimers in catalysis. *Chem. Rev.*, 101(9), 2991–3023. https://doi.org/10.1021/cr010323t

Baars, M.W. P. L., & Meijer, E. W., (2000). Host-guest chemistry of dendrimer molecules. *Top. Curr. Chem.*, 210, 131–182. https://doi.org/10.1007/3-540-46577-4_3

Camarada, M. B., (2017). PAMAM dendrimers as support for the synthesis of gold nanoparticles: Understanding the effect of the terminal groups. *J. Phys. Chem. A*, 121(42), 8124–8135. https://doi.org/10.1021/acs.jpca.7b08272

Cho, T., Yoon, C. W., & Kim, J., (2018). Repetitively coupled chemical reduction and galvanic exchange as a synthesis strategy for expanding applicable number of Pt atoms in dendrimer-encapsulated Pt nanoparticles. *Langmuir*, 34(25), 7436–7444. https://doi.org/10.1021/acs.langmuir.8b01169

Crespilho, F. N., Nart, F. C., Oliveira, Jr, Os. N., & Brett, C. M. A., (2007). Oxygen reduction and diffusion in electroactive nanostructured membranes (enm) using a layer-by-layer dendrimer-gold nanoparticle approach. *Electrochim. Acta*, 52(14), 4649–4653. https://doi.org/10.1016/j.electacta.2007.01.048

Deraedt, C., Ye, R., Ralston, W. T., Toste, F. D., & Somorjai, G. A., (2017). Dendrimer-stabilized metal nanoparticles as efficient catalysts for reversible dehydrogenation/hydrogenation of N-heterocycles. *J. Am. Chem. Soc.*, 139(49), 18084–18092. https://doi.org/10.1021/jacs.7b10768

Devadas, B., & Imae, T., (2016). Hydrogen evolution reaction efficiency by low loading of platinum nanoparticles protected by dendrimers on carbon materials. *Electrochem. Commun.*, 72, 135–139. https://doi.org/10.1016/j.elecom.2016.09.022

Devadas, B., Periasamy, A. P., & Bouzek K., (2021). A review on poly(amidoamine) dendrimer encapsulated nanoparticles synthesis and usage in energy conversion and storage applications. *Coord. Chem. Rev.*, 444, 214062. https://doi.org/10.1016/j.ccr.2021.214062

Dumitrescu, I., & Crooks, R. M., (2012). Effect of mass transfer on the oxygen reduction reaction catalyzed by platinum dendrimer encapsulated nanoparticles. *Proc. Nat. Acad. Sci.*, 109(29), 11493–11497. https://doi.org/10.1073/pnas.1201370109

Esumi, K., & Torigoe, K., (2001). Preparation and characterization of noble metal nanoparticles using dendrimers as protective colloids. In: Dékány, I. (ed.), *Adsorption and Nanostructure. Progress in Colloid and Polymer Science*, vol. 117, pp. 80–87, Springer. https://doi.org/10.1007/3-540-45405-5_15

Esumi, K., Isono R., & Yoshimura, T., (2004). Preparation of PAMAM- and PPI-metal (silver, platinum, and palladium) nanocomposites and their catalytic activities for reduction of 4-nitrophenol. *Langmuir*, 20(1), 237–243. https://doi.org/10.1021/la035440t

Esumi, K., Kameo, Azusa., Suzuki, A., & Torigoe K., (2001). Preparation of gold nanoparticles in formamide and N, N-dimethylformamide in the presence of poly(amidoamine) dendrimers with surface methyl ester groups. *Colloids Surf. A: Physicochem. Eng. Aspects*, 189, 155–161. https://doi.org/10.1016/S0927-7757(00)00811-6

Esumi, K., Nakamura, R., Suzuki, A., & Torigoe, K., (2000). Preparation of platinum nanoparticles in ethyl acetate in the presence of poly(amidoamine) dendrimers with a methyl ester terminal group. *Langmuir*, 16(20), 7842–7846. https://doi.org/10.1021/la0005006

Esumi, K., Suzuki A., Yamahira, A., & Torigoe, K., (2000). Role of poly(amidoamine) dendrimers for preparing nanoparticles of gold, platinum, and silver. *Langmuir*, 16(6), 2604–2608. https://doi.org/10.1021/la991291w

Fan, FR. F., Mazzitelli, C. L., Brodbelt, J. S., & Bard, A. J., (2005). Electrochemical, spectroscopic, and mass spectrometric studies of the interaction of silver species with poly(amidoamine) dendrimers. *Anal. Chem.*, 77(14), 4413–22. https://doi.org/10.1021/ac048133g

Fréchet, J. M. J., & Tomalia, D. A., (2001). *Dendrimers and other Dendritic Polymers*. Wiley Series in Polymer Science. Wiley.

Garcia, M. E., Baker, L. A. & Crooks, F. R. M., (1999). Preparation and characterization of dendrimer-gold colloid nanocomposites. *Anal. Chem.* 71, 256–258. https://doi.org/10.1021/ac980588g

Goubert-Renaudin, S. N., & Wieckowsk, A., (2011). Ni and/or Co nanoparticles as catalysts for oxygen reduction reaction (ORR) at room temperature. *J. Electroanal. Chem.*, 652, 44–51. https://doi.org/10.1016/j.jelechem.2010.11.022

Ilunga, A. K., & Meijboom, R., (2016). Catalytic oxidation of methylene blue by dendrimer encapsulated silver and gold nanoparticles. *J. Mole. Catal. A: Chem.* 411, 48–60. https://doi.org/10.1016/j.molcata.2015.10.009

Kalbasi, R. J., & Zamani, F., (2014). Synthesis and characterization of Ni nanoparticles incorporated into hyperbranched polyamidoamine-polyvinylamine/SBA-15 catalyst for simple reduction of nitro aromatic compounds. *RSC Adv.*, 4, 7444. https://doi.org/10.1039/C3RA44662J

Kim, Y.-G., Oh, S.-K. & Crooks, R. M., (2004). Preparation and characterization of 1-2 nm dendrimer-encapsulated gold nanoparticles having very narrow size distributions. *Chem. Mater. 16* (1), 167–172. https://doi.org/10.1021/cm034932o

Knecht, M. R., & Crooks, R. M., (2007). Magnetic properties of dendrimer-encapsulated iron nanoparticles containing an average of 55 and 147 atoms. *New J. Chem.*, 31, 1349–1353. https://doi.org/10.1039/B616471B

Knecht, M. R., Garcia-Martinez, J. C., & Crooks, R. M., (2006). Synthesis, characterization, and magnetic properties of dendrimer-encapsulated nickel nanoparticles containing. *Chem. Mater.*, 18(21), 5039–5044. https://doi.org/10.1021/cm061272p

Kottakkat, T., Sahu, A. K., Bhat, S. D., Sethuraman, P., & Parthasarathi, S., (2011). Catalytic activity of dendrimer encapsulated Pt nanoparticles anchored onto carbon towards oxygen reduction reaction in polymer electrolyte fuel cells. *Appl. Catal. B: Environ.*, 110, 178–185. https://doi.org/10.1016/j.apcatb.2011.08.041

Ledesma-Garcia, J., Barbosa, R., Chapman, T. W., Arriaga, L. G., & Godinez, L. A., (2008). Evaluation of assemblies based on carbon materials modified with dendrimers containing platinum nanoparticles for PEM-fuel cells. *Int. J. Hydr. Energy.* 34(4), 2008–2014. https://doi.org/10.1016/j.ijhydene.2008.11.106

Ledesma-Garcia, J., Garcia, I. L. E., Chapman, T. W., Arriaga, L. G., Baglio, V., Antonucci, V., Arico, A. S., Ornelas, R., & Godinez, L. A., (2010). Pt dendrimer nanocomposites for oxygen reduction reaction in direct methanol fuel cells. *J. Solid State Electrochem.*, 14(5), 835–840. https://doi.org/10.1007/s10008-009-0862-x

Ledesma-Garcia, J., Garcia, I. L. E., Rodriguez, F. J., Chapman, T. W., & Godinez L. A., (2008). Immobilization of dendrimer-encapsulated platinum nanoparticles on pretreated carbon-fiber surfaces and their application for oxygen reduction. *J. Appl. Electrochem.*, 38(4), 515–522. https://doi.org/10.1007/s10800-007-9466-2

Li, Y., & El-Sayed, M. A., (2001). The effect of stabilizers on the catalytic activity and stability of Pd colloidal nanoparticles in the Suzuki reactions in aqueous solution. *J. Phys. Chem.*, 105(37), 8938–8943. https://doi.org/10.1021/jp010904m

Liu, L., & Corma, A., (2018). Metal catalysts for heterogeneous catalysis: From single atoms to nanoclusters and nanoparticles. *Chem. Rev.*, 118(10), 4981–5079. https://doi.org/10.1021/acs.chemrev.7b00776

Manna, A., Imae T., Aoi K., Okada M., & Yogo, T., (2001). Synthesis of dendrimer-passivated noble metal nanoparticles in a polar medium: Comparison of size between silver and gold particles. *Chem. Materials*, 13, 1674–1681. https://doi.org/10.1021/cm000416b

Markowicz-Piasecka, M., & Olasik, E. M., (2016). Dendrimers in drug delivery, Nanobiomaterials in Drug Delivery, *William Andrew Pub.*, *2,* 39–74. https://doi.org/10.1016/B978-0-323-42866-8.00002-2

Mazumder, A., Davis, J., Rangari, V., & Curry, M., (2013). Synthesis, characterization, and applications of dendrimer-encapsulated zero-valent Ni nanoparticles as antimicrobial agents. *Nanomater*, 2013, 843709. https://doi.org/10.1155/2013/843709

Newkome, G. R., Moorefield, C. N., & Vogtle, F., (2001). *Dendrimers and Dendrons: Concepts, Syntheses, Applications.* Wiley. https://doi.org/10.1002/3527600612

Oh, S. K., Niu, Y., & Crooks, R. M., (2005). Size-selective catalytic activity of Pd nanoparticles encapsulated within end-group functionalized dendrimers. *Langmuir*, 21(22), 10209–13. https://doi.org/10.1021/la050524i

Ostojic, N., & Crooks, R. M., (2016). Electrocatalytic reduction of oxygen on platinum nanoparticles in the presence and absence of interactions with the electrode surface. *Langmuir*, 32(38), 9727–35. https://doi.org/10.1021/acs.langmuir.6b02578

Ottaviani, M. F., Bossmann, S., Turro N. J., & Tomalia D. A., (1994). Characterization of starburst dendrimers by the EPR technique. 1. Copper complexes in water solution. *J. Am. Chem. Soc.*, 116(2), 661–671. https://doi.org/10.1021/ja00081a029

Patala, R., Noh, J. H., & Meijboom, R., (2018). Determination of maximum loading capacity of polyamidoamine (PAMAM) dendrimers and evaluation of Cu55 dendrimer, encapsulated nanoparticles for catalytic activity. *Int. J. Chem. Kinet.*, 12(1), 693–704. https://doi.org/10.1002/kin.21193

Rahim, E. H., Kamounah, F. S., Frederiksen, J., & Christensen, J. B., (2001). Heck reactions catalyzed by PAMAM-dendrimer encapsulated Pd (0) nanoparticles. *Nano Lett.*, 1(9), 499–501. https://doi.org/10.1021/nl015574w

Salas, F. A., Gonzalez, R. I., Ríos, P. L., Duran, I. A., & Camarada, M. B., (2020). Effect of the generation of PAMAM dendrimers on the 2 stabilization of gold nanoparticles. *J. Chem. Inf. Model.* 60(6), 2966–2976. https://doi.org/10.1021/acs.jcim.0c00052

Scott, R. W. J., Ye, H., Henriquez, R. R., & Crooks, R. M., (2003). Synthesis, characterization, and stability of dendrimer-encapsulated palladium nanoparticles. *Chem. Mater.*, 15(20), 3873–3878. https://doi.org/10.1021/cm034485c

Shao, M., Chang, Q., Dodelet, J. P., & Chenitz, R., (2016). Recent advances in electrocatalysts for oxygen reduction reaction. *Chem. Rev.*, 116, 3594–3657. https://doi.org/10.1021/acs.chemrev.5b00462

Stofik, M., Stryhal, Z., & Maly, J., (2009). Dendrimer-encapsulated silver nanoparticles as a novel electrochemical label for sensitive immunosensors. *Biosensors Bioelectron.* 24(7), 1918–1923. https://doi.org/10.1016/j.bios.2008.09.028

Strasser, J. W., Hersbach, T. J. P., Liu, J., Lapp, A. S., Prof. Frenkel, A. I., & Crooks, R. M., (2021). Electrochemical cleaning stability and oxygen reduction reaction activity of 1-2 nm dendrimer-encapsulated Au nanoparticles. *ChemElectroChem*, 8, 2545–2555. https://doi.org/10.1002/celc.202100549

Sun, L., Ca, D. V., & Cox, J. A., (2005). Electrocatalysis of the hydrogen evolution reaction by nanocomposites of poly(amidoamine)-encapsulated platinum nanoparticles and phosphotungstic acid. *J. Solid State Electrochem.*, 9, 816–822. https://doi.org/10.1007/s10008-005-0008-8

Tomalia, B. L., & Balogh, L., (1998). Poly(amidoamine) dendrimer-templated nanocomposites. 1. synthesis of zerovalent copper nanoclusters. *J. Am. Chem. Soc.*, 120(29), 7355–7356. https://doi.org/10.1021/ja980861w

Tomalia, D. A., Baker, H., Dewald, J., Hall, M., Kallos, G., Martin, S., Roeck, J., Ryder, J., & Smith, P., (1985). New class of polymers: Starburst-dendritic macromolecules. *Polymer J.*, 17(1), 117–132. https://doi.org/10.1295/polymj.17.117

Tomalia, D. A., Baker, H., Dewald, J., Hall, M., Kallos, G., Martin, S., Roeck, J., Ryder, J., & Smith, P., (1986). Dendritic macromolecules: Synthesis of starburst dendrimers. *Macromolecules*, 19(9), 2466–2468. https://doi.org/10.1021/ma00163a029

Trindell, J. A., Clausmeyer, J., & Crooks, R. M., (2017). Size stability and H_2/CO selectivity for Au nanoparticles during electrocatalytic CO_2 reduction. *J. Am. Chem. Soc.*, 139(45), 16161–16167. https://doi.org/10.1021/jacs.7b06775

Wu, L., Zhang, X., & Tao, Z., (2012). A mild and recyclable nano-sized nickel catalyst for the Stille reaction in water. *Catal. Sci. Technol.*, 2, 707–710. https://doi.org/10.1039/C2CY00466F

Xu, D. M., (2007). Generational poly(ester-amine) dendrimers. *J. Appl. Polym. Sci.* 104(1), 422–426. https://doi.org/10.1002/app.25229

Xu, D. M., Zhang, K.D., & Zhu, X. L., (2007). Preparation of Ag Nanoparticles in the Presence of Low generational poly (ester-amine) dendrimers. *J. Appl. Polym. Sci.*, 104(1), 422–426. https://doi.org/10.1002/app.25229

Yamamoto, K., Imaoka, T., Tanabe, M., & Kambe, T., (2020). New horizon of nanoparticle and cluster catalysis with dendrimers. *Chem. Rev.*, 120(2), 1397–1437. https://doi.org/10.1021/acs.chemrev.9b00188

Ye, H., Crooks, J. A., & Crooks, R. M., (2007). Effect of particle size on the kinetics of the electrocatalytic oxygen reduction reaction catalyzed by Pt dendrimer-encapsulated nanoparticles. *Langmuir*, 23(23), 11901–11906. https://doi.org/10.1021/la702297m

Yu, T., Wang, W., Chen, J., Zeng, Y., Li, Y., Yang, G., & Li Y., (2012). Dendrimer-encapsulated Pt nanoparticles: An artificial enzyme for hydrogen production. *J. Phys. Chem.*, 116(19), 10516–10521. https://doi.org/10.1021/jp3021672

Zeng, F., & Zimmerman, S. C. (1997). Recognition to self-assembly. *Chem. Rev.*, 97(5), 1681–1712. https://doi.org/10.1021/cr9603892

Zeng, F., & Zimmerman, S. C., (1997). Dendrimers in supramolecular chemistry: From molecular Recognition to Self-Assembly. *Chem. Rev.*, 97(5), 1681–1712. https://doi.org/10.1021/cr9603892

Zhao, M., & Crooks R. M., (1999). Dendrimer-encapsulated Pt nanoparticles: Synthesis, characterization, and applications to catalysis. *Adv. Mater.*, 11(3), 217–220. https://doi.org/10.1002/(SICI)1521-4095(199903)11:3<217::AID-ADMA217>3.0.CO;2–7

Zhao, M., & Crooks, R. M., (1999). Homogeneous hydrogenation catalysis with monodisperse, dendrimer-encapsulated Pd and Pt nanoparticles. *Angew. Chem. Int. Ed. Engl.*, 38(3), 364–366. https://doi.org/10.1002/(SICI)1521-3773(19990201)38:3<364::AID-ANIE364>3.0.CO;2-L

Zhao, M., & Crooks, R. M., (1999). Intradendrimer exchange of metal nanoparticles. *Chem. Mater. Chem. Mater.*, 11(11), 3379–3385. https://doi.org/10.1021/cm990435p

Zhao, M., Sun Li., & Crooks, R. M., (1998). Preparation of Cu nanoclusters within dendrimer templates. *J. Am. Chem. SocietySoc.*, 120(19), 4877–4878. https://doi.org/10.1021/ja980438n

11 Organic Nanomaterials

Tanu Gupta, Amit Jaiswal, Ranjeet Kumar, and Ranjeet Kumar

11.1 INTRODUCTION

Nanomaterials are a class of materials with diameters ranging from 1 to 100 nm. In other words, a nanomaterial is described as "any external dimension in the nanoscale or having internal or surface structure in the nanoscale (length range about 1 nm to 100 nm)". It encompasses a broad range of multidisciplinary research and development activities that are expanding rapidly across the world (Enoki et al., 2013). Nanotubes, dendrimers, quantum dots, and fullerenes are examples of nanomaterials. Nanomaterials are gaining popularity owing to their unique optical characteristics, which have a substantial influence on a variety of sectors, including electronics, mechatronics, medicine, pharmaceuticals, ionic liquids, polymers, and many more. Nanoscale titanium dioxide is used in cosmetics, sunscreen, and self-cleaning windows; nanocoatings and nanocomposites are used in windows, sports equipment, bicycles, and automobiles; nanoscale silica is used as a filler in cosmetics and dental fillings; many more applications are also reported (Sayed et al., 2006).

11.2 SOURCES OF NANOMATERIAL

1. Engineered nanoparticles created by humans to have specific qualities such as colloidal or particulate materials can be formed using an appropriate approach (Kochmann et al., 2020).
2. It can be encountered accidentally as a byproduct of mechanical operations such as car engine exhausts, smelting, and home solid fuel burning and vaporization (Marinella et al., 2012; Daniel et al., 2009).
3. It may also be naturally derived, like foraminifera and viruses, from wax crystals such as nasturtium leaf, spider mite silk, tarantula blue color, certain butterfly wing scales, natural colloids, and horny materials (University of Western Ontario, 2013). Weathering processes of industrial effluent sites can produce natural nanomaterials (Bae et al., 2006). Organic nanomaterials have attracted a lot of interest and are being explored, with liposomes and polymersomes being used for scanning.

Application of organic nanomaterial: There are numerous applications of organic nanomaterials, which are as follows.

Biomedicine and Biomedical: Organic nanomaterials offer a wide range of applications in medicine and biomedicine, including biological mimetics, polymeric nanoconstructs as biomaterials, nanoscale microfabrication-based devices, and nanomachines (Juliano, et al., 2012). A large variety of therapeutic compounds have been

DOI: 10.1201/9781003481157-11

reported for use in the treatment of cancer, asthma, allergies, infections, diabetes, and pain, among other conditions (Donaldson et al., 2008).

i. **Cancer therapy**: Organic nanoparticles, such as peptide-based nanomaterials, have been created for use as pharmaceuticals in therapeutic strategies because they may target cancer cells while causing less harm to healthy organs (Li et al., 2014). (KLAKLAK)2 (KLAK) is a self-assembled peptide-based nanomaterial used in cancer treatment to prevent the development of aggressive tumors by disrupting the mitochondrial membrane (Wang et al., 2018). Wang and his colleagues created a library of self-assembled KLAK-based nanomaterials for cancer treatment (Wang et al., 2018). The Beclin1 peptide has been found to be a haploid-sufficient tumor suppressor; however, because of its enzymatic breakdown in the blood, it is poorly delivered to tumors. Wang et al. addressed this challenge by creating conjugated poly(-amino ester)s, which is a simple and supramolecular technique (P-Becl). As a result, the P-Becl nanoparticles increased MCF-7 cell cytotoxicity and demonstrated therapeutic benefits on breast cancer in vivo (Wang et al., 2015).

   Furthermore, photodynamic therapy-based cancer therapies have been carried out by employing organic nanomaterials to treat deeper tumors and improve immune response. Kuangda and colleagues created nanomaterials of chlorin-based organic frameworks that can encapsulate small molecule immunotherapeutic drugs and inhibit indoleamine 2,3-dioxygenase. Through photodynamic therapy, this nanostructure was able to induce an antitumor immune response while also delivering an inhibitor of indoleamine 2,3-dioxygenase to the target tumor (Lin et al., 2016).

   Photodynamic therapy still confronts a significant hurdle because of its oxygen-dependent feature, which restricts its clinical effectiveness for hypoxic tumors. Tian and colleagues solved the problem by employing covalent organic nanostructure, which was made with a donor-acceptor molecular heterostructure and may handle hypoxic-tumor photodynamic therapy in two ways: type I photodynamic therapy and type II photodynamic therapy. Under single wavelength irradiation, type I photodynamic therapy and type II photodynamic therapy capabilities may be reduced by reducing the oxygen-dependent aspect of photodynamic therapy and ensuring significant destroying properties on tumor cells in vivo (Figure 11.1; Cheng et al., 2019). Furthermore, Maksimenko et al. (2013) produced nanoassemblies of novel polyisoprenoyl prodrugs of gemcitabine to circumvent antitumor toxicity. The polyisoprenoyl gemcitabine nanoassemblies demonstrated considerable cytotoxicity in murine melanoma cell line B16F10 as well as many other human cell lines (Couvreur et al., 2013).

ii. **Diabetes treatments**: Nanomaterials have enhanced glucose monitoring, glucose detection, and insulin administration, all of which have been utilized to enhance diabetes therapy. Gu et al. (2015) investigated glucose sensors using nanoscale carbon nanostructure components, which improve glucose sensor sensitivity and temporal response and lead to in vivo glucose monitoring sensors. Furthermore, his group investigated nanotechnology methods to "closed-loop"

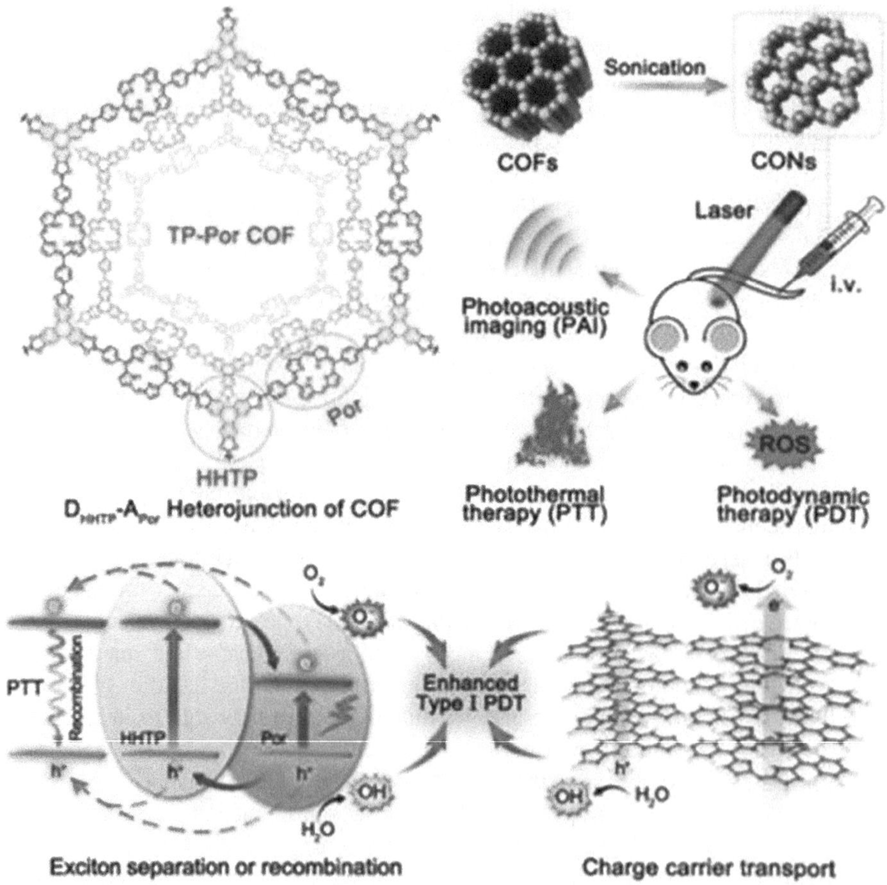

FIGURE 11.1 2D CON construction, in vivo tumor treatment, and the process of photodynamic therapy and type I photodynamic therapy creation are all depicted in this diagram. Reproduced by Cheng et al. (2019). Copyright 2019 American Chemical Society.

insulin administration systems that occur between blood glucose level monitoring. Ali and colleagues created gastro-resistant imine-linked-covalent organic conceptual model (nCOF) nanoparticles for oral insulin delivery to manage diabetes. The nanoparticles, which were made from layered nanosheets with insulin packed between the layers, showed insulin safety in digestive fluids in vitro as well as glucose-responsive release (Ali et al., 2021). Graphene-based organic nanoparticles have been developed for non-invasive sweat-based diabetic treatment (Kim et al., 2016). By reducing graphene oxide, Chang et al. (2013) showed glucose and hydrogen peroxide detection.

iii. **Pain treatments**: Organic nanoparticles can be used to relieve pain in many regions of the body. Due to switchable characteristics such as size, surface characteristics, ability to respond, circulation time, loading efficiency, and specifically target tissues (Almeida et al., 2012), FDA-approved organic nanomaterials such as liposomes (Turánek et al., 2012) have much application to the next generation of nanomaterials (2015). PEGylated liposomes have been utilized to increase the accumulation of zoledronic acid, allowing it to flow through the blood-brain barrier for effective pain relief (Maione et al., 2007, 2013). Garcia et al. (2011) produced a bupivacaine nanomaterial that was nanoencapsulated utilizing poly(lactic-co-glycolic acid) for testing in vitro and in vivo. Kim et al. (2018) employed (Poly(D,L-lactic-co-glycolic acid)) PLGA-encapsulated nanoparticles that contain p38 siRNA to inhibit microglial activation.

iv. **Asthma treatment**: Elhissi et al. created polyamidoamine (PAMAM) G3 dendrimers that operate as nanocarriers for pulmonary administration of the model moderately soluble anti-asthma medication beclomethasone dipropionate (BDP) via nebulization, resulting in decreased asthma symptoms (Elhissi et al., 2014). Telomere dendrimer has also been created as an effective nanocarrier for delivering dexamethasone directly into the lungs, lowering allergic pulmonary inflammation and eosinophils (Burt et al., 2004). Kumar et al. (2003) discovered that chitosan-IFN-pDNA nanoparticles can decrease allergic asthma airway hyperresponsiveness and lung histopathology. Furthermore, chitosan-IFN-pDNA nanoparticles may be used to treat allergic asthma by inhibiting the generation of pro-inflammatory mediators in lung OVA-specific CD8+ T lymphocytes.

v. **Treatment of allergies, infections, and other diseases**: 1-fluoro-2, 4-dinitrobenzene and 2-deoxyurushiol nanoparticles can be used to treat allergic skin (DeLouise et al., 2017). The majority of organic nanoparticle vaccines are being used to prevent infectious illness. Thomas et al. created nanoparticles with varying ratios of polylactic acid and poly(lactic-co-glycolic acid) to transfer hepatitis B surface antigen to the virus (Ahsan et al., 2011). Moon and his colleagues created delivery vehicles for VMP001, a plasmodium vivax malaria vaccine made up of poly(lactic-co-glycolic acid) conjugated malaria antigen encapsulated in lipid membrane (Irvine et al., 2012). A hybrid multilamellar vaccination particle has been made to produce long-lasting antigen-specific CD8+ and CD4+ T-cell responses (Moon et al., 2019).

   a. **Environmental improvement**: Continuous human activities are releasing harmful substances into the water, soil, and atmosphere, which is becoming a huge concern all over the world. As a result, many academics are concentrating on recent advancements in nanotechnology in order to deal with expanding hazardous wastes at a lower cost, with less energy, and with higher efficiency. Kümmerer et al. (2011) studied the biodegradability of carboxylated and non-carboxylated carbon nanotubes, fullerenes of various sizes (C60 and C70), and crystalline nanoparticles of cellulose and starch in the aquatic environment. According to Long et al. (2001), carbon nanotubes might be employed as an adsorbent for the removal of nitrous oxide, sulfur dioxide, and carbon dioxide at room temperature. Furthermore, Mochida et al. (1997) reported the oxidation of nitrous oxide to nitro on activated carbon fiber at ambient temperature. Hydrophobic, electrostatic, pi–pi, hydrogen, and covalent bonding sites exist in carbon nanotubes. As a result, organic molecules which have aromatic rings that include carbon-carbon double bonds can generate pi–pi interactions with carbon nanotubes (Yang et al., 2010). Organic molecules containing –COOH, –OH, and $-NH_2$ functional groups may contact through non-covalent interactions with the surface of graphitic carbon nanotube (Yang et al., 2008).

b. **Drugs delivery**: Organic nanoparticles have been designed to encapsulate and deliver therapies, allowing for drug delivery applications. Nanoparticles, liposomes, polymeric micelles, and prodrugs are some of the nano-scaled drug delivery technologies that have been created. (Zhang et al., 2017). Shen et al. (2021) employed pillar[5]arene-based host–guest complex formation to create supramolecular nanoparticles with increased Aggregation-Induced Emission (AIE) effect, which may also be used as fluorescent nanocarriers for imaging-directed drug administration. The targeted charge-reversal nanoparticles (TCRNs) for nuclear drug delivery were disclosed by Y. Shen (2007). TCRNs made with folic acid-functionalized poly(ε-caprolactone)-bl ock-polyethyleneimine/amide are negatively charged in neutral solution and at once soon converted to positively charged at pH 6 and substantially positively charged at pH 5, enhancing cellular absorption and directing TCRNs to the nucleus. Wang et al. (2019) demonstrated the light-triggered (Glutathione) GSH-responsive nanoparticles by proving the perallyloxycucurbit[6]uril-based nanoparticles for targeted drug administration in melanoma cells, which may be used for precisely delivering chemotherapeutic medicines into melanoma cells. Gong et al. (2021) and colleagues created a glutathione-activated carrier-free nanodrug of triptolide as an actionable drug carrier for tumor therapy. In this approach, carrier-free triptolide nanodrugs self-assemble to generate a carrier-free nanodrug that has a monodisperse spherical morphology and is stable under various physiological settings (pH, high salt, etc.). Organic nanomaterials, such as dendrimers, have been successfully demonstrated as dual drug delivery systems, glycosylated dendrimer-based targeted delivery systems, monoclonal antibody-targeted delivery systems, and FA-targeted delivery system (Shi et al., 2013).

c. **Automotive industry**: For better fuel demand and higher protection, catalyst supports, fuel additives, lubricating oils, super-duper anti-glare layers for windows, sensing, and electronics are some of the applications of organic nanomaterials in the automotive industry (Asmatulu et al., 2013). When nanographene platelets, also known as graphene, nanowire, or nanoprene, are applied to tire treads, they significantly improve resistance properties as well as cornering stability, driving response, and noise tolerance (Nuraje et al., 2010). CNTs, carbon black, graphene, and polytetrafluorethylene are placed in primers and top paints and then sprayed on the surface of automobiles to improve hardness, resistance to abrasion, damage tolerance, wearing, and subconscious properties (Tracton et al., 2006 and Moradiya et al., 2019). Graphene nanoparticles have been employed in automobile fluids and lubricants to improve mechanical qualities and provide cost savings.

d. **Solar energy**: Nowadays, solar energy harvesting and storage have attracted tremendous research efforts due to unique properties when the particle size reduced to the nanometer scale. Feng et al. (2015) developed a nano-template for solar energy fuels that is made up of azobenzene and graphene nanosheets that are covalently connected through methoxyl and carboxyl groups and have a large capacity and long-term storage due to intermolecular H-bonds. Guerrero et al. (2013) exhibited nanosize fullerene binding the interface at cathode contacts, resulting in greater electron selectivity and improved device performance in organic solar cells. Agbolaghi created core–mantle nanohybrids by grafting polyaniline onto multi-walled carbon nanotubes for solar cell and photovoltaic efficiency (Agbolaghi et al., 2020). Andersson 2018 and co-workers reported water-dispersed for the first time in NP-OPV research where O-xylene was used in the miniemulsion procedure to create photoactive nanoparticles. Investigation of PTNT:PC71BM nanoparticles could be useful in the future for up-scaling of OPV production.

e. **Optical application**: There are tremendous optical applications of organic nanomaterials. Yao et al. (2011) have synthesized the optical activity-mediated porphyrin-based nanoparticles. They discovered that the Soret band of porphyrin nanoparticles has a significant bathochromic shift when compared to the porphyrin monomer. In addition, Kimura et al. found that induced circular dichroism of porphyrin nanoparticles shows split Cotton effects in the Soret band region (Kimura et al., 2011). Liu and his co-workers have developed the AIE active fluorescent probe (TPFE-Rho) and fabricated AIE active NIR fluorescent organic nanoparticles for non-invasive in vivo tumor growth monitoring because it shows intense Near-Infrared (NIR) fluorescent emission and large Stokes shift (Liu et al., 2018). Ahn et al. (2018) and colleagues investigated the molecular shape-dependent luminescence activity of bent-shaped one-dimensional molecules (benzo[d]thiazol-2-yl)-6-substituted-naphthalen-2-ol and their 2-methoxy derivatives with an excited-state intramolecular proton transfer (ESIPT) property that is predicted to change their optical properties and can be used to label cells using confocal and two-photon microscopic imaging. Binary organic nanoparticles were successfully synthesized for three-photon imaging with brightest luminescence for improving the performance of multi-photon imaging techniques used in mouse brain vessel imaging, with a depth of 1.68 mm (Liu et al., 2020) (Figure 11.2). Schenning et al. demonstrated the co-assembly of synthetically accessible prefunctionalized conjugated oligomers to produce multifunctional fluorescent organic nanoparticles for multitargeting imaging.

f. **Electronic application**: Carbon nanotubes are widely used as organic nanomaterials in electronics in various fields such as semiconductor chips, lighting, and electric double-layer capacitors. In light, carbon nanotubes are used as active memory elements developed by Nantero

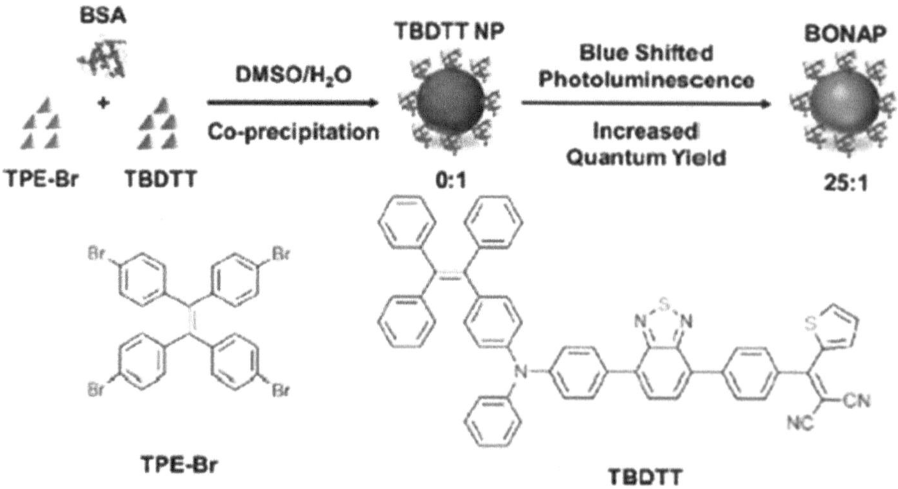

FIGURE 11.2 Molecular structures of binary organic nanoparticles. Reproduced by Liu et al. (2020). Copyright 2020 American Chemical Society.

where Intel is looking at the possible replacement of copper wires inside semiconductors using carbon nanotubes (Kunellos et al., 2006). The Electric Double-Layer Capacitors (EDLC) is applicable as a hybrid energy source for electric vehicles and portable electric devices by using carbon nanotubes.

Methods of synthesis of organic nanomaterials: Organic nanomaterials can be prepared using the following methods:

a. Emulsification
b. Nano-precipitation
c. Spray drying process

a. **Emulsification**: Methods for introducing better emulsions with nanosized particles to design organic nanoparticles have significantly evolved over the last decade as a result of technological advancements in emulsification equipment and the ongoing growth of low-energy stirring routes due to environmental limitations (Becher et al., 1965). For the creation of nanoorganic particles, many methods have been utilized. **(i) Solvent withdrawal causes precipitation**: Macromolecules dissolved in the dispersed medium of an emulsion can precipitate when the organic solvent is removed via evaporation of the solvent, solvent diffusion, or salting-out (Doelker et al., 1999). **(ii) Gelation of the emulsion droplets**: The nanoemulsion is broken down into polymer gel droplets, which are then broken down further into nanoparticles (Gohla et al., 2000). **(iii) Emulsion polymerization**: Emulsion polymerization techniques are divided into several categories, including traditional emulsion polymerization, surfactant-free emulsion polymerization, nano-emulsions, and micro-emulsion polymerizations, all of which have kinetically and thermodynamically distinct emulsion characteristics. Furthermore, interfacial polymerization is a highly helpful approach for the creation of nanocapsules (Allouche et al., 2013).

b. **Nano-precipitation**: Fessi and colleagues were the first to invent and patent the nano-precipitation process for preparing nanoparticles for drug administration, commonly known as the interfacial deposition method (Benita et al., 1989). On phase separation, the procedure needs the addition of two miscible solvents, which leads to the spontaneous creation of nanoparticles (Doelker et al., 2005). The Marangoni effect, governed by interfacial turbulences that occur at the interface of the solvent and the non-solvent and derived from intricate and cumulated processes, is responsible for fast nanoparticle production (Doelker et al., 1998). The nano-precipitation approach may be used with a variety of polymers, including poly(e-caprolactone), polylactide (Barratt et al., 1999), poly(lactide co-glycolide) (Desai et al., 2008), poly(hydroxyl butyrate) (Gurunathan et al., 2009), and even peptides (Orecchioni et al., 1998). The technique is most commonly used on non-polymeric substances such as cyclodextrin (Gulik et al., 1996) and pharmaceuticals (Cabane et al., 1997).

c. **Spray drying process**: Spray drying has been used to produce micro-sized organic nanoparticles and regulate features for biomedical applications, particularly in therapeutics (Vehring et al., 2008 and Schafroth et al., 2009). The process consists of four steps: liquid feedstock, atomization of the feed, evaporative mass transport of the solvent, and finally, isolation of the dried product from the gas (Masters et al., 1991). For example, Li et al. (2010) used a "nano spray dryer" to prepare several types of polymeric nanoparticles such as Arabic gum, whey protein, polyvinyl alcohol, modified starch, and malt-o-dextrin. Lee et al. (2011) used a similar approach to create bovine serum albumin nanoparticles in 2011.

11.3 CHARACTERIZATION OF ORGANIC NANOPARTICLES

a. **Atomic force microscopy (AFM)**: Quantitative information regarding nanoparticles, such as length, breadth, and height, and other physical qualities, including shape and surface roughness, may be obtained using AFM data. Particles with a height of 1 nm to 5 μm may be measured using AFM in a single scan, and AFM scanning is done with a physical probe in either direct or close contact (Rebouillat et al., 1998 and Gatenholm et al., 1999). Many more scientists are employing these techniques to characterize organic nanoparticles (Moribe et al., 2015; Mao et al., 2013; and Masuhara et al., 2008).

b. **Transmission electron microscopy (TEM)**: Transmission electron microscopy (TEM) is a tool for understanding nanostructured materials such as nanoparticles, nanotubes, bulk metallic, grapheme, grapheme oxide, and polymer nanocomposites. Furthermore, electron diffraction patterns of many materials are explained, and material structure is anticipated using electron diffraction pattern findings interpretation (Jafari et al., 2015). Prasant et al. (2017) produced and characterized the L-phenylalanine-tethered, naphthalene diimide-based organic nanoparticles via Transmission Electron Microscope (TEM). Many researchers have also created organic nanomaterials and described them using TEM (Thanh et al., 2018).

c. **Light scattering**: Scattering techniques are effective for nanoparticle characterization, and a variety of light scattering methods, including dynamic light scattering, static light scattering, and neutron scattering, have been described. Dynamic light scattering is a commonly used technique for determining the size of nanoparticles in colloidal solutions in the nano- and submicrometer ranges. Dynamic light scattering determines the nanoparticle hydrodynamic diameter in solution by measuring light scattering as a function of time, which is combined with the Stokes–Einstein assumption. A relatively low nanoparticle concentration is required in dynamic light scattering to avoid a multiple scattering effect (Wolfgang et al., 2015). Waclawik et al. (2014) used dynamic light scattering to determine the molecular thickness of thin organic shells on tiny inorganic cores. Static light scattering provides information on particle shape when combined with dynamic light scattering (Domingues et al., 2018).
d. **X-ray scattering**: X-ray scattering techniques, such as X-Ray Diffraction (XRD), are extensively used to characterize nanoparticles and offer data on crystalline size, crystalline structure, phase nature, and lattice parameters. The Scherrer equation is used to calculate the latter parameter of an X-ray diffraction measurement for a given sample (Ingham et al., 2015 and Warren et al., 1969). This equation is stated in terms of the diffracting angle 2 by $D = b/\text{FWHM}(2)$ cos (where it is provided in radians and is the X-ray wavelength) or in terms of Q by $D = b2/\text{FWHM}(Q)$ cos (where it is given in radians and is the X-ray wavelength) (Beyerlein et al., 2011).

11.4 CONCLUSIONS

In this chapter, we presented a detailed overview of organic nanoparticles such as application, synthesis, and characterization techniques. Because of their small size, nanoparticles have a great surface area, making them a good contender for a variety of significant applications, including biomedicine and biomedical which include cancer therapy, pain, asthma, diabetes treatments, allergies, infections, and other diseases. Furthermore, these are also applicable to environmental improvement, drugs delivery, automotive industry, solar energy, optical application, and electronic application. Due to these applications, it contributes to a healthy environment by delivering cleaner air and water, as well as renewable power for a great future. Furthermore, nanotechnology has attracted a lot of interest, with leading universities, corporations, and organizations investing more in research and development. Different synthesis methods have been reported for organic nanoparticles such as emulsification, nano-precipitation, and spray drying process. The form, structure, and characteristics of organic nanoparticles may be controlled using these several synthetic approaches. Organic nanoparticles have sizes ranging from a few nanometers to 500 nm, according to several characterization techniques such as AFM, TEM, light scattering, and XRD.

ACKNOWLEDGMENTS

The authors are grateful to CMP Degree College (a constituent college of the University of Allahabad in Prayagraj) for providing us with the required assistance and resources to conduct this research.

REFERENCES

Agbolaghi, S. (2020). Core-mantle-shell novel nanostructures for efficacy escalating in poly (3-hexylthiophene): Phenyl-C71-butyric acid methyl ester photovoltaics. *Carbon Letters*, *30*(1), 45–54.

Allouche, J. (2013). Synthesis of organic and bioorganic nanoparticles: An overview of the preparation methods. In: *Nanomaterials: A Danger or a Promise?*, 27–74. doi:10.1007/978-1-4471-4213-3_2.

Arpagaus, C., & Schafroth, N. (2009). Laboratory scale spray drying of biodegradablepolymers. In: *Respiratory Drug Delivery* (pp. 269–274).

Asadi A M., & Jafari E. M. (2015). Transmission electron microscopy as best technique for characterization in nanotechnology. *Synthesis and Reactivity in Inorganic, Metal-Organic, and Nano-Metal Chemistry*, *45*(3), 323–326.

Asahi, T., Sugiyama, T., & Masuhara, H. (2008). Laser fabrication and spectroscopy of organic nanoparticles. *Accounts of Chemical Research*, *41*(12), 1790–1798.

Asmatulu, R. (Ed.). (2013). *Nanotechnology Safety*. Newnes.

Asmatulu, R., Khan, S. I., Misak, H., & Nuraje, N. (2010). Graphene-based nanocomposite coating on fiber reinforced composites subjected to UV degradation. In: *SAMPE Fall Technical Conference* (pp. 17–20).

Becher, P. (1965). *Emulsions: Theory and Practice*, Reinhold Pub. Corp.

Bianco, A., Cheng H. M., Enoki T, et al. (2013). All in the graphene family-a recommended nomenclature for two-dimensional carbon materials. *Carbon*, *65*, 1–6

Bilati, U., Allémann, E., & Doelker, E. (2005). Development of a nanoprecipitation method intended for the entrapment of hydrophilic drugs into nanoparticles. *European Journal of Pharmaceutical Sciences*, *24*(1), 67–75.

Börås, L., & Gatenholm, P. (1999). Surface composition and morphology of CTMP fibers. https://doi.org/10.1515/HF.1999.031

Caraglia, M., Luongo, L., Salzano, G., Zappavigna, S., Marra, M., Guida, F., & Maione, S. (2013). Stealth liposomes encapsulating zoledronic acid: A new opportunity to treat neuropathic pain. *Molecular Pharmaceutics*, *10*(3), 1111–1118.

Carvalho, P. M., Felício, M. R., Santos, N. C., Gonçalves, S., & Domingues, M. M. (2018). Application of light scattering techniques to nanoparticle characterization and development. *Frontiers in Chemistry*, *6*, 237.

Cheng, Q., Li, S., Sun, C., Yue, L., & Wang, R. (2019). Stimuli-responsive perallyloxycucurbit [6] uril-based nanoparticles for selective drug delivery in melanoma cells. *Materials Chemistry Frontiers*, *3*(2), 199–202.

Damia, B., & Marinella, F. (2012). *Analysis and Risk of Nanomaterials in Environmental and Food Samples*. Elsevier

Deepak, V., Kalishwaralal, K., & Gurunathan, S. (2009). Purification, immobilization, and characterization of nattokinase on PHB nanoparticles. *Bioresource Technology*, *100*(24), 6644–6646.

DiSanto, R. M., Subramanian, V., & Gu, Z. (2015). Recent advances in nanotechnology for diabetes treatment. *Wiley Interdisciplinary Reviews: Nanomedicine and Nanobiotechnology*, *7*(4), 548–564.

Duclairoir, C., Nakache, E., Marchais, H., & Orecchioni, A. M. (1998). Formation of gliadin nanoparticles: Influence of the solubility parameter of the protein solvent. *Colloid and Polymer Science*, *276*(4), 321–327.

Durán-Lobato, M., Martín-Banderas, L., Gonçalves, L. M., Fernández-Arévalo, M., & Almeida, A. J. (2015). Comparative study of chitosan-and PEG-coated lipid and PLGA nanoparticles as oral delivery systems for cannabinoids. *Journal of Nanoparticle Research*, *17*(2), 1–17.

Egami, K., Higashi, K., Yamamoto, K., & Moribe, K. (2015). Crystallization of probucol in nanoparticles revealed by AFM analysis in aqueous solution. *Molecular Pharmaceutics*, *12*(8), 2972–2980.

Endo M., Hayashi T., Kim Y. A., & Muramatsu H. (2006). Development and application of carbon nanotubes. *Japanese Journal of Applied Physics*, *45*(6A): 4883–4892.

Eustis, S., & El-Sayed, M. A. (2006). Why gold nanoparticles are more precious than pretty gold: Noble metal surface plasmon resonance and its enhancement of the radiative and nonradiative properties of nanocrystals of different shapes. *Chemical Society Reviews*, *35*(3), 209–217.

Fan, Y., Stronsky, S. M., Xu, Y., Steffens, J. T., van Tongeren, S. A., Erwin, A., & Moon, J. J. (2019). Multilamellar vaccine particle elicits potent immune activation with protein antigens and protects mice against Ebola virus infection. *ACS Nano*, *13*(10), 11087–11096.

Farah, B., Nawel, K., Thirumurugan, P., Gobinda, D., Sudhir, K. S., Sneha, A. T., Fadia, B.-S., Jamie, W., Mohammed, A. A., Mostafa, K., Hassan, T., Renu, P., Ramesh, J., Nassima, M.-S., Felipe, G., & Ali, T. (2021). In vivo oral insulin delivery via covalent organic frameworks. *Chemical Science*, *12*, 6037–6047.

Fessi, H. P. F. D., Puisieux, F., Devissaguet, J. P., Ammoury, N., & Benita, S. (1989). Nanocapsule formation by interfacial polymer deposition following solvent displacement. *International Journal of Pharmaceutics*, *55*(1), R1–R4.

Garcia, X., Escribano, E., Domenech, J., Queralt, J., & Freixes, J. (2011). In vitro characterization and in vivo analgesic and anti-allodynic activity of PLGA-bupivacaine nanoparticles. *Journal of Nanoparticle Research*, *13*(5), 2213–2223.

Guerrero, A., Dorling, B., Ripolles-Sanchis, T., Aghamohammadi, M., Barrena, E., Campoy-Quiles, M., & Garcia-Belmonte, G. (2013). Interplay between fullerene surface coverage and contact selectivity of cathode interfaces in organic solar cells. *ACS Nano*, *7*(5), 4637–4646.

Ingham, B. (2015). X-ray scattering characterisation of nanoparticles. *Crystallography Reviews*, *21*(4), 229–303.

Jackson, J. K., Zhang, X., Llewellen, S., Hunter, W. L., & Burt, H. M. (2004). The characterization of novel polymeric paste formulations for intratumoral delivery. *International Journal of Pharmaceutics*, *270(1-2)*, 185–198.

Jatana, S., Palmer, B. C., Phelan, S. J., & DeLouise, L. A. (2017). Immunomodulatory effects of nanoparticles on skin allergy. *Scientific Reports*, *7*(1), 1–11.

Juliano, R. L. (2012). The future of nanomedicine: Promises and limitations. *Science and Public Policy*, *39*(1), 99–104.

Koudelka S, Turánek J. (2012). Liposomal paclitaxel formulations. *J Control Release*. *163*(3), 322–34.

Kumar, M., Kong, X., Behera, A. K., Hellermann, G. R., Lockey, R. F., & Mohapatra, S. S. (2003). Chitosan IFN-γ-pDNA nanoparticle (CIN) therapy for allergic asthma. *Genetic Vaccines and Therapy*, *1*(1), 1–10.

Kümmerer, K., Menz, J., Schubert, T., & Thielemans, W. (2011). Biodegradability of organic nanoparticles in the aqueous environment. *Chemosphere*, *82*(10), 1387–1392.

Lannibois, H., Hasmy, A., Botet, R., Chariol, O. A., & Cabane, B. (1997). Surfactant limited aggregation of hydrophobic molecules in water. *Journal de Physique II*, *7*(2), 319–342.

Lee, H., Choi, T. K., Lee, Y. B., Cho, H. R., Ghaffari, R., Wang, L., & Kim, D. H. (2016). A graphene-based electrochemical device with thermoresponsive microneedles for diabetes monitoring and therapy. *Nature Nanotechnology*, *11*(6), 566–572.

Lee, S. H., Heng, D., Ng, W. K., Chan, H. K., & Tan, R. B. (2011). Nano spray drying: A novel method for preparing protein nanoparticles for protein therapy. *International Journal of Pharmaceutics*, *403(1-2)*, 192–200.

Li, X., Anton, N., Arpagaus, C., Belleteix, F., & Vandamme, T. F. (2010). Nanoparticles by spray drying using innovative new technology: The Büchi Nano Spray Dryer B-90. *Journal of Controlled Release*, *147*(2), 304–310.

Liu, D., Du, J., Qi, S., Li, M., Wang, J., Liu, M., & Shen, J. (2021). Supramolecular nanoparticles constructed from pillar [5] arene-based host-guest complexation with enhanced aggregation-induced emission for imaging-guided drug delivery. *Materials Chemistry Frontiers*, *5*(3), 1418–1427.

Liu, M., Gu, B., Wu, W., Duan, Y., Liu, H., Deng, X., & Liu, B. (2020). Binary organic nanoparticles with bright aggregation-induced emission for three-photon brain vascular imaging. *Chemistry of Materials*, *32*(15), 6437–6443.

Long, R. Q., & Yang, R. T. (2001). Carbon nanotubes as a superior sorbent for nitrogen oxides. *Industrial & Engineering Chemistry Research*, *40*(20), 4288–4291.

Lu, K., He, C., Guo, N., Chan, C., Ni, K., Weichselbaum, R. R., & Lin, W. (2016). Chlorin-based nanoscale metal-organic framework systemically rejects colorectal cancers via synergistic photodynamic therapy and checkpoint blockade immunotherapy. *Journal of the American Chemical Society*, *138*(38), 12502–12510.

Luo, W., Feng, Y., Cao, C., Li, M., Liu, E., Li, S., & Feng, W. (2015). A high energy density azobenzene/graphene hybrid: A nano-templated platform for solar thermal storage. *Journal of Materials Chemistry A*, *3*(22), 11787–11795.

Maksimenko, A., Mougin, J., Mura, S., Sliwinski, E., Lepeltier, E., Bourgaux, C., & Couvreur, P. (2013). Polyisoprenoyl gemcitabine conjugates self assemble as nanoparticles, useful for cancer therapy. *Cancer Letters*, *334*(2), 346–353.

Masters, K. (1991). *The Spray Drying Handbook*. Longman Scientific Publication

Mochida, I., Kawabuchi, Y., Kawano, S., Matsumura, Y., & Yoshikawa, M. (1997). High catalytic activity of pitch-based activated carbon fibres of moderate surface area for oxidation of NO to NO_2 at room temperature. *Fuel*, *76*(6), 543–548.

Moon, J. J., Suh, H., Polhemus, M. E., Ockenhouse, C. F., Yadava, A., & Irvine, D. J. (2012). Antigen-displaying lipid-enveloped PLGA nanoparticles as delivery agents for a Plasmodium vivax malaria vaccine. *PloS One*, *7*(2), e31472.

Mourdikoudis, S., Pallares, R. M., & Thanh, N. T. (2018). Characterization techniques for nanoparticles: Comparison and complementarity upon studying nanoparticle properties. *Nanoscale*, *10*(27), 12871–12934.

Müller, R. H., Mäder, K., & Gohla, S. (2000). Solid lipid nanoparticles (SLN) for controlled drug delivery-a review of the state of the art. *European journal of pharmaceutics and biopharmaceutics*, *50*(1), 161–177.

Nantero (2006). NRAMTM.

Nasr, M., Najlah, M., D'Emanuele, A., & Elhissi, A. (2014). PAMAM dendrimers as aerosol drug nanocarriers for pulmonary delivery via nebulization. *International Journal of Pharmaceutics*, *461*(*1–2*), 242–250.

Nehilla, B. J., Bergkvist, M., Popat, K. C., & Desai, T. A. (2008). Purified and surfactant-free coenzyme Q10-loaded biodegradable nanoparticles. *International Journal of Pharmaceutics*, *348*(1–2), 107–114.

Pan, X., Sharma, A., Gedefaw, D., Kroon, R., De Zerio, A. D., Holmes, N. P., & Andersson, M. R. (2018). Environmentally friendly preparation of nanoparticles for organic photovoltaics. *Organic Electronics*, *59*, 432–440.

Peng CM, Donnet JB, Wang TK, et al. Surface treatment of carbon fibers. In: Donnet JB, Wang TK, Rebouillat S, Peng JCM (eds). *Carbon fibers*, 2nd ed. New York: Marcel Dekker, 1990, pp. 161–230.

Phys.org. (2013). Novel natural nanomaterial spins off from spider-mite genome sequencing. *Phys.Organic*.

Poland, C. A., Duffin, R., Kinloch, I., Maynard, A., Wallace, W. A., Seaton, A., & Donaldson, K. (2008). Carbon nanotubes introduced into the abdominal cavity of mice show asbestos-like pathogenicity in a pilot study. *Nature Nanotechnology*, *3*(7), 423–428.

Portela, C. M., Vidyasagar, A., Krödel, S., Weissenbach, T., Yee, D. W., Greer, J. R., & Kochmann, D. M. (2020). Extreme mechanical resilience of self-assembled nanolabyrinthine materials. *Proceedings of the National Academy of Sciences*, *117* (11), 5686–5693.

Pritam C., Krishnendu D., & Prasanta K. D. (2017). l-Phenylalanine-tethered, naphthalene diimide-based, aggregation-induced, green-emitting organic nanoparticles. *Langmuir*, *33*(18), 4500–4510.

Qi, G. B., Gao, Y. J., Wang, L., & Wang, H. (2018). Self-assembled peptide-based nanomaterials for biomedical imaging and therapy. *Advanced Materials*, *30*(22), 1703444.

Qiao, Z. Y., Lin, Y. X., Lai, W. J., Hou, C. Y., Wang, Y., Qiao, S. L., & Wang, H. (2016). A general strategy for facile synthesis and in situ screening of self-assembled polymer peptide nanomaterials. *Advanced Materials*, *28*(9), 1859–1867.

Qiu, L. Y., & Bae, Y. H. (2006). Polymer architecture and drug delivery. *Pharmaceutical Research*, *23*(1), 1–30.

Quintanar-Guerrero, D., Allémann, E., Fessi, H., & Doelker, E. (1999). Pseudolatex preparation using a novel emulsion-diffusion process involving direct displacement of partially water-miscible solvents by distillation. *International Journal of Pharmaceutics*, *188*(2), 155–164.

Quintanar-Guerrero, D., Allémann, E., Fessi, H., & Doelker, E. (1998). Preparation techniques and mechanisms of formation of biodegradable nanoparticles from preformed polymers. *Drug Development and Industrial Pharmacy*, *24*(12), 1113–1128.

Santra, M., Jun, Y. W., Reo, Y. J., Sarkar, S., Choi, W., Kwon, J. E., & Ahn, K. H. (2018). Exploration of molecular shape-dependent luminescence behavior: Fluorogenic organic nanoparticles based on bent shaped excited-state intramolecular proton-transfer dyes. *ACS Applied Bio Materials*, *1*(1), 136–145.

Saura, S., & Daniel, C. (2009). *Nanotoxicity: From in vivo and in vitro Models to Health Risks*. John Wiley & Sons.

Scardi, P., Leoni, M., & Beyerlein, K. R. (2011). On the modelling of the powder pattern from a nanocrystalline material. *Journal of Crystallography*, *226*, 924–933.

Seyler, I., Appel, M., Devissaguet, J. P., Legrand, P., & Barratt, G. (1999). Macrophage activation by a lipophilic derivative of muramyldipeptide within nanocapsules: Investigation of the mechanism of drug delivery. *Journal of Nanoparticle Research*, *1*(1), 91–97.

Shin, J., Yin, Y., Park, H., Park, S., Triantafillu, U. L., Kim, Y., & Kim, D. W. (2018). p38 siRNA-encapsulated PLGA nanoparticles alleviate neuropathic pain behavior in rats by inhibiting microglia activation. *Nanomedicine*, *13*(13), 1607–1621.

Shortell, M. P., Fernando, J. F., Jaatinen, E. A., & Waclawik, E. R. (2014). Accurate measurement of the molecular thickness of thin organic shells on small inorganic cores using dynamic light scattering. *Langmuir*, *30*(2), 470–476.

Skiba, M., Wouessidjewe, D., Puisieux, F., Duchêne, D., & Gulik, A. (1996). Characterization of amphiphilic β-cyclodextrin nanospheres. *International Journal of Pharmaceutics*, *142*(1), 121–124.

Thomas, C., Rawat, A., Hope-Weeks, L., & Ahsan, F. (2011). Aerosolized PLA and PLGA nanoparticles enhance humoral, mucosal and cytokine responses to hepatitis B vaccine. *Molecular Pharmaceutics*, *8*(2), 405–415.

Tracton, A. A. (Ed.). (2006). *Coatings Materials and Surface Coatings*. CRC Press.

Vehring, R. (2008). Pharmaceutical particle engineering via spray drying. *Pharmaceutical Research*, *25*(5), 999–1022.

Wang, K., Zhang, Z., Lin, L., Chen, J., Hao, K., Tian, H., & Chen, X. (2019). Covalent organic nanosheets integrated heterojunction with two strategies to overcome hypoxic-tumor photodynamic therapy. *Chemistry of Materials*, *31*(9), 3313–3323.

Wang, S., Sobczynski, D. J., Jahanian, P., Xhahysa, J., & Mao, G. (2013). A supra-monolayer nanopattern for organic nanoparticle array deposition. *ACS Applied Materials & Interfaces*, *5*(7), 2699–2707.

Wang, Y., Lin, Y. X., Qiao, Z. Y., An, H. W., Qiao, S. L., Wang, L., & Wang, H. (2015). Self assembled autophagy inducing polymeric nanoparticles for breast cancer interference in-vivo. *Advanced Materials*, *27*(16), 2627–2634.

Warren, B. E. (1969). *X-ray Diffraction*. Dover.

Wolfgang, M. (2015). Powerpoint presentation: Nanoparticle size analysis: A survey and review. In: *Nanomedicines Alliance*. https://pqri.org/wp-content/uploads/2015/09/03-Wolfgang-FDA-PQRI-Particle-Size-Presentation_Final.pdf

Wu, D., Gao, Y., Qi, Y., Chen, L., Ma, Y., & Li, Y. (2014). Peptide-based cancer therapy: Opportunity and challenge. *Cancer Letters*, *351*(1), 13–22.

Xia, Q., Chen, Z., Yu, Z., Wang, L., Qu, J., & Liu, R. (2018). Aggregation-induced emission-active near-infrared fluorescent organic nanoparticles for noninvasive long-term monitoring of tumor growth. *ACS Applied Materials & Interfaces*, *10*(20), 17081–17088.

Xu, P., Van Kirk, E. A., Zhan, Y., Murdoch, W. J., Radosz, M., & Shen, Y. (2007). Targeted charge-reversal nanoparticles for nuclear drug delivery. *Angewandte Chemie International Edition*, *46*(26), 4999–5002.

Yang, K. U. N., Wu, W., Jing, Q., & Zhu, L. (2008). Aqueous adsorption of aniline, phenol, and their substitutes by multi-walled carbon nanotubes. *Environmental Science & Technology*, *42*(21), 7931–7936.

Yang, K., & Xing, B. (2010). Adsorption of organic compounds by carbon nanomaterials in aqueous phase: Polanyi theory and its application. *Chemical Reviews*, *110*(10), 5989–6008.

Yao, H., Sasahara, H., & Kimura, K. (2011). Organic porphyrin nanoparticles with induced optical activity: Ion-based synthesis from achiral chromophore and chiral counterions. *Chemistry of Materials*, *23*(3), 913–922.

Yeh, T. Y., Wang, C. I., & Chang, H. T. (2013). Photoluminescent C-dots@ RGO for sensitive detection of hydrogen peroxide and glucose. *Talanta*, *115*, 718–723.

Li, Y., Zhou, L., Zhu, B., Xiang, J., Du, J., He, M., Zhang, P., Zeng, R., & Gong, P. (2021). A glutathione-activated carrier free nanodrug of triptolide as trackable drug delivery system for monitoring and improving tumor therapy. *Materials Chemistry Frontiers*. 5, 5312–5318.

Zhu, D., Wu, S., Hu, C., Chen, Z., Wang, H., Fan, F., & Zhang, L. (2017). Folate-targeted polymersomes loaded with both paclitaxel and doxorubicin for the combination chemotherapy of hepatocellular carcinoma. *Acta Biomaterialia*, *58*, 399–412.

Zhu, J., & Shi, X. (2013). Dendrimer-based nanodevices for targeted drug delivery applications. *Journal of Materials Chemistry B*, *1*(34), 4199–4211.

12 Polyoxometalate-Induced Nano-Engineered Composite Materials for Energy Storage

Battery Application

Shankab J. Phukan, Neeraj K. Sah, Dasnur Nanjappa Madhusudan, Ruchi Sharma, Rohit Sharma, Kamatchi Sankaranarayanan, Chandni Pathak, Manas Roy, Shyam S. Pandey, and rer. nat. Somenath Garai

12.1 INTRODUCTION

Electrochemical energy conserver plays a crucial function in resolving energy scarcity and environmental pollution hazards (Tarascon & Armand, 2011). In the current scenario, harvesting and storage of energy are considered as two prime factors in which we need to concentrate. The storage of harvested green energy from solar, thermal, hydrothermal sources is achieved by energy storage devices, which involves the conversion of electrical or non-mechanical energy into chemical energy or potential energy. The main advantages of the energy storage devices are that they improve reliability and resilience and integrate diverse resources (Liu et al., 2018). In most cases, energy storage devices (mainly batteries) are closed systems which implies that energy storage and conversion take place in the same compartment. The battery consists of one or more electrochemical cells, and the electrical energy is generated by converting chemical energy via redox processes at the anode and cathode. Generally, batteries are of two kinds, *viz.* primary and secondary batteries: primary batteries are used just once and then discarded, whereas secondary batteries can be recharged, which means by reversing current flow through the cell (Winter & Brodd, 2004). The quantity of electrical charge provided by a battery at a particular rated voltage is referred to as its capacity. The C-rate measures how quickly a battery is charged or drained; if the C-rate is low, then the battery capacity is high and *vice versa*. The process of production of a battery is controlled by mainly four factors:

DOI: 10.1201/9781003481157-12

cost, performance, reliability and its appearance. Alkaline battery is a general example of primary battery, and common examples of secondary batteries are lead-acid batteries and lithium-ion batteries (LIBs). Conventional batteries consist of hazardous metals like mercury, cadmium and lead as the electrode or electrolyte materials; therefore, it is necessary to dispose these batteries safely when it is discharging to avoid damage to the environment. Furthermore, battery discharge is also an exothermic process that releases a substantial amount of heat is dissipated which can also be considered as a limiting factor for battery performances.

Very recently, the researchers have concentrated more on integrated energy storage systems than conventional devices due to their multiform functions like micro-electromechanical systems and self-powering systems. Wang et al. (2014) recently created a flexible, fiber-based integrated system that recognizes both flexible energy storage as well as flexible optoelectronic detection systems simultaneously on an isolated fiber device.

It is quite impossible to anticipate which of the energy storage devices will play a major role in the near future. Still, a lot of drawbacks exist and the development of some energy storage devices lies in its infant stage due to lack of resources, environmental issues, etc. Sustainable attempts should be taken to improve the renewable resources of energy production to reduce the emissions and environmental impacts. No doubt, future research on energy storage devices offers opportunities for innovation and advancements in the energy storage process. In these circumstances, polyoxometalates (POMs) have been recognized as one of the promising electrode materials for battery applications.

Polyoxometalates are well-organized inorganic nanoclusters made up of early transition metals with generally high oxidation states and oxide ligands. An early transition metal-based compound possesses a higher overall charge, distinctive crystal shapes and thermal stability; owing to these features, POMs have encrypted their name as one of the potent alternative electrode materials for electrochemical rechargeable batteries and supercapacitors (SCs). POM-based electrode materials gained more attention due to their facile synthetic routes and significantly minimized some of the trending issues associated with metal-oxide-based electrode materials, such as rapid capacity decay, volume change and phase transition (Huang et al., 2020). Advantageously, POMs are based on inexpensive and easily available precursors for large-scale synthesis with intriguing properties like high negative charges, tunable sizes and structural diversity. POMs, which may be combined with different polymers or carbon-based nanostructures, are extremely attractive candidates for hybrid energy storage devices. Consequently, Wang et al. have utilized electropolymerization to implant a lacunary Keggin anion $[SiW_{11}O_{39}]^{8-}$ and poly(3,4-ethylenedioxythiophene) [PEDOT] composite on an ITO electrode to construct a traditional dye-sensitized solar cell (DSSC) counter electrode (Yuan et al., 2013). It has been demonstrated (Alaaeddine et al., 2014) that the Dawson anion $[P_2W_{18}O_{62}]^{6-}$ can be used as an electroactive contact between an indium tin oxide electrode and a bulk heterojunction of the conductive polymer in an arrangement of stacked organic solar cells while poly(3-hexylthiophene) (P3HT) and [6,6]-phenyl-C_{61}-butyric acid methyl ester (PCBM) are used to fabricate the bulk heterojunction.

It has been observed that POM-based composites have some limitations in generating photocurrent due to the following reasons: (a) Surface area: POM and hybrid POM layer-by-layer (LBL) film photo anodes, unlike nano-crystalline semiconductor films, do not create large surface area nanostructured films. (b) Poor electron diffusion and film conductivity. (c) Film crystallinity: after cycling electrochemically in acids, POM thin films with Ru (II) or Os(II) complexes seem to crystallize; however, the effect on charge transfer in the emerging film is unknown. The POM hybrid materials that are electrostatically assembled are often non-crystalline. (d) Limitation of dye-cells: even though $[Ru(bpy)_3]^{2+}$ and $[Ru(bpy)_2(PVP)_{10}]^{2+}$ have shown to be suitable model compounds for sensitive POM photocurrent production, their absorption cross-sections are rather small (Walsh et al., 2016). However, luminescent sensor arrays, ferrioxalate–POM actinometry and ECL, all benefit from the utilization of hybrid POM photochemical technologies. To overcome the above-mentioned confines of POM-based composites in energy harvesting and storing applications, researchers are pushing the POM chemistry boundaries in search of appropriate companions for the metal-oxide clusters to revolutionize the portable energy device sector. Some of the ground-breaking experiments relating to this particular domain are discussed in this chapter.

This chapter primarily emphasizes the vintage points of extensive presentations of POM-based nanostructured composite systems for energy storage. An attempt has been made to deliver an outline of probable resolutions of challenges to energy storage and supercapacitor applications by POM-based composite systems along with its structural aspects. Instead of conventional energy storage systems, the virtues of incorporating POM-based nanostructured composite systems are emphasized and the future challenges are manifested in this chapter.

12.2 STRUCTURAL ASPECTS OF POM

Polyoxometalates (POMs) are anionic metal-oxo clusters with the formula $[M_xO_y]^{n-}$, which are composed of one or more transition metals with a high valency (M = V, Mo, W, Nb) connected by oxo-ligands, formed through a unique self-assembly mechanism, beginning with basic building blocks such as MoO_4^{2-} or VO_4^{3-}, mostly in acidic medium (Herrmann et al., 2015). POM clusters have molecular diameters ranging from 1 to 10 nanometers and are mostly anionic in nature. This combination can produce homonuclear POM anions up to the nuclearity of 36 without metal *d*-electron while the nuclearity of the cluster can be enhanced up to a value of 368 (characterizable by single-crystal XRD) upon reduction and further acidification. The structural diversity of POMs is due to the large possible variety of combinations of the fundamental metal-oxide building components from the dynamic library (Herrmann et al., 2015; Long et al., 2010; Miras et al., 2012) as illustrated in Figure 12.1, exhibiting different components of the POM clusters.

The numerous types of POM species may be classified into three groups: (a) heteropolyanions: these clusters are constituted by the central heteroatom X and addenda metals M around X; they are most decorated in terms of POM research, the classic Keggin ($XM_{12}O_{40}$) and Wells-Dawson ($X_2M_{18}O_{62}$) anions are of particular attention (Long et al., 2010). The addenda atom is generally Mo, W, V and Nb in high oxidation states, but the heteroatom can be any of over 60 elements, comprising mostly the non-metals or some of the transition metals (Liu & Tang, 2010). (b) Isopolyanions: these are built by bottom-up approach through charge-driven self-assembly processes mostly in the acidic medium through corner/edges or even face sharing of the tetrahedral or mostly octahedral building blocks. They usually possess a higher

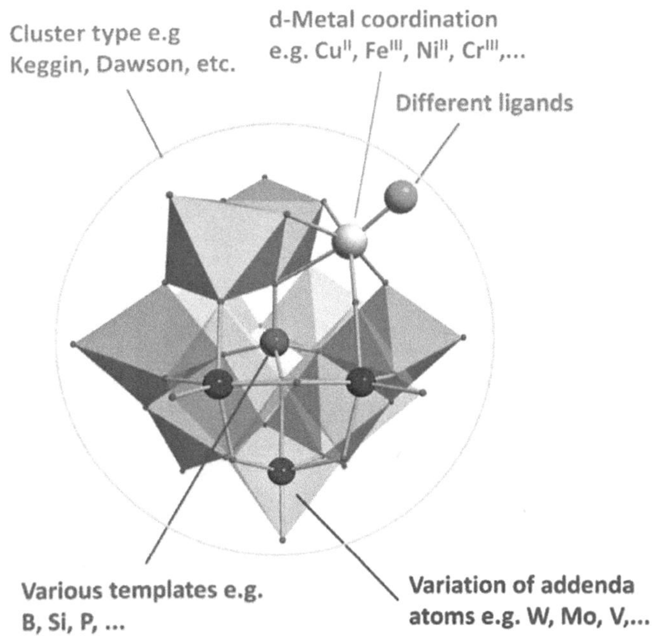

FIGURE 12.1 Constituents of a polyoxometalate cluster which are susceptible to deliberated chemical modifications. Reprinted from Herrmann et al. (2015). Copyright 2015, with permission from *Royal Society of Chemistry.*

cluster charge and therefore are rather less stable compared to the heteropolyanions, which makes them more desirable as intermediate constituents for the formation of larger assemblies. The core heteroatom is absent from these molecules, which have a similar metal-oxide framework to the heteropolyanions (Genovese & Lian, 2017). When an oxometalate ion is protonated under certain conditions (such as pH, concentration and temperature), the MO_4^{2-} units poly-condenses, resulting in the development of more complex structures known as polyanions. The isopolyanion (IPA) has the generic formula $M_nO_{(4n-m)}^{(2n-m)-}$ while the condensation happens between similar building species. However, when numerous oxoanions condense about a core heteroatom (X = Si, P, As, Ge) or other metal atoms, a heteropolyanion (HPA) with the common formula $X_sM_nO_m^{y-}$ is formed (Ammam, 2013). (c) Macro-molecular polymolybdate clusters: the newest branch of POM clusters to be investigated are the molybdenum blue and molybdenum brown type reduced POM clusters, which are comprised of very high-nuclearity metal-oxide cluster anions like $[Mo_{154}]$ or $[Mo_{132}]$ clusters that may be generated in symmetrical ring type or spherical conformations (Liu & Tang, 2010; Long et al., 2010).

The Keggin ion system, Dawson complex, Lindqvist ion, Anderson complex, sandwich structures and Presslyer forms are the most prevalent POM structures, as depicted in Figure 12.2. Covalent functionalization of classic POM structures has recently been identified by Proust et al., Izzet et al. and Izarova et al., resulting in new classes of hybrid POM species (Izarova et al., 2010; Proust et al., 2012).

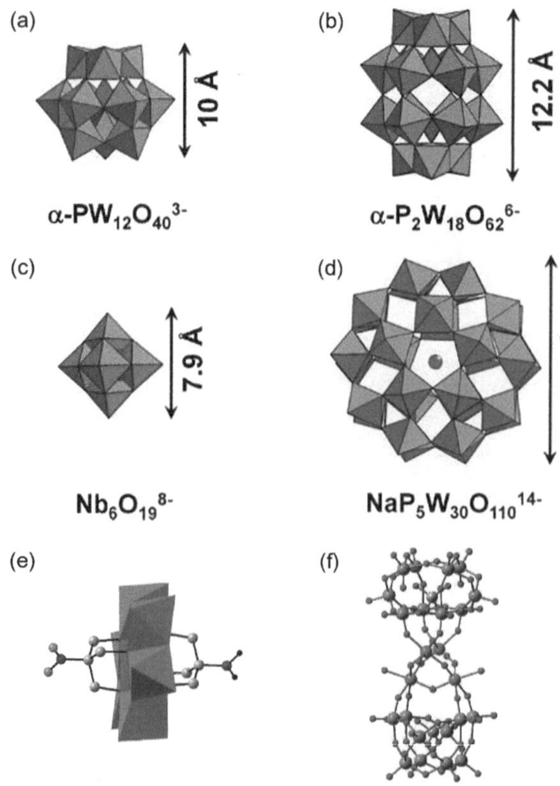

FIGURE 12.2 Pictorial depiction of different polyoxometalates. (a) The Keggin form; (b) Dawson form; (c) Lindqvist form; (d) a view down the molecular C_5 axis of a Presslyer structure containing Na^+. Reprinted from Wang and Weinstock (2010). Copyright, with permission from *Royal Society of Science;* (e) Anderson System. Reprinted from Song et al. (2008). Copyright 2010, with permission from *American Chemical Society; and (f)* "Sandwich" anion $[Ru_4(\mu\text{-}O)_4(\mu\text{-}OH)_2(H_2O)_4(\gamma\text{-}SiW_{10}O_{36})_2]^{10-}$, displaying Ru(IV) centers fitted in the middle of two $\{SiW_{10}O_{36}\}$ parts. Reprinted from Sartorel et al. (2008) Copyright 2008, with permission from *American Chemical Society.*

12.2.1 Lindqvist

The Lindqvist is best described as a six-edge octahedron (Figure 12.3) with a general representation of $[M_6O_{19}]^{n-}$ (where M can be Mo and W for n = 2 and V, Nb, Ta for n = 8). The solid structure of the Lindqvist ion salt varies as a function of the cluster's protonation and linkage to the group-I cations. Generally, the NbO_6(octahedral) polyoxoniobates consist of a longer central Nb–O_c bond length of 2.3–2.5 Å which is in trans position with respect to a shorter Nb–O_t bond length of about 1.75–1.80 Å, while the four Nb–O_b bridging bonds vary within the range of 2 Å; as a consequence of that, the bond length of other Nb–O_b bonds of the NbO_6H decrease approximately to 1.90–1.95 Å (Nyman et al., 2006).

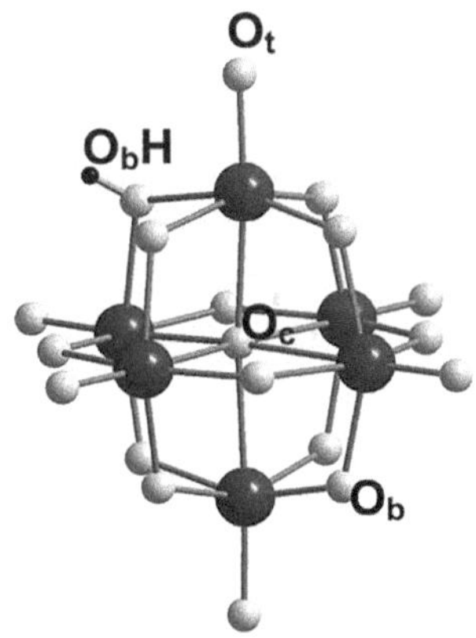

FIGURE 12.3 Representation of Lindqvist cluster $[HNb_6O_{19}]$ in Ball & Stick representation, correspondingly, signifying the different oxygen sites: O_c, O_b, O_t and O_bH as central, bridging, terminal and protonated bridging oxygen, respectively. Reprinted from Herrmann et al. (2015). Copyright 2015, with permission from the *Royal Society of Chemistry*.

12.2.2 Keggin

Keggin structure is one of the most vastly studied structural forms of heteropoly nanoacids. Keggin POM-cluster anions can be generally represented by $[XM_{12}O_{40}]^{n-}$ resulting from 12 octahedrally coordinated metal atoms (M = V, Nb, Ta, Mo and W) arranged around a tetrahedral heteroatom (X = P, Ar, Si, Ge, *etc.*), with an M:X ratio of 12:1, as suggested by Pauling in 1929, but the confirmation comes from Keggin only in 1933 (Tarascon & Armand, 2011) who explored the framework of $[PW_{12}O_{40}]^{3-}$ by employing single-crystal X-ray diffraction. The Keggin can also be realized from the $\{\mu_3\text{-O}\}$ coordination of the building $\{M_3O_{13}\}$ units around central heteroatoms to yield the general formula $(XO_4)M_{12}O_{36}$ (Figure 12.4).

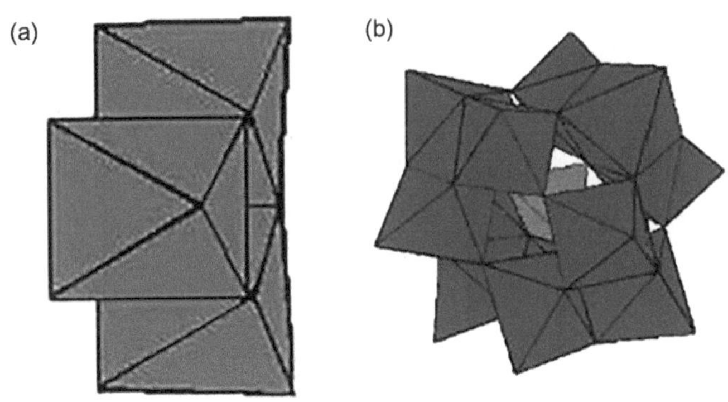

FIGURE 12.4 (a) Representation of three octahedrons to form a deltahedral trimetallic cluster of $\{M_3O_{13}\}$- moiety. (b) General representation of the Keggin cluster. Reprinted from Ammam (2013). Copyright 2013, with permission from the *Royal Society of Science*.

The existence of four C_3-axis of symmetry in the Keggin structure and their rotation give rise to various isomers. There are five theoretically possible structural isomers, but researchers are able to identify, synthesize and isolate frequently three of them. These isomers of α-Keggin sometimes are also termed as the rotational isomers by Baker–Figgis, designated as α, β, γ, δ and ε. The α-structure is designated as the original Keggin structure and all other isomers can be derived from it by successive rotation of 60° for the $\{M_3O_{13}\}$ moiety. These structures are later confirmed by various researchers (Keggin, 1933; Kobayashi & Sasaki, 1975; Pope et al., 1983). The symmetry of α, β, γ, δ and ε is T_d, C_{3V}, C_{2V}, C_{3V} and T_d, respectively.

The electronic structure, atomic structure and the electrostatic interaction between these units are tuned as a function of rearrangement of the triads: $\alpha < \beta < \gamma < \delta < \varepsilon$, representing the trend of increasing stability among the fully oxidized Keggin anions, but this trend is not found to be compatible for the reduced clusters (Gumerova & Rompel, 2018). There are various ways by which this trend can be explained; first, we will try to understand via the cooperative effect of all the moieties; as one of the $\{M_3O_{13}\}$ moiety rotates, it prompts the other contingent moieties to rotate exposing the heteroatoms. The rotation by π/3 degree of three such subunits is known to create a nucleophilic void, which can be a potential attacking center for the *Lewis* acids like Al^{3+}, and this structural engineering of the moieties finds applications in Al batteries.

Secondly, both DFT calculations and molecular orbital approach have concreted the above results. The LUMO of the β-form lies below the LUMO of the α-form, reflecting that the $2e^-$ or $4e^-$ reduced species prefers to exist in its β isomeric form. Because of its low-lying (stable) LUMO levels, the reduced form of the γ isomer is also stable, obviously not to the extent of stability of β isomers. The δ and ε isomers are unstable compared to α or β isomers and can only be found to be stable (Zhou et al., 2013) and obtainable when present in a reduced state while bound to a transition metal ion. To stabilize isomers other than α and β, we require to put them in reducing ambiance, roughly 4–8 electrons are required to stabilize the γ, 8–11 electrons for the δ and 10–14 electrons for the ε-isomer. Tézé et al. exhibited that the $M_{12}O_{36}$ framework has a certain amount of rotated triads which are directly linked to the oxidizing character of Keggin POMs, that is the ability to reduce itself rises in the order $\alpha < \beta < \gamma$ (Tézé et al., 1996). Comparing oxidized and reduced Keggin POMs, the oxidized form is more prone toward addenda atom substitution and the reduced form is more resistant toward the same owing to the stabilization by the delocalization of the "blue" electrons. However, metal/metal-oxo cations such as {Ni(II) or V(V/IV)O} can be used to cap the highly charged ε-Keggin anions, thereby affording the "pseudo-Keggin" structures (López & Poblet, 2004).

12.2.3 Lacunary Keggin

The term lacunary basically means "missing fragments", sometimes called defect structures. POMs that have metal-oxo octahedra (one or more) as shown in Figure 12.5 missing from a general skeleton are termed as "lacunary". The reaction of alkaline solutions containing Keggin-kind POMs under fixed and optimum

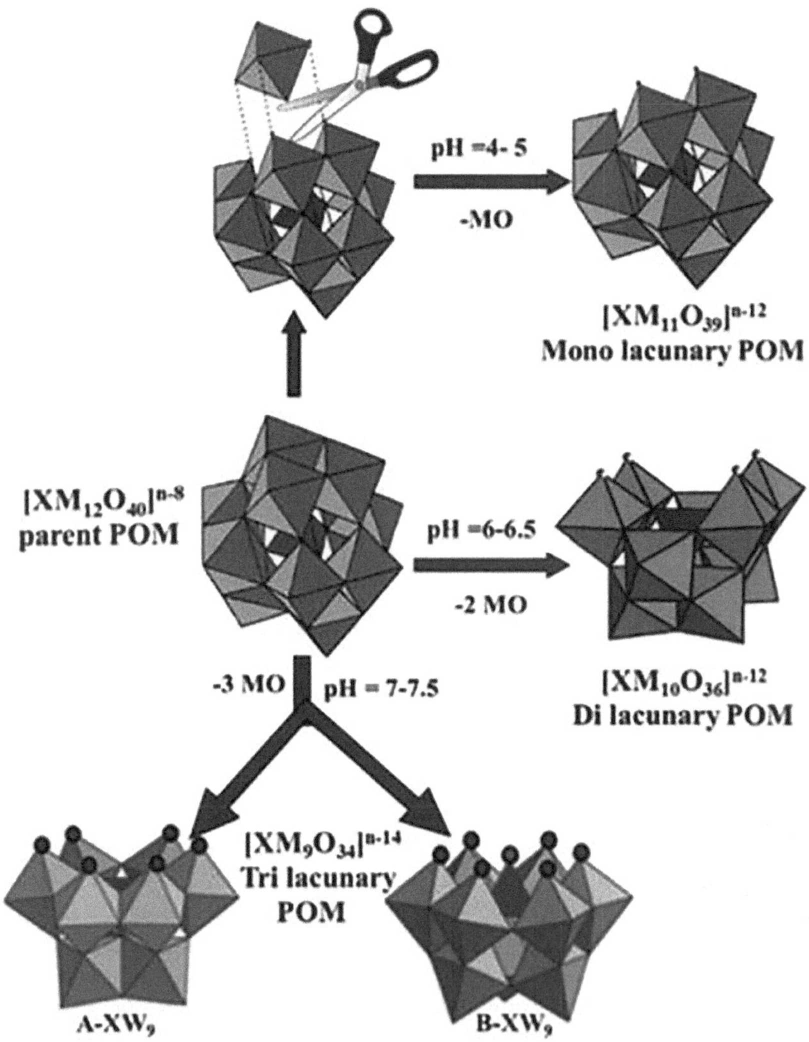

FIGURE 12.5 Various lacunary POM frameworks obtained from the parent Keggin structure, where X = P or Si, M = Mo or W and n indicates charge on X. Reprinted from Patel et al. (2016). Copyright 2016, with permission from *Taylor & Francis*.

experimental environments (*e.g.*, temperature, pH and concentration) results in the depletion of metal centers to produce "lacunary" POMs species wherein $[XM^{VI}{}_{11}O_{39}]^{(n+4)-}$ and $[XM^{VI}{}_{10}O_{36}]^{(n+5)-}$ are mono- or di-lacunary POMs, respectively, which are formed by removal of several MO_x units. This gives lacunary POM species of which the framework still belongs to the Keggin sequence (Ammam, 2013). These reactive species have their unique reactivity and stability patterns and these vacancies can be filled by apposite linker atoms.

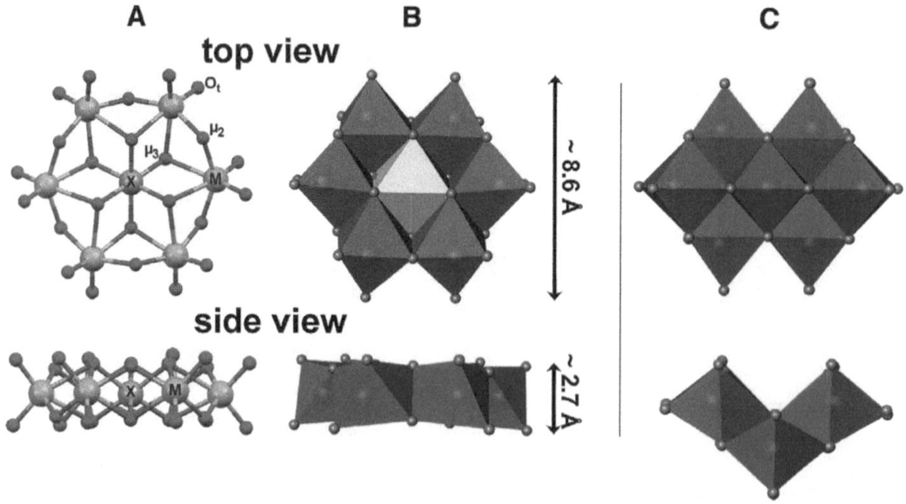

FIGURE 12.6 (a and b) Representation of Anderson-Evans polyhedral POM $[XM_6O_{24}]^{n-}$ where heteroatom X (gray sphere/gray octahedron) is generally a first transition metal, the addenda atom M (green spheres/octahedra) represents W(VI) or Mo(VI) and the three different modes of oxo-linkage (red spheres) shown as μ_3 (triply bridged between X and two M) and μ_2 (doubly bridged between two M) and O_t (terminal oxygen atoms). (c) Representation of the bent isomer $[M_7O_{24}]^{n-}$ [51] *e.g.*, $[Mo_7O_{24}]^{6-}$ [52] (c bottom). Reprinted from Blazevic and Rompel (2016). Copyright 2016, with permission from *Elsevier.*

12.2.4 Anderson

The structural topology of Anderson-type POMs consists of a central octahedral unit (XO_6) surrounded by six edges sharing MO_6 octahedral units (Wu et al., 2021) with an octahedral central atom (Al) and tetrahedrally linked heteroatoms (P or Si) leading to the D_{3d} symmetry (Pope et al., 1983). General representation of Anderson-type POMs suggests $[H_a(XO_6)\ M_6O_{18}]^{n-}$, where ($0\leq a \leq 6$), ($2\leq n \leq 8$), X is the central heteroatom and M is the addenda atom (Mo^{VI} or W^{VI}) (Blazevic & Rompel, 2016; Pratt III et al., 2013) as shown in Figure 12.6.

The Anderson-type POMs can be classified into two categories: the non-protonated species containing highly oxidized heteroatom (Te^{VI}, I^{VII}) with the common formula $[X^{n+}\ M_6O_{24}]^{(12-n)-}$ (Perloff, 1970; Pratt III et al., 2013) and the protonated species consisting of heteroatom present in low oxidation state (Cr^{III}, Fe^{III}) with the common formula $[X^{n+}(OH)_6M_6O_{18}]^{(6-n)-}$ (Bergamo & Sava, 2007). The protonated species contain six μ_3-O atoms attached with six protons, surrounding the heteroatom (Pratt III et al., 2013).

12.2.5 Well-Dawson

The Dawson POM is the result of combination of two lacunary Keggin monomers $[XW_9O_{34}]^{z-}$ with dual structural arrangements: di-metallic groups (M_2O_{10}) formed *via* the condensation of two octahedrons where metal atom is at the center and

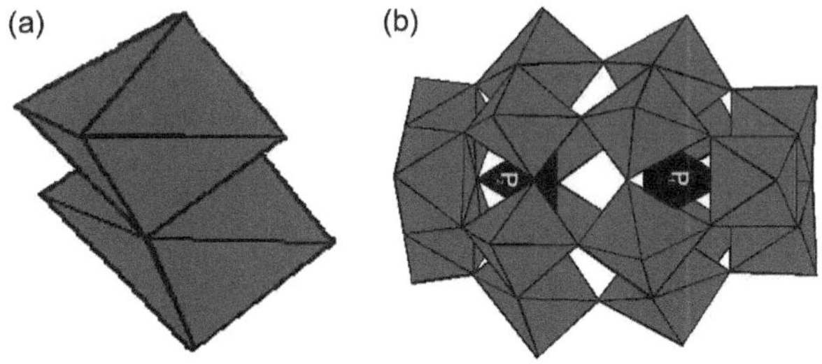

FIGURE 12.7 (a) Union of two octahedra to form bi-metallic moiety of M_2O_{10}. (b) Pictorial representation of the Dawson POM structure. Reprinted from Ammam (2013). Copyright 2013, with permission from the *Royal Society of Science.*

oxygen atoms are at vertices (Figure 12.7) and trimetallic groups (M_3O_{13}) formed by the combination of three octahedrons WO_6 (already mentioned in Figure 12.3a of the Keggin structure). The general representation of Wells-Dawson is given by $[(XO_4)_2M_{18}O_{54}]^{n-}$ (where M can be Mo or W and X is an element of the main group) and the primitive Wells-Dawson skeleton contains two tetrahedral anions, *e.g.* PO_4^{3-}, AsO_4^{3-}, SO_4^{2-}, *etc.* The metal atoms can occupy mainly two locations in the Dawson structure: (a) 3*2 metal atoms occupy the apical position whereas (b) the equatorial position corresponds to 6*2 metal atoms. Various symmetry components, such as planes, centers and several axes, result in a large number of isomers (Ammam, 2013), while Contant et al. identified six feasible isomers in the Dawson form: α, β, γ, α*, β*, γ* (Contant & Thouvenot, 1993). The α-isomer contains two core heteroatoms which are coordinated by nine octahedral $\{XM_9O_{31}\}$ units connected through O atoms (Figure 12.8a). The rotation around the C_3 axis of α-isomer by π/3 generates the β-isomer (Figure 12.8b); formation of the β-isomer results in the dissipation of the symmetry plane, dividing the Dawson molecule and forming two identical halves

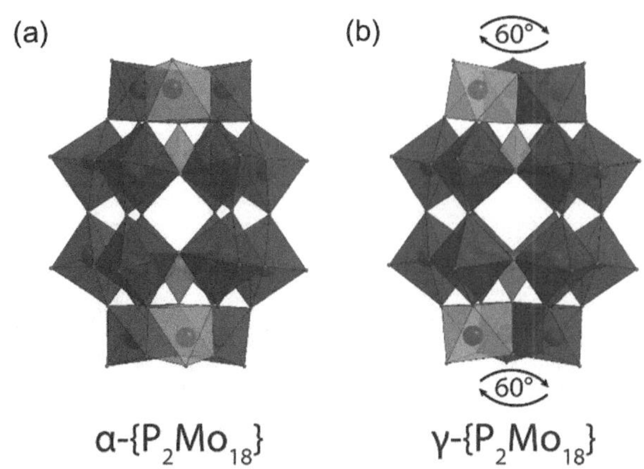

FIGURE 12.8 Polyhedral representation of single-crystal XRD structures of Dawson structural isomers (a) α-$\{P_2Mo_{18}\}$ and (b) γ-$\{P_2Mo_{18}\}$. A pair of Mo polyhedra are represented in green to visualize the cap rotation difference between isomers. Reprinted from Janusson et al. (2019). Copyright 2019, with permission from the *Royal Society of Science.*

yielding symmetry C_{3V}. The second rotation around the axis of symmetry C_3 by $\pi/3$ results in the γ isomer. This brings back the Dawson unit's symmetry plane with the existence of an inversion center present between two $[XW_9O_{34}]^{Z-}$ segments; similar rotation of α, β and γ yields the isomers α*, β* and γ*.

12.2.6 Presslyer POMs

The Presslyer anion is generally represented as $[X^{n+}P_5W_{30}O_{110}]^{(15-n)-}$ and is best known as the smallest POM with an internal cavity (Figure 12.12d) that allows cation exchange with the solution (Fernández et al., 2007).

12.2.7 Sandwich

The newly invented "sandwich" complexes result from the blend of lacunary entities: XW_9, X_2W_{15} and H_4XW_{15} for the Keggin, Dawson symmetrical and Dawson dissymmetrical. Figure 12.9a represents the sandwich structure where four transition metals in an octahedral environment are sandwiched between two tri-lacunary Keggin moieties of general formula $\{XM_9\}$. Figure B can be described as the sandwich structure where four transition metals in an octahedral environment are sandwiched between two tri-lacunary Wells-Dawson moieties of general formula$\{X_2M_{15}\}$.

The inclusion of two, three, or four transition metal ions at the connection point of two lacunary Keggin or two Dawson polyoxometalates generates a "sandwich-type complex" hybrid, which improves the structural stability. Two sandwich complexes constructed on Keggin and Dawson groups are represented in Figure 12.9a and b. Weakley et al., in 1973, reported the first synthesized sandwich POM, $K_{10}Co_4(H_2O)_2(PW_9O_{34})_2$ which has been fabricated by a combination of two Keggin moieties PW_9O_{34} by four co-centers (Weakley et al., 1973). During the 1980s–1990s, several researchers have worked extensively and investigated different sandwich-based Keggin, and the numbers dramatically rose between the 1990s and the early 2000s.

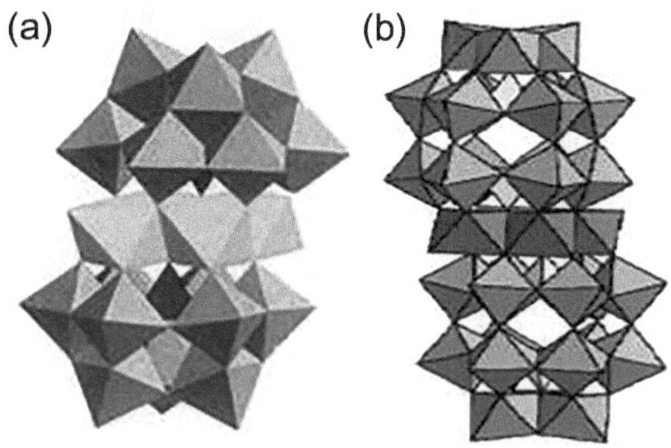

FIGURE 12.9 Structure of metal-oxide heterocycle sandwiched between (a) two Lacunary Keggin moieties and (b) two Lacunary Dawson moieties. Reprinted from Ammam (2013). Copyright 2013, with permission from the *Royal Society of Science*.

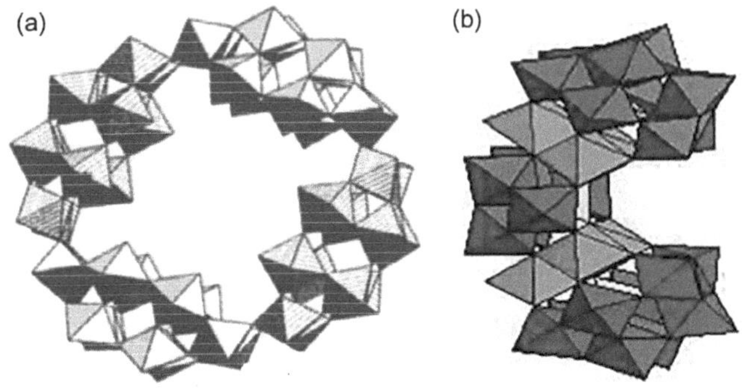

FIGURE 12.10 Structure of (a) Crown and Wheel POMs. (b) Banana structure. Reprinted from Ammam (2013). Copyright 2013 with permission from the *Royal Society of Science*.

12.2.8 Wheel and Banana

Another newly discovered POM structural species contains clusters such as the "crown" or "wheel". It was initially synthesized by Contant et al. in 1985 (Contant & Teze, 1985), and one of the examples is $K_{28}Li_5H_7P_8W_{48}O_{184}$ (Figure 12.10a). Mbomekalle et al. in the early 2000s have synthesized the "banana" structure (Mbomekalle et al., 2003) where two of the Keggin units, $XW_9M_3O_{40}$ or $XW_9M_2M'O_{40,}$ are linked with the XW_6O_{16}unit to give a banana-shaped POM (Figure 12.10b).

12.2.9 The Giant Big Wheel or the Bielefeld Wheel

Molybdenum blue solutions are known to exist in nature from the ancient times. First, Scheele and then Berzelius tried to isolate some solid out of this blue water and proposed the first formula $Mo_5O_{14}{\cdot}nH_2O$. The problem was elucidated by A. Müller and his coworkers in 1995 by the isolation of single crystals of the blue compound, finally formulated as $\{Mo_{154}\} \equiv [Mo_{154}O_{448}(NO)_{14}H_{14}(H_2O)_{70}]^{28-}$, which was considerably highlighted by several research journals as "Big Wheel Rolls Back Molecular Frontier", Mo_{154} is called the big wheel and Mo_{176}, the Bielefeld wheel.

The wheels of the Mo_{154} type having an average exterior diameter of 3.4 nm and a height of 1.5 nm are more compact than the Mo_{176}-type wheels, whose external diameter is about 4.1 nm and height 1.3 nm as shown in Figure 12.11. Mo_{176}-type wheels are more prone to hydrophilic solvent due to their exposed flat hydrophilic surface toward hydrophilic agents like water than Mo_{154}-type wheels. In contrast to nano-Keplerates, the wheel-shaped giant molecule is also soluble in organic solvents like acetonitrile, acetone or alcohols because of their lower charge resulting from the protonation (Müller & Serain, 2000). Each of the 14 or 16 incomplete cubane type-$\{Mo_5O_6\}$ chambers in the large wheel-shaped clusters Mo_{154} or Mo_{176} has two delocalized 4*d* electrons. There are 28 electrons in the Mo_{154} cluster while the Mo_{176} cluster has 32 electrons but their relative degrees of reduction are the same.

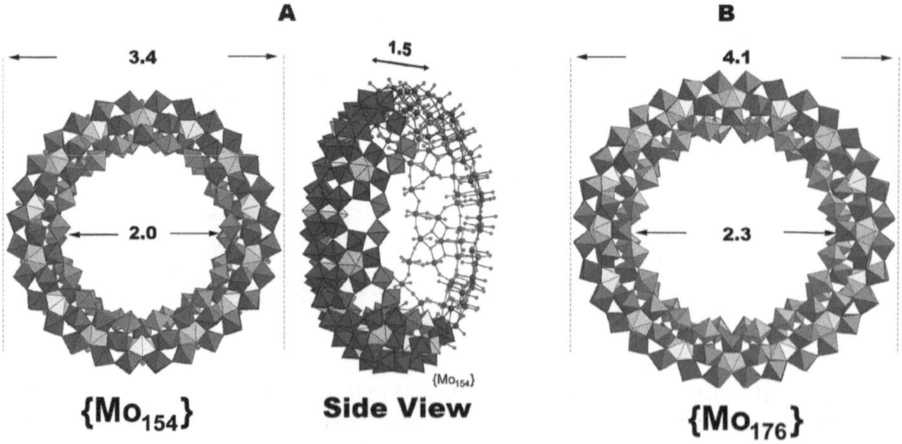

FIGURE 12.11 Dimensional comparison of the molybdenum wheels of (a) Mo_{154} and (b) Mo_{176}-type. All measurements are in nanometers.

The Mo_{154} and Mo_{176} wheels are classified as "reduced protonated hydrated molybdenum oxides" and represented by the chemical formula $[(MoO_3)_{11n}H_n(H_2O)_{5n}]^{n-}$ where n = 14 or 16 repeating units of $[(MoO_3)_{11n}H_n(H_2O)_{5n}]^{n-}$ correspondingly. The equations can alternatively be written as $[\{Mo_8\}_n\{Mo_2\}_n\{Mo_1\}_n]^{n-}$ in terms of their constituents and the lacunary type wheel structure with missing $\{Mo_2\}$ and $\{Mo_1\}$ also known (Pakulski et al., 2019). The missing $\{Mo_2\}$-type units can be replaced by magnetically active metal ions like Ce^{3+}, Pr^{3+} and Eu^{3+} while the $\{Mo_1\}$-type nicks can be substituted by electrochemically important Cu^{2+} (D. Zhong et al., 2011).

12.2.10 Nanohedgehog or the Blue Lemon

The blue-colored Mo_{368} cluster with a 6-nm exterior diameter and 4-nm height is similar in shape to the hedgehog or lemon which signifies its nicknames like the "Nanohedgehog" or the "Blue Lemon" (Müller et al., 2002). The giant cluster consists of six different subunits of the dynamic molybdate library and has a positive curvature with D_{8d} symmetry of its central surface while the two capping regions show a negative curvature with C_{4v} symmetry, thus demonstrating interesting surface symmetry breaking shown in Figure 12.12. This container cluster contained a huge cavity of 4.0×2.5 nm inside which was capable of encapsulating about 400 water molecules.

12.2.11 Icosahedral Inorganic Superfullerenes or the Keplerates

The self-assembly of the $\{Mo^{VI}(Mo^{VI})_5O_{21}(H_2O)_6\}^{6-}$-type pentagonal and $\{Mo^{V}_2O_4(H_2O)_2\}^{2+}$-type dumbbell-shaped dinuclear building blocks in the carboxylate buffer solution leads to the formation of icosahedral-shaped $[\{Mo^{VI}(Mo^{VI})_5O_{21}(H_2O)_6\}_{12}\{Mo^{V}_2O_4(RCO_2)\}_{30}]^{42-}$ nanocluster. The basic exciting chemical and structural features of this nano-Keplerate include the following:

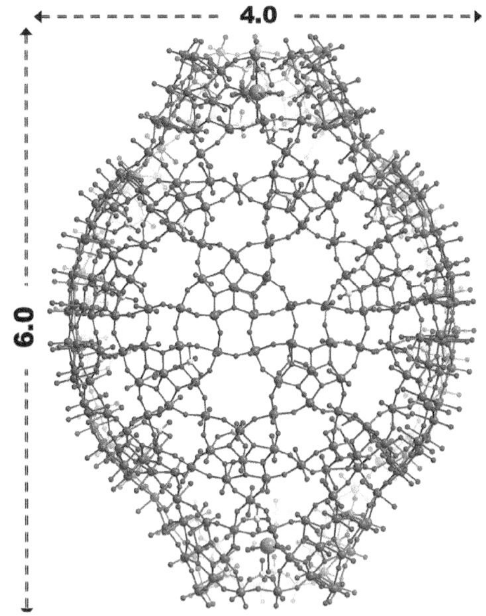

FIGURE 12.12 Polyhedral depiction of the nanohedgehog cluster $[H_{16}Mo_{368}O_{1032}(H_2O)_{240}(SO_4)_{48}]^{48-}$, demonstrating the building components of {(Mo)Mo_5}, {Mo_2} and {Mo_1} units sequentially. All measurements are in nanometers.

1. The capsule is highly charged and hence is soluble in the universal solvent water but also can be soluble in organic solvents by cation exchange or by the application of reverse micelle scaffold of the surfactants (Volkmer et al., 2000).
2. The interior of the capsule has 30 easily derivatizable bidentate ligands coordinated to the 60 Lewis acidic Mo^V centers. By altering the pH of the solution, the ligands can be replaced from hydrophobic to hydrophilic ones, thereby changing the interior cavity environment from hydrophobic to hydrophilic ones (Müller et al., 2006).
3. The cluster contains 20 well-defined {Mo_9O_9} "crown ether"-type pores (outer aperture diameter 1.1 nm (Figure 12.13a) while inner aperture diameter is about 0.4 nm) which can be reversibly blocked/plugged and opened prior to the entry of the cations ("molecular lockgates").
4. In the presence of adequate receptor sites, the encapsulation of a wide range of cations can also be readily achieved. The entry of the cations inside the capsule depends not only on their electrochemical gradient but also on the size and ease of their partial dehydration, necessary for passing the "cone-shaped"-{Mo_9O_9}-pores. This presents the basis of fixation of the cations in different depths of the pores, leading to their virtual separation at the nanoscale ("nano-ion chromatograph"). Surprisingly, these pores even allow the entry of larger substrates due to the breathing vibration of the capsule in solution, which changes the pore size (Ziv et al., 2009).

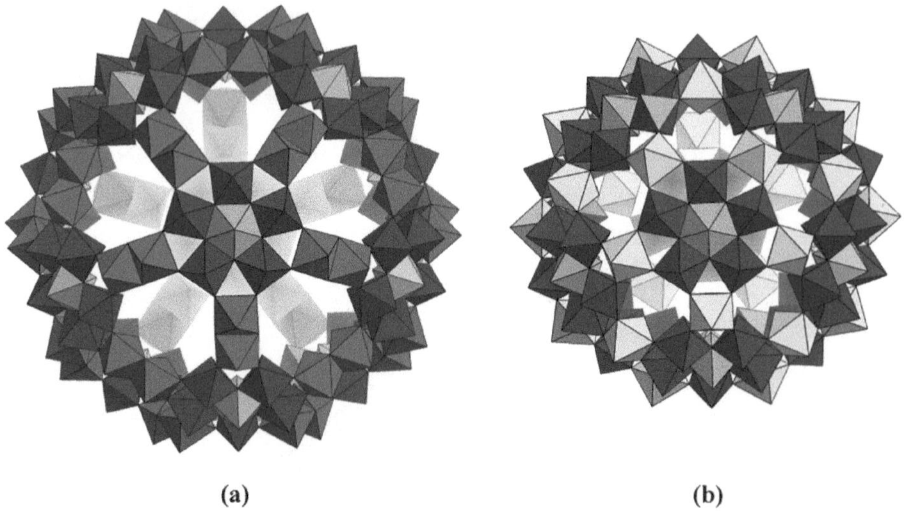

FIGURE 12.13 Overview of the (a) larger, (b) smaller Keplerate formed by the combination of dinuclear (shown in grey) and mononuclear (shown in white) linkers with the $M(M_5)$ (M=Mo/ W) pentagonal units (black). Reprinted from Koegerler et al. (2010). Copyright 2010, with permission from the *Royal Society of Science*.

As a general phenomenon, the introduction of preformed $\{Mo^{V}{}_2O_4(H_2O)_2\}^{2+}$ or $\{Mo^{V}{}_2O_2S_2(H_2O)_2\}^{2+}$ building blocks into the dynamic library of molybdates or tungstates at pH=4 results in the evolution of $\{M_{72}Mo_{60}O_{372}\}^{n-}$ or $\{M_{72}Mo_{60}S_{60}O_{312}\}^{n-}$ (M=Mo/W and the number of cavity interior ligands present determines n). Similarly, the presence of certain trivalent metal ions, *e.g.*, Fe^{3+}, Cr^{3+} or oxo-species like $(VO)^{2+}$ in an acidified aqueous solution of molybdate or tungstate at pH≈2 leads to the evolution of the smaller Keplerate of the type $\{M_{72}M'_{30}\}^{n-}$ (Figure 12.13b: M=Mo/W and M' =Fe^{III}, Cr^{III} or VO^{II}; n depends on the type and the number of cavity internal ligands present and also the presence of extra Mo fragments inside the cavity), where the M' atoms span the vertices of an icosidodecahedron. These $\{M_{72}M'_{30}\}$-clusters also have 20 $\{M_3M'_3O_6\}$-pores (pore outer aperture 0.9 pm) which can be plugged by ammonium, partially aquated potassium, barium or gadolinium ions. Although the smaller Keplerates are mostly neutral, they tend to behave as "polyprotic nanoacids" due to higher *Lewis* acidity of the M' centers (when M' =Fe/ Cr) by deprotonation of water molecules coordinated to M' centers.

12.3 APPLICATION OF POM AS BATTERY MATERIALS

The nominal "Baghdad battery", a vessel perceived on an archeological uncovering in a remote region near Baghdad, which had imputations with the Persian civilization, was assumed to be the oldest electrochemical battery where its invention was dated back to the first century BC (Dubpernell & Westbrook, 1978). Moreover, now it is accepted without any exception that the electrochemical cell was introduced to the modern civilization by the efforts consigned by two Italian pioneers: Luigi Galvani of the University of Bologna, Italy, and Alessandro Volta of the University of Pavia, Italy, toward the end of the eighteenth century. The lineage of the lithium-ion (Li-ion) battery (LIB) was established when Akira Yoshino developed a battery prototype in 1985, which was founded upon the research initiated by M. Stanley Whittingham, Koichi Mizushima, John Goodenough and Rachid Yazami during the decade of 1970–1980. Henceforth, the succeeding generation of commercial LIBs emerged as a result of the teamwork of Sony and Asahi Kasei, with the leadership of Yoshio Nishi in 1991.

Being the lightest metal, lithium showcases a high specific capacity value (3.86 Ah g^{-1}) and a lesser value of electrode potential (–3.04 V *vs.* SHE), incorporating as a perfect material for anode construction for high-energy batteries. The reduction potential of Li^+ to Li resides above the LUMO of the mostly familiar non-aqueous electrolytes; as a consequence of which a phenomenon of continuous reduction of electrolytes takes place, except that an inactivating solid electrolyte interface (SEI) is formed. The damage-responsive SEI restores in a non-uniform manner upon the Li surface, contributing to the formation and also the growth of "dendrites" (Figure 12.14).

This very phenomenon of dendrite growth renders the battery susceptible to electrical short circuit and even explosion/fire.

The fabrication of LIBs with two distinctive intercalation hosts has been recommended by Armand to bypass the safety arguments (Armand, 1980; Zhang et al., 2020). In the intercalation of graphite with Li^+ (Besenhard, 1976) on the layered

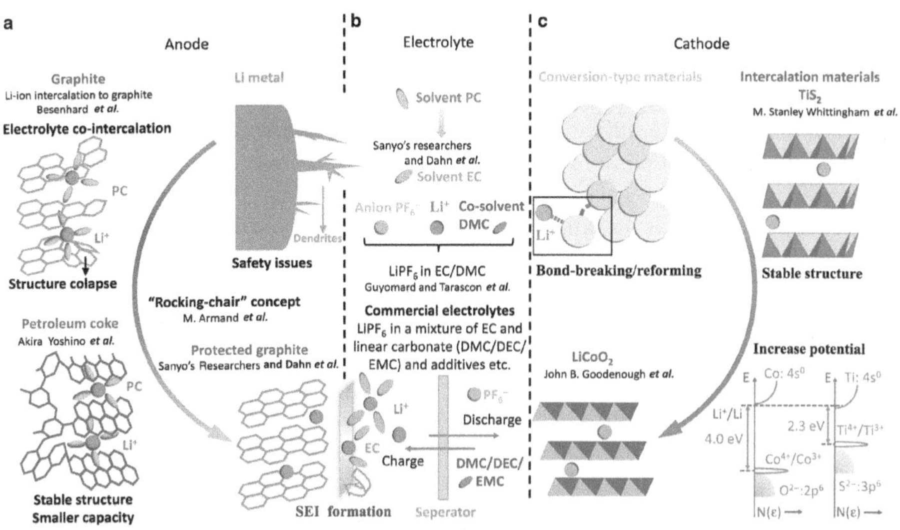

FIGURE 12.14 The development of (a) anode materials, including lithium metal, petroleum coke and graphite, (b) electrolytes with the solvent propylene carbonate (PC), a mixture of ethylene carbonate (EC) and at least one linear carbonate selected from dimethyl carbonate (DMC), diethyl carbonate (DEC), ethyl methyl carbonate (EMC) and many additives, (c) cathode materials, including conversion-type materials, intercalation materials titanium disulfide (TiS_2) and lithium cobalt oxide ($LiCoO_2$). Reprinted Xie and Lu (2020). Copyright 2020, with permission from *Nature*.

architecture of graphite, the p_z orbitals (half-filled) lying perpendicularly to the graphitic planes cooperate with the 2s orbitals of Li to restrict the expansion of volume and subsequent dendrite growth. The petroleum coke, a low graphitized form of carbon, which has been discovered by Akira Yoshino, is capable of facilitating low potential intercalation of Li^+ (~0.5 V relative to Li^+/Li) without going through any structural deformation (Yoshino et al., 1987). Although the current density of the system has been constrained by the amorphous character of petroleum coke, it has been commercially established as the very first intercalation anode for the LIBs, thanks to its cycling stability. Most recently, to resolve the issue of irregular Li-ion distribution across the electrode surface, which in turn results in the dendrite formation, Keggin-type POMs were employed to act as an "ionic sponge" in the LIBs. The results of the addition of a Keggin-type cluster with mixed metallic centers of molybdenum and vanadium, $H_5PMo_{10}V_2O_{40}$ ($PMo_{10}V_2$), to a general electrolyte, were reported (Zhong et al., 2022), which is capable of forming a lithium-rich layer and can effectively tackle the Li–dendrite growth. The Li-rich layer constrained the dendrite formation by filling up the Li-ion vacancies that occur on the surface of the metal anode, and the strength of the electric field was consistently established across the surface of the anode. An electrolyte's working domain is dependent upon the respective positions of LUMO and HOMO, where the position of LUMO should be higher than the anodic electrochemical potential and that of HOMO should reside lower than the cathodic electrochemical potential. Nevertheless, the SEI layer is only stable on the surface of anode and cathode if the above-mentioned condition is inverted. Though the consequences of structural damage of graphite have been reported by the propylene carbonate (PC), the electrolyte system containing the solvent propylene carbonate (PC) is more favored than that of ethylene carbonate (EC) due to the lower melting temperature (−48.8°C) of PC than compared to that of EC (36.4°C) (Winter et al., 2018). A fortunate lithiation of graphite through electrochemical procedure in EC-based electrolyte (Fujimoto et al., 1997) and the suppression of graphite depilation by EC, owing to the generation of SEI sacrificial layer (Fong et al., 1990), has been reported, which has initialized the application of graphite as anodic material for LIBs. The electrolyte system of $LiPF_6$ in EC/DMC formulated in 1993 has emerged as an apposite charge mediator for various applications owing to its upgraded oxidation stability. This is one of the most favorable electrolytes to date, pertaining to threefold superior energy density to lithium cobalt oxide-based LIBs as compared to its first generation (Figure 12.15).

Metal chalcogenides, owing to their layered structural arrangement, provide novel Li-ion storage mechanisms without exhibiting significant structural deformities. Whittingham and Gamble (1975) at Exxon investigated titanium disulfide (TiS_2) and reported that the material has the capability to chemically intercalate Li^+ over the entire stoichiometric window, showcasing minimal expansion of the crystal lattice. TiS_2 has been employed as a cathode for batteries since 1973 and has exemplified a 2.5-V cell in 1976 (Whittingham, 1976). Goodenough et al. have proposed three categories of oxide species for cathodic applications in battery: the $LiCoO_2$, reported in 1979 (Mizushima et al., 1980), can versatilely accumulate and deliver Li^+ at a potential value of more than 4.0 V *vs.* Li+/Li and has authorized a 4.0 V rechargeable battery coupled with lithium metal anode. However, the lithium ions residing in the tetrahedral voids in the spinel structure of $LiMn_2O_4$, interestingly, can deliver

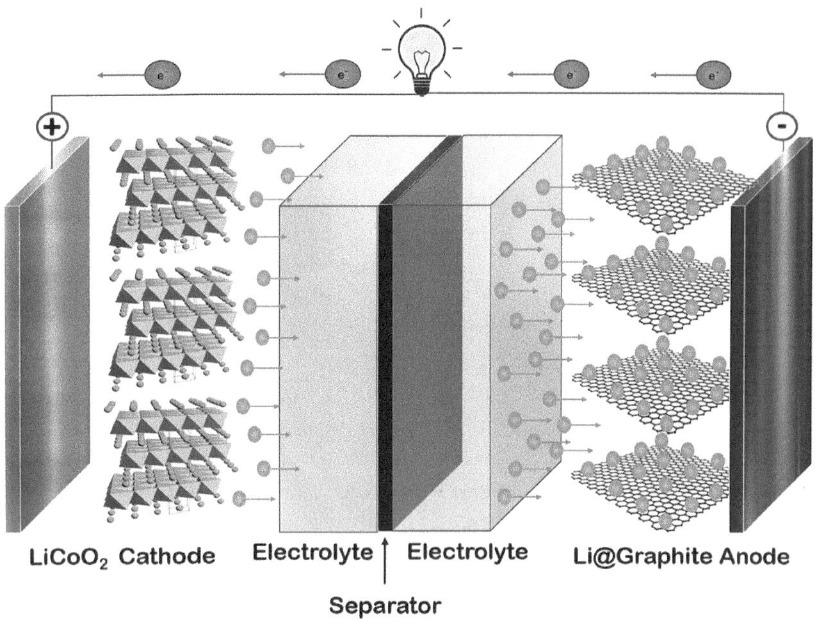

FIGURE 12.15 Construction of lithium cobalt oxide as cathode material and Li@graphite anode for the next-generation battery applications.

an electrochemical potential of ~4.0 V *vs.* Li^+/Li with additional cost-efficiency benefits. However, the disintegration of Mn-sites in the presence of H^+ ions restrained the application of the material (Thackeray et al., 1983). Whereas the polyanion oxide species of $Li_xFe_2(XO_4)_3$ (X = S, Mo, W, *etc.*) has been reported to put forward a higher value of cell voltage in contrast to the conventional oxides (Manthiram & Goodenough, 1987). The strength of the Fe–O bond has weakened due to the presence of the covalency of the X–O bond in the polyanion oxide species through the inductive effect which results in the diminished redox energy of Fe^{2+}/Fe^{3+} and hence maximized the redox potential.

The lithium-ion (Li-ion) battery (LIB) is the most familiar power device primarily used in laptops, smartphones, tablets, other rechargeable devices, *etc.* The omnipotent contribution of these energy devices has even been exercised to the electrical vehicle sectors as well as to the military-grade and aerospace applications; specifically, the combination of unconventional energy sectors like photovoltaics and wind power is yet to be explored to the fullest. Functionally, batteries are classified into two major types: primary battery (disposal type) and secondary battery (rechargeable type), and the classification of batteries based on the working mechanism involves electrolyte materials like aqueous electrolyte battery and non-aqueous electrolyte battery (Yoshino, 2012). The LIBs are non-aqueous electrolytic batteries that are being massively used in the off-grid type power supply systems, distinctively integrated with solar home systems (SHS) in underdeveloped areas. The major components associated with LIBs are cathode, anode, separator and a supporting solution through which the transportation of the Li ions occurs, from cathode to anode and *vice versa* throughout the exercise of charging and discharging, respectively.

The materials that are conventionally employed in the fabrication of anodes of LIBs are hard carbon, graphitic carbon, synthetic graphite, metallic lithium, tin-based alloys, silicon-based materials and lithium titanate, while V_2O_5, lithium-ion phosphate, FeS_2, oxide of manganese, nickel, cobalt with lithium counter cations and electronic conducting polymers are often used for the fabrication of cathode, and finally, the electrolyte materials involved are $LiAsF_6$, $LiPF_6$, $LiCF_3SO_3$ and $LiClO_4$ (Figure 12.16). Metallic lithium has proven to be a more vital anode material than any other metals owing to its lower reduction potential of order 3.04 V *vs.* primary reference electrode (standard hydrogen electrode, SHE), and the Li batteries possess a higher storage capacity of 3,860 mAh·g^{-1} or 2,061 mAh·cm^{-3} (Lin et al., 2017). Several obstacles are there on the path of constructing benign battery cells due to the high reactivity of lithium, which eventually can be surpassed through the utilization of the layered compounds which can adequately sandwich the Li atoms; for example, metallic lithium can be blended with graphite to construct less reactive and highly effective anodes.

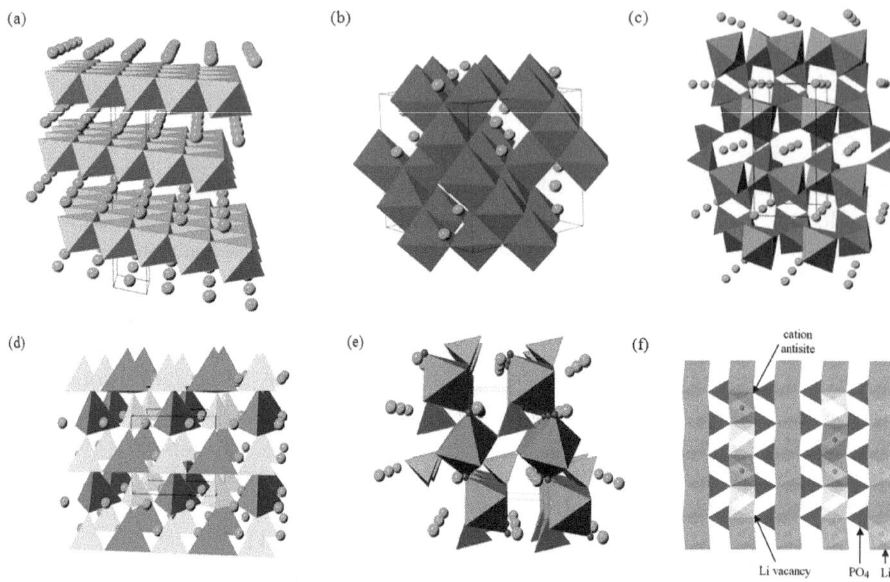

FIGURE 12.16 Representation of crystal moieties of cathode materials for LIBs (a) layered α-$LiCoO_2$; (b) cubic $LiMn_2O_4$ spinel; (c) olivine-structured $LiFePO_4$; (d) β_{II}-Li_2FeSiO_4; and (e) tavorite-type $LiFeSO_4F$. Li ions are depicted as grey spheres, CoO_6 octahedra in light grey; MnO_6 octahedra in black, Fe–O polyhedra in black, PO_4 tetrahedra in dark grey, SiO_4 tetrahedra in white, SO_4 tetrahedra in gray, fluoride ions in black. (f) Schematic of a structural plane of $LiFePO_4$ containing two neutral defect cluster arrangements comprised of two antisite defects (Fe on Li sites). Reprinted Islam and Fisher (2014). Copyright 2010, with permission from the *Royal Society of Science.*

The challenges for efficient anode construction have also been attempted by reacting silicon tetrachloride with magnesium powder to produce porous Si materials (Han et al., 2018) which showcased a reversible competence of 1,125 mAh·g^{-1} at 1 A·g^{-1} after the completion of 100 cycles of Li^+-battery. Owing to the unique hallmarks of the polymers with conducting type properties like agile weightage, flexibility, cost-effectiveness and renewability, these materials also stand as potent candidates for battery electrodes. The phenomena of internal transportation of charge carriers within the conjugated polymers introduce a few undesired inadequacies of low immovability, low reversibility and slope energy; however, significant efforts have been put forward *via* the installation of a functional group (carbonyl) to mitigate the problems associated with the conducting polymers (Tang et al., 2018). In another study, in an anode system of multiwalled carbon nanotubes (MWCNTs) coated with SnO_2, synthesized by a conventional method of wet synthesis, it was found that the external faces of the MWCNTs containing SnO_2 eventually triggered unstable behaviors during the process of charging and discharging (Kim et al., 2018). While further experimenting with carbon nanotubes, an effective material has been synthesized where the architecture of the composite represents a binary structure of flexible single-walled carbon nanotubes (SWNTs) encased around graphene foam (Ren et al., 2018). The resultant anode material composite had a reversible capacity of 606 mAh·g^{-1} and an elevated specific competence of 953 mAh·g^{-1} after the expiration of 1,000 cycles and had attained an exceptional capacity of 90% over 1,000 cycles at 1 A·g^{-1}.

The conventional cathodes used in the LIBs assembly consist of phosphate-based materials or oxides; however, to achieve elevated lithium-ion mobility which will eventually speed up the process of charging and discharging and enhance the diffusivity of the lithium ions, doping of potassium and sulfur has been observed in a spinal lithium-manganese composite. A favorite composite which is also used extensively as a material for cathodes of non-rechargeable LIBs is iron disulfide (FeS_2). By making use of *Auricularia auricula*, a recent report has been published: where adopting the carbonization-sulfuration process, a hybrid cathode of nanocomposites of biomass-carbon@FeS_2 has been fabricated for rechargeable LIBs (Xu et al., 2018), where the renewable and green biomass of *Auricularia auricula* functions as a source of carbon. With a boosted theoretical capacity of around 440 mAh·g^{-1}, V_2O_5 is also a potent candidate for cathode material composite for LIBs; however, owing to its destitute conductivity, the process of lithiation/delithiation is skeptically afflicted. As a solution, the V_2O_5 complexes have been coated with nitrogen-doped carbon (NCNPs-V_2O_5) by the *in-situ* hydrothermal treatment cooperated with organic solvent; as a result, enhancements in electrical conductivity along with the diffusion of Li ions have been observed (Cao et al., 2018). Meanwhile, lithium cobalt oxide ($LiCoO_2$) has its name carved as one of the most well-known metal-oxide composites for its phenomenal electrochemical performances. With an exfoliated structure containing close-packed oxygen ions networks where the Li^+ and Co^{3+} ions rest in octahedral holes, the diffusion of Li^+ ions is highly facilitated, whereas the amount of energy that can be stored in a battery is estimated upon the allowance of mass transport of lithium and the number of lithium ions present in the material which is confined in octahedral sites. Researchers have proposed that magnesium and phosphorus-coated $LiCoO_2$ cathodes may boost durability, efficacy, and substantial power density of LIB. For instance, Kalluri et al. (Kalluri et al. 2017) asserted that Mg, P-coated $LiCoO_2$ offers intensified cycle life and augmented power compactness around 400 Wh·Kg^{-1} also with an upgraded loading density of 3.6 g/cc^{-1}, along with a capability of 112 mAh·g^{-1} at 10°C, which is 14 times higher than uncoated $LiCoO_2$ at 4.35 V.

Even though the mechanization of LIBs has passed through several improvements for the applications in electric vehicles (EVs), the accessible batteries lack some of the particular features to fulfill the demand for energy consumption in EVs. The major cause for this occurrence is the consumption of energy in a non-monotonic fashion integrated with repeated switching during the process of battery discharging, which is also threatening to the battery itself (Kouchachvili et al., 2018). Unbalanced high-temperature action and declined performance of charging and discharging at low temperatures are two additional major issues associated with LIBs, which can be attributed predominantly to the elevated impedance of the SEI layer, reduction of reaction kinetics and low diffusion of Li ions (Lei et al., 2018). So, in order to detach the above-mentioned complications, scientists and researchers have been continuously putting their efforts to fabricate hybrid electrodes with precise modifications employed with implicit composites.

Polyoxometalates (POMs), owing to their distinctive, reversible and stable multi-electron redox abilities, are also getting recognition as novel electrode materials for LIBs. POMs can acquire a great number of electrons and are able to execute multi-electron redox processes: thanks to the transition metal ions which hold high oxidation states, *e.g.*, the Keggin-type $[PMo_{12}O_{40}]^{3-}$ $\{PMo_{12}\}$ can be dubbed as an electron "reservoir or nano-sponge" due to its capability to uptake higher number of electrons (max. up to 24). However, because of their moderated surface area, profound solubility and semiconducting nature, the POMs find suitable places to be applied in the LIBs; hence, it is often beneficial to link the POMs with electrically conductive components for the sake of constructing effective electrodes for LIBs. The most prominent companion of the POMs, in the development of constructing effective electrode materials for the LIBs applications, is the carbon nanotubes (CNTs) which are one of the profound materials with high thermal and electrical conductivity along with a "huge" activated surface area. A nano-hybrid composite system has been designed where the clusters of POMs (Keggin-type $\{PMo_{12}\}$) were absorbed on to the active surfaces of SWNTs without significant structural alteration/decomposition (Kawasaki et al., 2011). For the molecular cluster batteries (MCBs), induced with nano-hybrid POM/SWNT cathode, the quantification of charging and discharging phenomena indicated an elevated battery capacity measurement of the order of 320 mAh·g^{-1} and rapid charging and discharging action in contrast to those of the microcrystalline POM MCBs. In 2014, an initiatory step was taken by a research team of Prof. K. Awaga to covalently bind nanocomposites of amine-functionalized $SiW_{11}O_{39}$ $\{SiW_{11}\text{-}NH_2\}$ on CNTs to design novel CNTs-$\{SiW_{11}\}$-nanocomposite materials for the purpose of operating it as an anode material for LIBs (Chen et al., 2014). The subsequent anode material has been scrutinized at a current density of 0.5 mA·cm^{-2} where it displayed an intense discharge capacity of 650 mAh·g^{-1} while maintaining its stability up to 100 cycles. The emergence of the sidewall defects embodied on the walls of the CNTs, which is a consequence of the covalently established $\{SiW_{11}\}$, enhanced the electrochemical performances of the CNTs-$\{SiW_{11}\}$-nanocomposites and also diminished the diffusion length to permit effective transfer of electron density. In 2015, another venture with SWNT was explored where the carboxylic acid-functionalized SWNTs are covalently grafted *via* amide bond linkage with redox-active Anderson-type POM clusters, $[MnMo_6O_{24}]^{9-}$, organically modified with tris-($NH_2C(CH_2OH)_3$) moieties, to be utilized as electrode materials in LIBs (Ji et al., 2015). The modified composite system displayed improved electrochemical performances such as higher discharge capacities, extended cycling stability and minimized electrochemical impedance, which are justified by the phenomenon of chemisorption of the POM clusters on the SWNT walls rather than the sole physisorption circumstances. On account of their potent electrochemical characteristics such as flexibility, lightweight, multi-functionalities and straightforward operational abilities, the conjugative composites of conductive polymers have been recognized as one of the foremost candidates and also a sound collaborative confidante of POMs for LIB applications. An ordered *state-of-the-art* amalgamated nanocomposite material, synthesized by the fusion of Keggin-type POM, MWNT and polyaniline (PANI), has come into existence through a facile stepwise modification method for their implementation in LIBs (Hu et al., 2017). A refined discharge rate, capacities and comprehensively elongated stability have been detected, which are due to the synergistic effects between the $PMo_{12}O_{40}$, PANI and MWNTs, *i.e.* strong π-π interactions between PANI and CNTs, whereas strong electrostatic interactions facilitate the linking of POMs with PANI. In 2015, the cluster of polyoxovanadate ($Li_7[V_{15}O_{36}(CO_3)]$), which is capable of withstanding multielectron reduction processes without any deformation in its cluster structure, has been generalized as a novel category of exclusive cathode material for the future greater energy as well as power density LIBs. The rapid diffusion of lithium ions accompanied with profound electronic conductivity of the polyoxovanadate authorized accelerated charging and discharging phenomena in the resultant batteries (Chen et al., 2015). Whenever a class of POM is functionalized as a cathode material for LIBs, certain fragilities like lower electronic conductivity, high solubility in electrolyte and reactivity of surface obstruct the electrochemical performances. To overcome these drawbacks, a binary scheme has been proposed recently, which includes applying a coating of the conducting polymer, polypyrrole (PPy), on a material of POMOF, which has been synthesized *via* functionalization of the nitrogenous ligand (1, 10-phenanthroline monohydrate) onto the MOFs to construct the desired composite Cu-POMOF@PPy (Han et al., 2021). The effective application coating of PPy bestowed additional capacity, upgraded the electronic conductivity and reduced the volume contrast of Cu-based POMOF during cycling, whereas the entrapment of the POMs in MOFs terminated the suspension of POMs in the electrolyte and thereby enhanced the electrochemical stability. In 2018, two novel Wells-Dawson type POMOFs with various helix channels have been explored based upon the self-adjustive coordination nature of $\{P_2W_{18}\}$ polyanion (Yang et al., 2018), where twofold helical channels are isolated in the case of hemispherical coordination of $\{P_2W_{18}\}$ clusters with Ag^+ through 1, 2, 4-triazole ligand (POMOF-1). As a result, the formation of fourfold helical channels has been observed for the clockwise coordination of $\{P_2W_{18}\}$ clusters with Cu (POMOF-2). The implementation of the resultant composites as anode materials for the LIBs has been altogether effectual as facile access is provided to the lithium ions by the fourfold helical channels confining the Wells-Dawson type $\{P_2W_{18}\}$ polyanions. To tackle the general shortcomings of the contemporary electrode materials, the implantation of organic entities into the POMs also provides a worthwhile strategy for the enhancement of electronic conductivities of POMs, owing to the elevated metal-to-ligand charge transfer through d-π electronic interactions. Recognizing this characteristic, a composite of covalently grafted imido-hexamolybdate containing the Lindqvist structure has been investigated (Khan et al., 2014), where the imido-hexamolybdates displayed a coulombic efficiency ~100% along with an inceptive discharge capacity of ~1,678 $mAh{\cdot}g^{-1}$ with a retention capacity up to 85% through 100 cycles, which is due to the presence of electron-withdrawing group (–SCN) in Mo_6-SCN. With the abundancy of electroactive centers and ion-exchange channels with high surface area, a three-dimensional structurally integrated mesoporous nano-assembly of nickel cobalt sulfide/polyaniline @polyoxometalate/reduced graphene oxide ($NiCo_2S_4$/PANI@POM/rGO) was reported in 2022, which was originally synthesized through a facile route of oxidative polymerization followed by a hydrothermal method (Gautam et al., 2022). A profound reversible capacity of order 735.5 $mA{\cdot}h\ g^{-1}$ (0.1 $A{\cdot}g^{-1}$ after 200 cycles) along with a retention capacity (0.161% decay/per cycle at 0.5 $A{\cdot}g^{-1}$ for 1,000 cycles) and rate capability (355.6 $mAh{\cdot}g^{-1}$ at 2 $A{\cdot}g^{-1}$) was recorded by electrochemical analysis of the hybrid composite, employed as anode material for LIBs.

The latest domain of research on POMs as electrode materials for LIBs includes the application of Mo-based composites such as small Keplerates $\{Mo_{72}Fe_{30}\}$ and Giant Wheels $\{Mo_{154}\}$, as the Mo-based POMs display elevated electrochemical performances in contrast to other POM electrode materials. The Giant Wheel $\{Mo_{154}\}$ structural composite is one of the most captivating Mo-based metal-oxide clusters with an open host skeleton, which can be conventionally fabricated by metal-oxygen flexible coordination modes. The methods of ligand substitution, implantation of structural vacancies or generation of defects and integration of hetero elements in the architecture of $\{Mo_{154}\}$ can result in various effective prospects to achieve profound electrochemical performance. The "Giant Wheel" has been applied as a cathode material in magnesium rechargeable batteries (MRBs) with the prospect to achieve high specific current density and capacitance for multivalent ion batteries (Fan et al., 2020); the $\{Mo_8\}$ core units can harbor approximately up to 28 electrons. These kinds of core units are proven to be good hosts for multivalent ions as they have the capability to dispense the Mg^{2+} ions inside the local crystal structure. An introductory attempt has been taken where the proton conductivity has been measured for the spherical $\{Mo_{72}Fe_{30}\}$ composite *aka* "small Keplerate", which is trusted to inaugurate effective pathways for augmented electrochemical performances (Tandekar et al., 2021). The stability of the small Keplerate cluster in various cycles of proton conductivity, which is dependent upon low-temperature sensitivity and humidity tolerance, presents alluring aspects. The deprotonation of Fe-bonded water molecules results in the development of (Fe-OH) species, which leads to the formation of Fe-O-Fe linkage and thereby results in the further extension of $\{Mo_{72}O_{30}\}$ into an extended structure in the orthorhombic crystal. In 2020, small Keplerate was initially introduced as a high-energy anode material for LIBs (Huang et al., 2020), where it was found that a single molecule of $\{Mo_{72}O_{30}\}$ can hold up to 377 electrons. The resultant anode material interacts with Li^+ through diffusion processes and surface-capacitive reactions, and the Keplerate material exhibited an elevated capacity of around 1,250 mA $h{\cdot}g^{-1}$ at 100 $mA{\cdot}g^{-1}$ with a retention capacity up to 92% after 100 cycles. The electrochemical studies performed till now on the new generation of POM nanocomposites have convinced the scientific society about their intensified attributions toward energy storage applications. However, there are still plenty of nano-sized POM composites, *e.g.*, $\{Mo_{248}\}$ or the "big wheel" structure and the "Blue lemon or Hedgehog structure", that are still to be explored for energy storage applications.

Through disposing sufficient oxide ligands *via* the pyrolysis process, POM composites are capable of organizing porous architectures and can provide multiple metal sources, for which it has proven to be a beneficial scheme to designate POMs as the precursors for the fabrication of electrode materials. With implementing modifications like combining external species (*e.g.*, CNTs, POPs, graphene, MOFs, *etc.*) and employing heteroatoms, the electrochemical performances of the POMs can be boosted to new desired stages; however, there still remain some objections and arguments in the course of determining comprehensive synthetic routes and establishing connections between electrochemical properties and architecture, which will confidently be explored in future research.

12.4 CONCLUSION AND FUTURE PROSPECT

The construction of compact, eco-friendly, cost-effective and facilely tunable electronic materials with comparably higher power density and longer cycling life, which can be employed as a charge storage device, is an exacting task, considering the current circumstances of energy requirement. Owing to their superlative molecular level redox effects, distinctive structural flexibility and favorable aspects of physicochemical effects, POMs have materialized as prospective materials for energy storage applications. However, POMs deteriorate in some order in the energy storage applications as these metal-oxide clusters suffer from high solubility, lower conductive properties and other operational issues. So, it is a major urgency to reorient the POM clusters with other inorganic/organic species, metal-doping or framework assembly, to boost their electrochemical properties in order to employ them as electrode materials for energy storage devices.

This chapter is embellished with demonstrations of various structural prospects of different types of POMs with a nuclearity ranging from 6 to 368, as well as the cooperation of supplementary potent composites with POMs to furnish hybrid electrode materials for battery applications. A distinct contemplation is executed on the upgraded performance of these energy harvesting systems in association with literature-documented reference schemes. Additionally, each of the cooperative composite of POMs, employed as electrode materials, shows exclusive interactions, which were also explored. The carbon nanocomposites enable rapid reversible electrochemical redox reactions, and augment intrinsic redox activities of POM-based composites are insight. These POM-based nanocomposite materials have the ability to boost the capability of discharge of LIBs. Meanwhile, the companionship of conjugated system of conducting polymers with POMs was also elaborated, where we observed the linking of POMs with conducting polymers through strong electrostatic interactions. As a result of this synergistic interaction, a refined discharge rate, capacities and comprehensively elongated stability have been detected. These kinds of collaborative organizations manifest themselves as storage reservoirs for the charge carriers; thus, lithium ions for batteries and electrons in the case of supercapacitors produce perfect environments for charge storage devices. The fabrication of POM-based composite systems for the charge storage devices and other energy applications is still in an inaugural stage, and significant modifications are obligatory for its industrial applications reconditioning.

ACKNOWLEDGMENTS

Mr. Phukan, Mr. Sah and Dr. Garai are thankful to DST IC#3 Projects (DST/TM/EWO/MI/CCUS/15 & DST/TMD/HFC/2K18/06) while Dr. Garai is also thankful to SERB Project (ECR/2018/000254), BHU-IOE-SEED Grant and SPARC Project (SPARC/2018–2019/P124/SL) for kind financial help. Mr. Sah is thankful to UGC-CSIR for the JRF fellowships. Dr. Sankaranarayanan thanks DST, Govt. of India, for the Inspire grant (IFA13-PH-82) and SERB-SRG grant No. SRG/2020/001894. The authors are also thankful to their respective institutions/universities/departments as affiliated herewith for basic research and infrastructural facilities.

REFERENCES

Alaaeddine, M., Zhu, Q., Fichou, D., Izzet, G., Rault, J. E., Barrett, N., Proust, A., & Tortech, L. (2014). Enhancement of photovoltaic efficiency by insertion of a polyoxometalate layer at the anode of an organic solar cell. *Inorganic Chemistry Frontiers*, 1(9), 682–688.

Ammam, M. (2013). Polyoxometalates: Formation, structures, principal properties, main deposition methods and application in sensing. *Journal of Materials Chemistry A*, 1(21), 6291–6312.

Armand, M. B. (1980). "Intercalation electrode" in materials for advanced batteries. In: *NATO Conference Series, Series VI: Material Science*, 145.

Bergamo, A., & Sava, G. (2007). Ruthenium complexes can target determinants of tumour malignancy. *Dalton Transactions*, 13, 1267–1272.

Besenhard, J. O. (1976). The electrochemical preparation and properties of ionic alkali metal-and NR4-graphite intercalation compounds in organic electrolytes. *Carbon*, 14(2), 111–115.

Blazevic, A., & Rompel, A. (2016). The Anderson-Evans polyoxometalate: From inorganic building blocks via hybrid organic-inorganic structures to tomorrows "Bio-POM." *Coordination Chemistry Reviews*, 307, 42–64.

Cao, L., Kou, L., Li, J., Huang, J., Yang, J., & Wang, Y. (2018). Nitrogen-doped carbon-coated V_2O_5 nanocomposite as cathode materials for lithium-ion battery. *Journal of Materials Science*, 53(14), 10270–10279.

Chen, J.-J., Symes, M. D., Fan, S.-C., Zheng, M.-S., Miras, H. N., Dong, Q.-F., & Cronin, L. (2015). High-performance polyoxometalate-based cathode materials for rechargeable lithium-ion batteries. *Advanced Materials*, 27(31), 4649–4654.

Chen, W., Huang, L., Hu, J., Li, T., Jia, F., & Song, Y.-F. (2014). Connecting carbon nanotubes to polyoxometalate clusters for engineering high-performance anode materials. *Physical Chemistry Chemical Physics*, 16(36), 19668–19673.

Contant, R., & Teze, A. (1985). A new crown heteropolyanion $K_{28}Li_5H_7P_8W_{48}O_{184}.92H_2O$: Synthesis, structure, and properties. *Inorganic Chemistry*, 24(26), 4610–4614.

Contant, R., & Thouvenot, R. (1993). A reinvestigation of isomerism in the Dawson structure: Syntheses and ^{183}W NMR structural characterization of three new polyoxotungstates $[X_2W_{18}O_{62}]^{6-}$ (X= PV, AsV). *Inorganica Chimica Acta*, 212(1–2), 41–50.

Dubpernell, G., & Westbrook, J. H. (1978). *Proceedings of the Symposium on Selected Topics in the History of Electrochemistry*.

Fan, X., Garai, S., Gaddam, R. R., Menezes, P. V., Dubal, D. P., Yamauchi, Y., Menezes, P. W., Nanjundan, A. K., & Zhao, X. S. (2020). Uncovering giant nanowheels for magnesium ion-based batteries. *Materials Today Chemistry*, 16, 100221.

Fernández, J. A., López, X., Bo, C., de Graaf, C., Baerends, E. J., & Poblet, J. M. (2007). Polyoxometalates with internal cavities: Redox activity, basicity, and cation encapsulation in $[X^{n+}P5W_{30}O_{110}]^{(15-n)-}$Preyssler Complexes, with X= Na^+, Ca^{2+}, Y^{3+}, La^{3+}, Ce^{3+}, and Th^{4+}. *Journal of the American Chemical Society*, 129(40), 12244–12253.

Fong, R., Von Sacken, U., & Dahn, J. R. (1990). Studies of lithium intercalation into carbons using nonaqueous electrochemical cells. *Journal of the Electrochemical Society*, 137(7), 2009.

Fujimoto, M., Yoshinaga, N., Ueno, K., Furukawa, N., Nohma, T., & Takahashi, M. (1997). *Lithium Secondary Battery*. Google Patents.

Gautam, J., Liu, Y., Gu, J., Ma, Z., Dahal, B., Chishti, A. N., Ni, L., Diao, G., & Wei, Y. (2022). Three-dimensional nano assembly of nickel cobalt sulphide/polyaniline@ polyoxometalate/reduced graphene oxide hybrid with superior lithium storage and electrocatalytic properties for hydrogen evolution reaction. *Journal of Colloid and Interface Science*, 614, 642–654.

Genovese, M., & Lian, K. (2017). Polyoxometalates: Molecular metal oxide clusters for supercapacitors. In: *Metal Oxides in Supercapacitors* (pp. 133–164), Edited by DP Dubal and P Gomez-Romero, Elsevier.

Gumerova, N. I., & Rompel, A. (2018). Synthesis, structures and applications of electron-rich polyoxometalates. *Nature Reviews Chemistry*, 2(2), 1–20.

Han, Y., Lin, N., Xu, T., Li, T., Tian, J., Zhu, Y., & Qian, Y. (2018). An amorphous Si material with a sponge-like structure as an anode for Li-ion and Na-ion batteries. *Nanoscale*, 10(7), 3153–3158.

Han, Z., Li, X., Li, Q., Li, H., Xu, J., Li, N., Zhao, G., Wang, X., Li, H., & Li, S. (2021). Construction of the POMOF@ polypyrrole composite with enhanced ion diffusion and capacitive contribution for high-performance lithium-ion batteries. *ACS Applied Materials & Interfaces*, 13(5), 6265–6275.

Herrmann, S., Ritchie, C., & Streb, C. (2015). Polyoxometalate-conductive polymer composites for energy conversion, energy storage and nanostructured sensors. *Dalton Transactions*, 44(16), 7092–7104.

Hu, J., Jia, F., & Song, Y.-F. (2017). Engineering high-performance polyoxometalate/PANI/MWNTs nanocomposite anode materials for lithium ion batteries. *Chemical Engineering Journal*, 326, 273–280.

Huang, B., Yang, D.-H., & Han, B.-H. (2020). Application of polyoxometalate derivatives in rechargeable batteries. *Journal of Materials Chemistry A*, 8(9), 4593–4628.

Huang, S.-C., Lin, C.-C., Hsu, C.-T., Guo, C.-H., Chen, T.-Y., Liao, Y.-F., & Chen, H.-Y. (2020). Keplerate-type polyoxometalate $\{Mo_{72}Fe_{30}\}$ nanoparticle anodes for high-energy lithium-ion batteries. *Journal of Materials Chemistry A*, 8(41), 21623–21633.

Islam, M. S., & Fisher, C. A. (2014). Lithium and sodium battery cathode materials: Computational insights into voltage, diffusion and nanostructural properties. *Chemical Society Reviews*, 43(1), 185–204.

Izarova, N. V., Vankova, N., Heine, T., Biboum, R. N., Keita, B., Nadjo, L., & Kortz, U. (2010). Polyoxometalates made of gold: The polyoxoaurate $[Au^{III}{}_4As^{V}{}_4O_{20}]^{8-}$. *Angewandte Chemie International Edition*, 49(10), 1886–1889.

Izzet, G., Ishow, E., Delaire, J., Afonso, C., Tabet, J.-C., & Proust, A. (2009). Photochemical activation of an azido manganese-monosubstituted Keggin polyoxometalate: On the road to a Mn (V)- nitrido derivative. *Inorganic Chemistry*, 48(24), 11865–11870.

Janusson, E., de Kler, N., Vilà-Nadal, L., Long, D.-L., & Cronin, L. (2019). Synthesis of polyoxometalate clusters using carbohydrates as reducing agents leads to isomer-selection. *Chemical Communications*, 55(41), 5797–5800.

Ji, Y., Hu, J., Huang, L., Chen, W., Streb, C., & Song, Y.-F. (2015). Covalent attachment of Anderson-type polyoxometalates to single-walled carbon nanotubes gives enhanced performance electrodes for lithium ion batteries. *Chemistry-A European Journal*, 21(17), 6469–6474.

Kalluri, S., Yoon, M., Jo, M., Park, S., Myeong, S., Kim, J., Dou, S. X., Guo, Z., & Cho, J. (2017). Surface engineering strategies of layered $LiCoO_2$ cathode material to realize high-energy and high-voltage Li-ion cells. *Advanced Energy Materials*, 7(1), 1601507.

Kawasaki, N., Wang, H., Nakanishi, R., Hamanaka, S., Kitaura, R., Shinohara, H., Yokoyama, T., Yoshikawa, H., & Awaga, K. (2011). Nanohybridization of polyoxometalate clusters and single-wall carbon nanotubes: Applications in molecular cluster batteries. *Angewandte Chemie*, 123(15), 3533–3536.

Keggin, J. F. (1933). Structure of the molecule of 12-phosphotungstic acid. *Nature*, 131(3321), 908–909.

Khan, R. N. N., Mahmood, N., Lv, C., Sima, G., Zhang, J., Hao, J., Hou, Y., & Wei, Y. (2014). Pristine organo-imido polyoxometalates as an anode for lithium ion batteries. *RSC Advances*, 4(15), 7374–7379.

Kim, S. H., Lee, J. Y., & Yoon, Y. S. (2018). Effect of composite structure on capacity instability of SnO_2-Coated multiwalled carbon nanotube composite anode. *Journal of Alloys and Compounds*, 742, 542–548.

Kobayashi, A., & Sasaki, Y. (1975). The crystal structure of α-Barium 12-tungstosilicate, α-$Ba_2SiW_{12}O_{40}\cdot 16H_2O$. *Bulletin of the Chemical Society of Japan*, 48(3), 885–888.

Koegerler, P., Tsukerblat, B., & Mueller, A. (2010). Structure-related frustrated magnetism of nanosized polyoxometalates: Aesthetics and properties in harmony. *Dalton Transactions*, 39(1), 21–36.

Kouchachvili, L., Yaïci, W., & Entchev, E. (2018). Hybrid battery/supercapacitor energy storage system for the electric vehicles. *Journal of Power Sources*, 374, 237–248.

Lei, Z., Zhang, Y., & Lei, X. (2018). Temperature uniformity of a heated lithium-ion battery cell in cold climate. *Applied Thermal Engineering*, 129, 148–154.

Lin, D., Liu, Y., & Cui, Y. (2017). Reviving the lithium metal anode for high-energy batteries. *Nature Nanotechnology*, 12(3), 194–206.

Liu, J., Wang, J., Xu, C., Jiang, H., Li, C., Zhang, L., Lin, J., & Shen, Z. X. (2018). Advanced energy storage devices: Basic principles, analytical methods, and rational materials design. *Advanced Science*, 5(1), 1700322.

Liu, S., & Tang, Z. (2010). Polyoxometalate-based functional nanostructured films: Current progress and future prospects. *Nano Today*, 5(4), 267–281.

Long, D.-L., Tsunashima, R., & Cronin, L. (2010). Polyoxometalates: Building blocks for functional nanoscale systems. *Angewandte Chemie International Edition*, 49(10), 1736–1758.

López, X., & Poblet, J. M. (2004). DFT study on the five isomers of PW12O403−: Relative stabilization upon reduction. *Inorganic Chemistry*, 43(22), 6863–6865.

Manthiram, A., & Goodenough, J. B. (1987). Lithium insertion into Fe2(MO4)3 frameworks: Comparison of M= W with M= Mo. *Journal of Solid State Chemistry*, 71(2), 349–360.

Mbomekalle, I. M., Keita, B., Nierlich, M., Kortz, U., Berthet, P., & Nadjo, L. (2003). Structure, magnetism, and electrochemistry of the multinickel polyoxoanions $[Ni_6As_3W_{24}O_{94}(H_2O)_2]^{17-}$·$[Ni_3Na(H_2O)_2(AsW_9O_{34})_2]^{11-}$, and $[Ni_4Mn_2P_3W_{24}O_{94}(H_2O)_2]^{17-}$. *Inorganic Chemistry*, 42(17), 5143–5152.

Miras, H. N., Yan, J., Long, D.-L., & Cronin, L. (2012). Engineering polyoxometalates with emergent properties. *Chemical Society Reviews*, 41(22), 7403–7430.

Mizushima, K., Jones, P. C., Wiseman, P. J., & Goodenough, J. B. (1980). LixCoO_2 (0< x<-1): A new cathode material for batteries of high energy density. *Materials Research Bulletin*, 15(6), 783–789.

Müller, A., Beckmann, E., Bögge, H., Schmidtmann, M., & Dress, A. (2002). Inorganic chemistry goes protein size: A Mo_{368} nano-hedgehog initiating nanochemistry by symmetry breaking. *Angewandte Chemie International Edition*, 41(7), 1162–1167.

Müller, A., & Serain, C. (2000). Soluble molybdenum blues "des pudels kern." *Accounts of Chemical Research*, 33(1), 2–10.

Müller, A., Zhou, Y., Bögge, H., Schmidtmann, M., Mitra, T., Haupt, E. T., & Berkle, A. (2006). "Gating" the pores of a metal oxide based capsule: After initial cation uptake subsequent cations are found hydrated and supramolecularly fixed above the pores. *Angewandte Chemie International Edition*, 45(3), 460–465.

Nyman, M., Alam, T. M., Bonhomme, F., Rodriguez, M. A., Frazer, C. S., & Welk, M. E. (2006). Solid-state structures and solution behavior of alkali salts of the $\{Nb^{6}O^{19}\}^{8-}$ Lindqvist Ion. *Journal of Cluster Science*, 17(2), 197–219.

Pakulski, D., Gorczyński, A., Czepa, W., Liu, Z., Ortolani, L., Morandi, V., Patroniak, V., Ciesielski, A., & Samorì, P. (2019). Novel Keplerate type polyoxometalate-surfactant-graphene hybrids as advanced electrode materials for supercapacitors. *Energy Storage Materials*, 17, 186–193.

Patel, A., Narkhede, N., Singh, S., & Pathan, S. (2016). Keggin-type lacunary and transition metal substituted polyoxometalates as heterogeneous catalysts: A recent progress. *Catalysis Reviews*, 58(3), 337–370.

Perloff, A. (1970). Crystal structure of sodium hexamolybdochromate (III) octahydrate, $Na_3(CrMo_6O_{24}H_6).8H_2O$. *Inorganic Chemistry*, 9(10), 2228–2239.

Pope, M. T., Jeannin, Y., & Fournier, M. (1983). *Heteropoly and Isopoly Oxometalates* (Vol. 8). Springer.

Pratt III, H. D., Hudak, N. S., Fang, X., & Anderson, T. M. (2013). A polyoxometalate flow battery. *Journal of Power Sources*, 236, 259–264.

Proust, A., Matt, B., Villanneau, R., Guillemot, G., Gouzerh, P., & Izzet, G. (2012). Functionalization and post-functionalization: A step towards polyoxometalate-based materials. *Chemical Society Reviews*, 41(22), 7605–7622.

Ren, J., Ren, R.-P., & Lv, Y.-K. (2018). A New anode for lithium-ion batteries based on single-walled carbon nanotubes and graphene: Improved performance through a binary network design. *Chemistry-An Asian Journal*, 13(9), 1223–1227.

Sartorel, A., Carraro, M., Scorrano, G., Zorzi, R. D., Geremia, S., McDaniel, N. D., Bernhard, S., & Bonchio, M. (2008). Polyoxometalate embedding of a tetraruthenium (IV)-oxo-core by template-directed metalation of $[\gamma\text{-}SiW_{10}O_{36}]^{8-}$: A totally inorganic oxygen-evolving catalyst. *Journal of the American Chemical Society*, 130(15), 5006–5007.

Song, Y.-F., Long, D.-L., Kelly, S. E., & Cronin, L. (2008). Sorting the assemblies of unsymmetrically covalently functionalized Mn-Anderson polyoxometalate clusters with mass spectrometry. *Inorganic Chemistry*, 47(20), 9137–9139.

Tandekar, K., Singh, C., & Supriya, S. (2021). Proton conductivity in $\{Mo_{72}Fe_{30}\}$-type keplerate. *European Journal of Inorganic Chemistry*, 2021(8), 734–739.

Tang, M., Li, H., Wang, E., & Wang, C. (2018). Carbonyl polymeric electrode materials for metal-ion batteries. *Chinese Chemical Letters*, 29(2), 232–244.

Tarascon, J.-M., & Armand, M. (2011). Issues and challenges facing rechargeable lithium batteries. In *Materials for sustainable energy: A collection of peer-reviewed research and review articles from Nature Publishing Group* (pp. 171–179). Edited by Vincent Dusastre, World Scientific.

Tézé, A., Canny, J., Gurban, L., Thouvenot, R., & Hervé, G. (1996). Synthesis, structural characterization, and oxidation- reduction behavior of the γ-isomer of the dodecatungstosilicate anion. *Inorganic Chemistry*, 35(4), 1001–1005.

Thackeray, M. M., David, W. I. F., Bruce, P. G., & Goodenough, J. B. (1983). Lithium insertion into manganese spinels. *Materials Research Bulletin*, 18(4), 461–472.

Volkmer, D., Du Chesne, A., Kurth, D. G., Schnablegger, H., Lehmann, P., Koop, M. J., & Müller, A. (2000). Toward nanodevices: Synthesis and characterization of the nanoporous surfactant-encapsulated keplerate $(DODA)_{40}(NH_4)_2[(H_2O)n\subset Mo1_{32}O_{372}(CH_3COO)_{30}(H_2O)_{72}]$. *Journal of the American Chemical Society*, 122(9), 1995–1998.

Walsh, J. J., Bond, A. M., Forster, R. J., & Keyes, T. E. (2016). Hybrid polyoxometalate materials for photo (electro-) chemical applications. *Coordination Chemistry Reviews*, 306, 217–234.

Wang, X., Liu, B., Liu, R., Wang, Q., Hou, X., Chen, D., Wang, R., & Shen, G. (2014). Fiber-based flexible all-solid-state asymmetric supercapacitors for integrated photodetecting system. *Angewandte Chemie*, 126(7), 1880–1884.

Wang, Y., & Weinstock, I. A. (2010). Cation mediated self-assembly of inorganic cluster anion building blocks. *Dalton Transactions*, 39(27), 6143–6152.

Weakley, T. J., Evans, H. T., Showell, J. S., Tourné, G. F., & Tourné, C. M. (1973). 18-Tungstotetracobalto (II) diphosphate and related anions: A novel structural class of heteropolyanions. *Journal of the Chemical Society, Chemical Communications*, 4, 139–140.

Whittingham, M. S. (1976). Electrical energy storage and intercalation chemistry. *Science*, 192(4244), 1126–1127.

Whittingham, M. S., & Gamble Jr, F. R. (1975). The lithium intercalates of the transition metal dichalcogenides. *Materials Research Bulletin*, 10(5), 363–371.

Winter, M., Barnett, B., & Xu, K. (2018). Before Li ion batteries. *Chemical Reviews*, 118(23), 11433–11456.

Winter, M., & Brodd, R. J. (2004). What are batteries, fuel cells, and supercapacitors? *Chemical Reviews*, 104(10), 4245–4270.

Wu, P., Wang, Y., Huang, B., & Xiao, Z. (2021). Anderson-type polyoxometalates: From structures to functions. *Nanoscale*, 13(15), 7119–7133.

Xie, J., & Lu, Y.-C. (2020). A retrospective on lithium-ion batteries. *Nature Communications*, 11(1), 1–4.

Xu, X., Meng, Z., Zhu, X., Zhang, S., & Han, W.-Q. (2018). Biomass carbon composited FeS_2 as cathode materials for high-rate rechargeable lithium-ion battery. *Journal of Power Sources*, 380, 12–17.

Yang, X.-Y., Li, M.-T., Sheng, N., Li, J.-S., Liu, G.-D., Sha, J.-Q., & Jiang, J. (2018). Structure and LIBs anode material application of novel wells-dawson polyoxometalate-based metal organic frameworks with different helical channels. *Crystal Growth & Design*, 18(9), 5564–5572.

Yoshino, A. (2012). The birth of the lithium-ion battery. *Angewandte Chemie International Edition*, 51(24), 5798–5800.

Yoshino, A., Sanechika, K., & Nakajima, T. (1987). *Secondary Battery*. Google Patents.

Yuan, C., Guo, S., Wang, S., Liu, L., Chen, W., & Wang, E. (2013). Electropolymerization polyoxometalate (POM)-doped PEDOT film electrodes with mastoid microstructure and its application in dye-sensitized solar cells (DSSCs). *Industrial & Engineering Chemistry Research*, 52(20), 6694–6703.

Zhang, H., Li, C., Eshetu, G. G., Laruelle, S., Grugeon, S., Zaghib, K., Julien, C., Mauger, A., Guyomard, D., & Rojo, T. (2020). From solid-solution electrodes and the rocking-chair concept to today's batteries. *Angewandte Chemie*, 132(2), 542–546.

Zhong, D., Sousa, F. L., Müller, A., Chi, L., & Fuchs, H. (2011). A nanosized molybdenum oxide wheel with a unique electronic-necklace structure: STM study with submolecular resolution. *Angewandte Chemie*, 123(31), 7156–7159.

Zhong, Y., Su, Y., Huang, P., Jiang, Q., Lin, Y., Wu, H., Hensen, E. J., Abdelkader, A. M., Xi, K., & Lai, C. (2022). Polyoxometalate ionic sponge enabled dendrite-free and highly stable lithium metal anode. *Small Methods*, 6(3), 2101613.

Zhou, B., Sherriff, B. L., Taulelle, F., & Wu, G. (2003). Nuclear magnetic resonance study of Al: Si and F: OH order in zunyite. *The Canadian Mineralogist*, 41(4), 891–903.

Ziv, A., Grego, A., Kopilevich, S., Zeiri, L., Miro, P., Bo, C., Müller, A., & Weinstock, I. A. (2009). Flexible pores of a metal oxide-based capsule permit entry of comparatively larger organic guests. *Journal of the American Chemical Society*, 131(18), 6380–6382.

Index

Note: **Bold** page numbers refer to tables and *italic* page numbers refer to figures.

FSC
www.fsc.org
MIX
Papier | Fördert
gute Waldnutzung
FSC® C083411

Kolibri 360 GmbH
Ferdinand-Jühlke-Straße 7
99095 Erfurt, Deutschland
produktsicherheit@kolibri360.de